NEUROMETHODS ☐ 31

G Protein Methods and Protocols

NEUROMETHODS

Program Editors: Alan A. Boulton and Glen B. Baker

NEUROMETHODS ☐ 31

G Protein Methods and Protocols

Role of G Proteins in Psychiatric and Neurological Disorders

Edited by

Ram K. Mishra
McMaster University, Hamilton, Canada

Glen B. Baker
University of Alberta, Edmonton, Canada

and

Alan A. Boulton
University of Saskatchewan, Saskatoon, Canada

Humana Press ✳ **Totowa, New Jersey**

© 1997 Humana Press Inc.
999 Riverview Drive, Suite 208
Totowa, New Jersey 07512

This publication is printed on acid-free paper. ∞
ANSI Z39.48-1984 (American National Standards Institute)
Permanence of Paper for Printed Library Materials.

For additional copies, pricing for bulk purchases, and/or information about other Humana titles, contact Humana at the above address or at any of the following numbers: Tel.: 973-256-1699; Fax: 973-256-8341; E-mail: humana@mindspring.com, or visit our Website: http://humanapress.com

Cover illustration: Fig. 1 in "G Protein-Coupled Melatonin Receptors" by Lennard P. Niles.

Cover design by Patricia F. Cleary.

Photocopy Authorization Policy:

ISBN 0-89603-490-9
ISSN 0893-2336

Printed in the United States of America. 10 9 8 7 6 5 4 3 2 1

1999/1041

Preface

The G proteins are a family of structurally homologous, plasma membrane-associated guanine-nucleotide-binding proteins. These proteins play an integral role in the transduction of extracellular signals through second messenger systems. As such, G proteins affect a wide variety of intracellular biochemical reactions by regulating the concentration of second messengers in cells.

G proteins are heterotrimeric, consisting of α, β, and γ polypeptide chains, with G protein specificity largely determined by the α-subunit. Molecular cloning of G protein subunits has revealed 23 distinct α-subunits, encoded by 17 different genes. Based on functional measures, G proteins are generally classified into three major categories: the G_s family, which is stimulatory for adenylyl cyclase; the G_i family, which is inhibitory for adenylyl cyclase; and the G_q family, which stimulates phospholipases (Birnbaumer and Birnbaumer, 1995). Alternatively, on the basis of sequence homology, G proteins can be subdivided into four categories: G_s, G_i, G_q, and G_{12}.

Although the α-subunits are specific for a given G protein, the β- and γ-subunits share considerable homology with each other. As previously mentioned, it has generally been assumed that the diversity of the α-subunits provides the essential specificity for G protein receptor interactions. However, there is also evidence suggesting that the structural diversity of the β- and γ-subunits may also contribute to the specificity of these interactions. On the basis of molecular cloning studies, 5 β-subunits and 10 γ-subunits have been described. These two subunits can be regarded as one functional subunit, considering their tight association. The five β-subunits are highly homologous, and display a characteristic feature of repetitive tryptophan-aspartate (WD) motif,

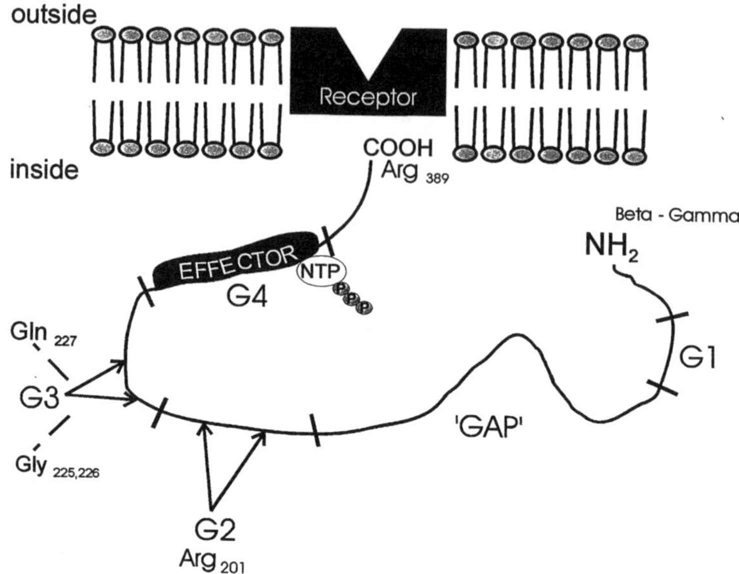

Fig. 1. A hypothetical model of G protein α-subunit (modified from Gordeladze et al., 1994). The figure displays various functions and the spatial interrelationships within the α-subunit. The C-terminus is involved in receptor binding, whereas the N-terminal and probably some other stretch of the subunit are involved in its anchoring to the β γ-subunit complex. Various other regions, G-1, G-2, G-3, and G-4, are important for the binding of GTP. The GTPase activity and the GDP/GTP exchange reside in G-2, G-3 regions. NTP stands for nucleotide triphosphate. Amino acids denoted with a number represent their corresponding position in the sequence. Mutations in these positions have led to the identification of their functional role.

whereas the γ-subunits display significantly more diverse primary structures (Simon et al., 1991; Helper and Gilman, 1992; Ray et al., 1995; Guderman et al., 1996).

G proteins cycle between a GDP-bound inactive form (which maintains the high-affinity agonist-binding state of the receptor) and a GTP-bound active form (low-affinity agonist binding state of the receptor). On agonist binding, the cell surface receptor catalyzes the GDP/GTP exchange at the α-subunit, and this active form goes on to stimulate effector pathways in transmembrane signaling. The salient features of α-subunit signaling are displayed in Fig. 1.

Table 1
G Proteins in Disease States

Disease states	Subtypes				References
	$G_{s\alpha}$	$G_{i\alpha}$	G_o	β_γ	
Congestive heart failure	↓80%	↑	ND	ND	Horn et al., 1988
	↓	↑	—	—	Horn et al., 1995
Dilated cardiomyopathy	ND	↑37%	ND	ND	Bohm et al., 1990
Heart failure	ND	↑	ND	ND	Neumann et al., 1988
Left ventricle failure	↓50%	ND	ND	ND	Longabaugh et al., 1988
Heart—septic shock	ND	↑65%	ND	ND	Bohm et al., 1995
Schizophrenia	—	↓	↓	ND	Okada et al., 1991
Depression	↑	NC	—	—	Manji et al., 1995
	↑	↑FC	↓	—	Lesch et al., 1992; Young et al., 1994
Cocaine addiction	NC	↓NAC	↓NAC	NC, NAC	Self et al., 1994
	—	Low levels VTA, NAC, LC	Low levels VTA, NAC, LC	—	Manji, 1992
	↑30–70% (hypo)	NC	↑20% clostrum, NE	NC	Parolaro et al., 1993
Morphine tolerance	↑>50% (striatal neurons)	↓16%	—	—	Van Vliet et al., 1993

[a]Abbreviations: ND, not determined; NC, no change; FC, frontal cortex, NAC, Nucleus accumbens, VTA, ventral tegmental area; LC, locus ceruleus; NE, nucleus endopiriform.

Changes in G Protein Subunits in Diseases

In view of the critical and important role carried out by G proteins in signal transduction, it is very likely that qualitative or quantitative changes in G proteins could have significant effects on intracellular effectors. It has been recently recognized by many investigators that the simple estimation of receptor numbers in disease states may not provide a complete perspective of receptor activity (Brodde and Michel, 1989). Indeed, the modification of signal transduction pathways could provide important clues to the pathophysiology of many disease conditions. For example, it has been demonstrated that in experimentally induced diabetes, the gene expression of the $G_{i\alpha}$-subunit is significantly reduced (Gawler et al., 1987). Similarly, in patients with type IA pseudohypoparathyroidism, reduced levels of G_s protein and mRNA levels have been reported (Carter et al., 1987). In congestive heart failure, the levels of G_s in lymphocyte membranes are significantly reduced, and this reduction can be fully restored by therapeutic agents, such as angiotensin-converting enzyme inhibitor (Horn et al., 1988). A summary of alterations in G protein levels in various diseases is presented in Table 1. Although in the past decade significant advances have been made toward unraveling the role of neurotransmitter receptors in various psychiatric and neurological disorders, the role of second messenger systems in these disease states is still poorly understood. The purpose of *G Protein Methods and Protocols* is to provide novel information and techniques pertaining to the study of G proteins and their role in CNS functions.

References

Birnbaumer, L. and Birnbaumer, M. (1995) Signal transduction by G-proteins. *J. Recept. Signal Trans. Res.* **15,** 213–252.

Bohm, M., Gierschik, P., Pieske, B., Schnabel, P., Ungerer, M., and Erdmann, E. (1990) Increase of Gi alpha in human hearts with dilated but not ischemic cardiomyopathy. *Circulation* **82,** 1249–1265.

Brodde, O. E. and Michel, M. C. (1989) Disease states can modify both receptor number and signal transduction pathways. *TIPS* **10,** 383, 384.

Carter, A., Bardin, C., Collins, R., Simons, C., Bray, P., and Spiegel, A. (1987) Reduced expression of multiple forms of the alpha subunit of the stimulatory GTP binding protein in pseudohypoparathyroidism type Ia. *Proc. Natl. Acad. Sci. USA* **84,** 7266–7269.

Gawler, D. J., Milligan, G., Spiegel, A. M., Unson, C. G., and Houslay, M. D. (1987) Abolition of the expression of inhibitory guanine nucleotide regulatory protein Gi activity in diabetes. *Nature* **327,** 229–232.

Gordeladze, J. O., Johansen, P. W., Paulson, R. H., Paulssen, E. J., and Gautvik, K. M. (1994) G-proteins: implications for pathophysiology and disease. *Eur. J. Endocrinol.* **131,** 557–574.

Guderman, T., Kalbrenner, F., and Schultz, G. (1996) Diversity and selectivity of receptor G-protein interaction. *Ann. Rev. Pharmacol. Toxicol.* **36,** 429–459.

Helper, J. R. and Gilman, A. G. (1992) G-proteins. *Trends Biochem. Sci.* **17,** 383–387.

Horn, E. M., Corwin, S. J., Steinberg, S. F., Chow, Y. K., Neuberg, G. W., Cannon, P. J., Powers, E. R., and Bilezikian, J. P. (1988) Reduced lymphocyte stimulatory guanine nucleotide regulatory protein and beta-adrenergic receptors in congestive heart failure and reversal with angiotensin converting enzyme inhibitor therapy. *Circulation* **78,** 1373–1379.

Horn, E. M., Kukin, M. L., Neuberg, G. W., Goldsmith, R. L., McCarty, M., Gratch, M., Medina, N., Yushak, M., and Packer, M. (1995) Lymphocyte G-proteins reflect response to treatment in congestive heart failure. *Am. Heart J.* **129,** 98–106.

Lesch, K. P., Hough, C. J., Aulakh, C. S., Wolozin, B. L., Tolliver, T. J., Hill, J. L., Akiyoshi, J., Chuang, D., and Murphy, D. L. (1992) Fluoxetine modulates G-protein alpha-s, alpha-q and alpha-12 subunit mRNA expression in rat brain. *Eur. J. Pharmacol. Mol. Pharmacol.* **227,** 233–237.

Longabaugh, J. P., Batner, D. E., Batner, S. F., and Homcy, C. J. (1988) Decreased stimulatory guanosine triphosphate binding protein in dogs with pressure overload left ventricular failure. *J. Clin. Invest.* **81,** 420–424.

Manji, H. K. (1992) G-proteins: implications for psychiatry. *Am. J. Psychiatry* **149,** 746–760.

Manji, H. K., Chen, G., and Shimon, H. (1995) Guanine nucleotide binding proteins in bipolar affective disorders: effect of long-term lithium treatment. *Arch. Gen. Psychiatry* **52,** 135–144.

Neumann, J., Schmitz, W., Scholz, H., von Meyerinck, L., Doring, V., and Kalmar, P. (1988) Increase in myocardial Gi-proteins in heart failure. *Lancet* **2,** 936, 937.

Okada, F., Crow, T. J., and Roberts, G. W. (1990) G proteins (Gi, Go) in the basal ganglia of control and schizophrenic brain. *J. Neural Transm. [GenSect]* **79,** 227–234.

Parolaro, D., Rubino, T., Gori, E., Massi, P., Bendotti, C., Patrini, G., Marcozzi, C., and Parenti, M. (1993) In situ hybridization reveals specific increases in G-alpha-s and G-alpha-o mRNA in discrete brain regions of morphine-tolerant rats. *Eur. J. Pharmacol. Mol. Pharmacol.* **244,** 211–222.

Ray, K., Kunsch, C., Bonnes, L. M., and Robishaw, J. D. (1995) Isolation of cDNA clones encoding eight different human G-protein g subunits, including three novel forms designated the g4, g10 and g11 sunbunits. *J. Biol. Chem.* **270,** 21,765–21,771.

Self, D. W., Terwilliger, R. Z., Nestler, E. J., and Stein, L. (1994) Inactivation of G_i and G_o proteins in nucleus accumbens reduces both cocaine and heroin reinforcment. *J. Neurosci.* **14,** 6239–6247.

Simon, M. I., Strathmann, M. P., and Gautam, N. (1991) Diversity of G-proteins in signal transduction. *Science* **252,** 802–808.

Young, L. T., Li, P. P., Kamble, A., Siu, K. P., and Warsh, J. J. (1994) Mononuclear levels of G-proteins in depressed patients with bipolar disorder or major depressive disorders. *Am. J. Psychiatry* **151,** 594–596.

Contents

Contributors

PAUL R. ALBERT • *Neuroscience Research Institute, University of Ottawa, Ontario, Canada*

RAJINDER PAL BHULLAR • *Department of Oral Biology, University of Manitoba, Winnipeg, Manitoba, Canada*

STEVEN R. CHILDERS • *Department of Physiology and Pharmacology, Bowman Gray School of Medicine, Winston-Salem, NC*

WILLARD J. COSTAIN • *Department of Psychiatry, McMaster University, Hamilton, Ontario, Canada*

IVONE GOMES • *Department of Biochemistry, All India Institute of Medical Sciences, New Delhi, India*

JAN O. GORDELADZE • *Instituttgruppe for Medisinske Basalfag, Institutt for Medisinsk Biokjemi, Universitetet I Oslo, Norway*

SURESH K. GUPTA • *Department of Psychiatry, McMaster University, Hamilton, Ontario, Canada*

PER WIIK JOHANSEN • *Instituttgruppe for Medisinske Basalfag, Institutt for Medisinsk Biokjemi, Universitetet I Oslo, Norway*

RODNEY L. JOHNSON • *Department of Medicinal Chemistry, University of Minnesota, Minneapolis, MN*

YASUO KAJIMOTO • *Department of Psychiatry and Neurology, Kobe University School of Medicine, Kobe, Japan*

NOBURO KITAMURA • *Department of Psychiatry, Shinko Hospital, Kobe, Japan*

PAOLA M. C. LEMBO • *Department of Pharmacology and Therapeutics, McGill University, Montréal, Quebec, Canada*

ANNA LORENZEN • *Pharmakologisches Institut, Universität Heidelberg, Germany*

HUSSEINI K. MANJI • *Department of Psychiatry and Behavioral Neurosciences, Wayne State University School of Medicine, Detroit, MI*

ERIC R. MARCOTTE • *Department of Psychiatry and Biomedical Sciences, McMaster University, Hamilton, Ontario, Canada*

RAM K. MISHRA • *Department of Psychiatry, McMaster University, Hamilton, Ontario, Canada*

STEPHEN J. MORRIS • *Neuroscience Research Institute, University of Ottawa, Ontario, Canada*

ERIC J. NESTLER • *Division of Molecular Psychiatry, Department of Psychiatry, Yale University, New Haven, CT*

LENNARD P. NILES • *Department of Biomedical Sciences, McMaster University, Hamilton, Ontario, Canada*

NAOKI NISHINO • *Department of Psychiatry and Neurology, Kobe University School of Medicine, Kobe, Japan*

HYMAN B. NIZNIK • *Clarke Institute of Psychiatry, University of Toronto, Ontario, Canada*

FUMIHIKO OKADA • *Okada Research Institute, Sapporo, Japan*

ZDENEK B. PRISTUPA • *Clarke Institute of Psychiatry, University of Toronto, Ontario, Canada*

DANA E. SELLEY • *Department of Physiology and Pharmacology, Bowman Gray School of Medicine, Winston-Salem, NC*

PUSHKAR SHARMA • *Laboratory of Neurochemistry, National Institute of Neurological Disease and Stroke, National Institutes of Health, Bethesda, MD*

SHAIL K. SHARMA • *Department of Biochemistry, All India Institute of Medical Sciences, New Delhi, India*

YUTAKA SHIRAI • *Department of Psychiatry and Neurology, Kobe University School of Medicine, Kobe, Japan*

OSAMU SHIRAKAWA • *Department of Psychiatry and Neurology, Kobe University School of Medicine, Kobe, Japan*

LAURA J. SIM • *Department of Physiology and Pharmacology, Bowman Gray School of Medicine, Winston-Salem, NC*

ROGER K. SUNAHARA • *Department of Pharmacology, University of Texas South Western Medical Center, Dallas, TX*

JUN-FENG WANG • *Department of Psychiatry and Biomedical Sciences, McMaster University, Hamilton, Ontario, Canada*

KATHERINE L. WIDNELL • *Division of Molecular Psychiatry, Department of Psychiatry, Yale University, New Haven, CT*

HIDEO YAMAMOTO • *Department of Psychiatry and Neurology, Kobe University School of Medicine, Kobe, Japan*

CHANG-QING YANG • *Department of Psychiatry and Neurology, Kobe University School of Medicine, Kobe, Japan*

L. TREVOR YOUNG • *Department of Psychiatry and Biomedical Sciences, McMaster University, Hamilton, Ontario, Canada*

In Vitro Autoradiographic Localization of Receptor-Stimulated [^{35}S]GTPγS Binding in Brain

Laura J. Sim, Dana E. Selley,
and Steven R. Childers

Introduction

Neurotransmitter receptors contain two main functional components: a ligand binding domain, which specifically recognizes the neurotransmitter, and a signaling component, which translates the binding of the neurotransmitter (or its agonists) into a physiological response. Radioligand binding assays represented a breakthrough in understanding the properties of these receptors in brain at the level of the ligand binding site, and the development of techniques like in vitro autoradiography of receptor binding allowed high-resolution analysis of the neuroanatomical localization of these receptors. However, these methodologies provide no information regarding the signal transduction component of these receptors, and in particular, cannot provide a true picture of the biological activity of receptors and of the efficacy of neurotransmitters and their agonists to produce a biological response. Even more recent methodological developments, like *in situ* hybridization of receptor mRNA, provide excellent localization of receptor gene expression, but also give little information about receptor coupling to intracellular signaling mechanisms. In order to address this question, techniques must be

From: *Neuromethods, Vol. 31: G Protein Methods and Protocols*
Ed: R. K. Mishra, G. B. Baker, and A. A. Boulton Humana Press Inc.

developed to allow for in vitro localization of receptor-mediated signal transduction.

If agonist efficacy were controlled primarily by a complex interplay between different intracellular effector mechanisms, many steps removed from the receptor, there would be little opportunity to develop methods to localize receptor signaling mechanisms. Fortunately for the superfamily of receptors coupled to G proteins, agonist efficacy is determined primarily at the level of the transducer itself, i.e., at the point at which receptors activate the α-subunits of G proteins to bind guanosine triphosphate (GTP). As reviewed below, studies over the past 20 years have developed techniques to measure this activation process, including assays of GTP hydrolysis, guanine nucleotide release, and binding of GTP analogs. In most receptor systems studied, the abilities of agonists and partial agonists to activate these in vitro systems correlate closely with agonist efficacies in more traditional pharmacological and physiological assays. In the same way, the potencies of specific antagonists to block agonist activation of G proteins also correlate well with their potencies in blocking agonist effects in vivo. Unfortunately, most of the methods used to measure receptor-mediated G protein activation in brain combine a relatively high technical complexity with a relatively low signal, which does not allow for their straightforward application to neuroanatomical localization. However, the discovery that receptor activation of G proteins could be measured with a potent GTP analog, [35S]guanylyl-5′-O-(γ-thio)-triphosphate (GTPγS), in the presence of excess guanosine diphosphate (GDP), has provided a unique opportunity to develop such methods. In vitro autoradiography of agonist-stimulated [35S]GTPγS binding in brain sections provides a method of localizing G protein-coupled receptors that is both faster and more versatile than traditional in vitro autoradiography of radioligand binding sites, since any number of unlabeled agonists can be used to localize receptors with one radioligand, [35S]GTPγS. More importantly, it provides a method to localize receptors not only by their ability to bind ligands, but also according to their ability to activate an intracellular signal transducer.

Agonist-Stimulated Labeling of G Protein

Mechanism of Receptor G Protein Activation

G protein-coupled receptors represent a major superfamily of receptors in brain: at least 80% of neurotransmitters and hormones produce their biological effects via G protein-coupled receptors (Birnbaumer et al., 1990). Several classes of G proteins have been identified, including G_s, which stimulates adenylyl cyclase and is a substrate for ADP-ribosylation by cholera toxin, and G_i/G_o, which inhibit adenylyl cyclase, modulate ion channel activity, and are ADP-ribosylated by pertussis toxin (PTX). The cycle of receptor G protein activation has been characterized and reviewed elsewhere (Gilman, 1987; Birnbaumer et al., 1990). Each G protein is a heterotrimer composed of three subunits, α, β, and γ. The α-subunit binds and hydrolyzes GTP, whereas the βγ dimer does not bind guanine nucleotides. Binding of agonist produces an activated receptor, which changes the conformation of the G protein, increasing the affinity of the α-subunit for GTP and decreasing its affinity for GDP. GTP binding activates the α-subunit, and decreases its affinity for βγ, causing βγ to dissociate from α. The receptor and α then dissociate, enabling α to interact with effectors, which include adenylyl cyclase, ion channels, and phospholipases A_2 and C (Hildebrandt et al., 1983; Gilman, 1987; Birnbaumer et al., 1990; Brown and Birnbaumer, 1990; Childers, 1991). Studies have shown that in addition to involvement in receptor coupling to the G protein (Fung, 1983; Florio and Sternweis, 1989), βγ subunits may also regulate adenylyl cyclase and potassium channels (Tang and Gilman, 1992; Clapham and Neer, 1993; Taussig et al., 1994). The α-subunit spontaneously inactivates by intrinsic GTPase activity that hydrolyzes GTP to GDP, increasing the affinity of α for βγ. The α-subunit reassociates with βγ, completing the cycle. The reaction is catalytic, since each receptor can activate multiple G proteins, which can result in tremendous amplification of the receptor signal into an intracellular response (Asano et al., 1984; Gierschik et al., 1991).

Historical Review

Receptor Localization

The first studies on receptor localization were performed by injecting radiolabeled ligands in vivo, allowing a brief survival time, then removing brains, and autoradiographically visualizing the radioactive signal (Pert et al., 1975; Kuhar and Yamamura, 1975). This method has several disadvantages, including high nonspecific binding, restriction to examining only one ligand per brain, and the use of large amounts of radioligand. The development of in vitro receptor autoradiography (Young and Kuhar, 1979; Herkenham and Pert, 1980), which was based on receptor binding assays in membranes (Pert and Snyder, 1973), solved many of these problems and is still used in receptor localization studies. In addition to the ability to localize receptors in an anatomically specific manner, this technique allows the examination of multiple receptor systems in the same brain. Furthermore, the assay is not limited to radioligands that are stable for in vivo procedures, and allows in vitro optimization of assay conditions for specific receptors by manipulation of buffer composition and incubation time and temperature. However, receptor radioligand autoradiography is limited to the detection of ligand binding sites and provides no information about the functional activity of receptors.

Detection of Agonist-Stimulated G Protein Activity

Receptor activation of G proteins in isolated membranes has traditionally been measured by receptor-stimulated low K_m GTPase activity (Cassel and Selinger, 1976; Koski and Klee, 1981; Selley and Bidlack, 1992). However, this method can be misleading, since factors that affect GTP hydrolysis and not G protein activation will affect the results (Selley et al., 1993). The GTPase assay is also a rather indirect measure of G protein activity, since it actually measures hydrolysis, which is the inactivation reaction of the G protein cycle. Receptor-stimulated [^{35}S]GTPγS binding is a more direct assay of receptor activation of G proteins, since it measures the activation reaction of the cycle, i.e., the exchange of bound GDP for GTP (or [^{35}S]GTPγS). This technique has been used in biochemical studies of purified and reconstituted systems, where receptors and G proteins are assayed in solution or reconstituted in phospholipid vesicles (Asano et al., 1984; Kurose et al., 1986; Florio and Sternweis, 1989). More recently, the [^{35}S]GTPγS technique has been applied to isolated cardiac (Hilf et al., 1989), brain (Lorenzen et al., 1993), and cultured cell (Lazareno et al., 1993; Traynor and Nahorski, 1995) membranes to measure specific receptor-stimulated G protein activation.

The stimulation of [^{35}S]GTPγS binding to G proteins by receptor agonists is based on the G protein activation cycle discussed above (Fig. 1). Initially, a large excess of GDP is added to the assay to shift the G protein into the inactive state. [^{35}S]GTPγS and the desired agonist are then added to activate the G protein coupled to the receptor of interest. Receptor activation decreases the α-subunit's affinity for GDP relative to GTP, so that the G protein binds GTP or [^{35}S]GTPγS. In vivo, the α-subunit GTPase hydrolyzes GTP back to GDP; however, [^{35}S]GTPγS is resistant to hydrolysis and remains bound. The increase in bound [^{35}S]GTPγS induced by agonist can then be measured using autoradiography or liquid scintillation spectrophotometry. Although first developed for measuring agonist-induced stimulation of [^{35}S]GTPγS binding in isolated membranes, these same principles can be applied (with several major technical changes) to brain sections to localize receptor activity.

Pharmacological Specificity

Since [^{35}S]GTPγS autoradiography depends on agonist–receptor interactions, several pharmacological criteria must be met to verify the specificity of the assay. First, agonist-stimulated [^{35}S]GTPγS binding must be blocked by the

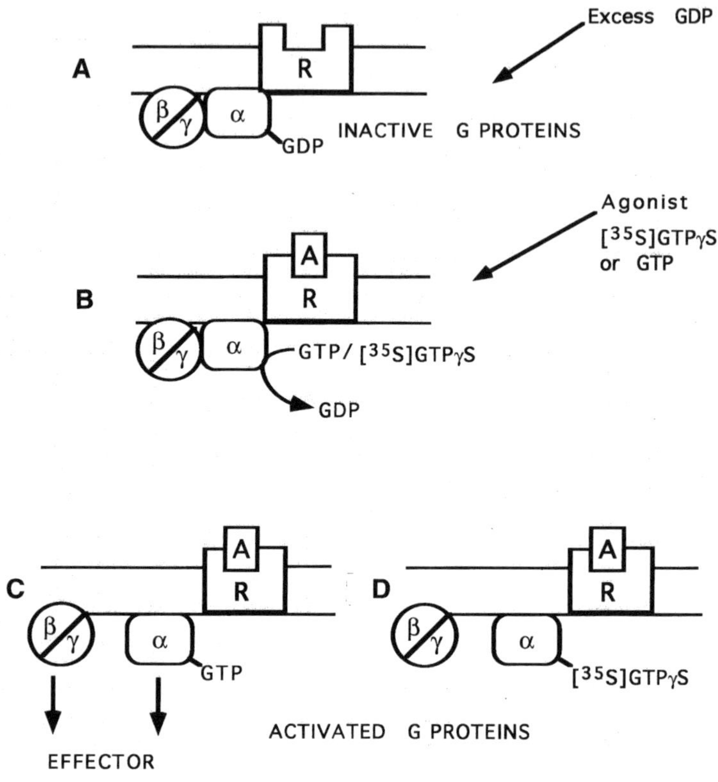

Fig. 1. Schematic diagram of the mechanism of agonist-stimulated [35S]GTPγ S binding. (1) Addition of excess GDP shifts the G proteins into the inactive state. (2) Agonist binds to the receptor, which increases the affinity of the G protein for GTP or [35S]GTPγ S. (3) Activation of the G protein by GTP activates effectors. (4) Activation of the G protein by [35S]GTPγ S provides labeled G protein.

appropriate antagonist (Fig. 2). For example, in the μ opioid receptor system, stimulation of [35S]GTPγ S binding by the μ agonist DAMGO was blocked by the opioid antagonist naloxone. Similarly, WIN 55212-2-stimulated [35S]GTPγ S binding via cannabinoid receptors was blocked by the cannabinoid antagonist SR 141716A (Sim et al., 1995). Second, agonist-stimulated [35S]GTPγ S binding must also be concentration-dependent and saturable with regard to agonist. This

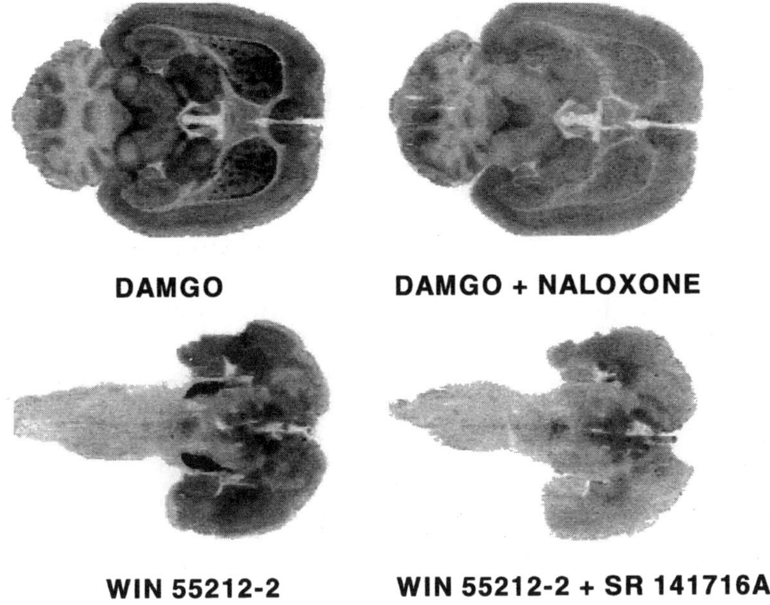

Fig. 2. Autoradiograms demonstrating antagonist reversal of μ opioid- (**top**) and cannabinoid- (**bottom**) stimulated [³⁵S]GTPγS binding. μ Opioid- (using 3 μM DAMGO as an agonist) stimulated [³⁵S]GTPγS binding is found in the caudate-putamen, thalamus, and periaqueductal gray, and is eliminated by the μ antagonist naloxone (0.3 μM). In a more ventral section, cannabinoid-stimulated [³⁵S]GTPγS binding (using 1 μM WIN 55212-2 as an agonist) is seen in the substantia nigra and is blocked by the addition of the cannabinoid antagonist SR 141716A (0.3 μM).

has been demonstrated in both isolated membranes and tissue sections (Sim et al., 1995). Finally, since agonist-stimulated [³⁵S]GTPγS binding results from receptor activation of G proteins, the anatomical distribution of agonist-stimulated [³⁵S]GTPγS binding should correlate with that of receptors. This has been true for the receptor systems examined thus far: cannabinoid, GABA$_B$, and opioid (Goodman et al., 1980; Herkenham and Pert, 1982; Chu et al., 1990; Herkenham et al., 1991; Jansen et al., 1992). However, since the receptor–G protein interaction is catalytic, the relative levels of agonist-stimulated [^{35}S]GTPγS binding may not correlate with receptor density (*see* Role of Coupling Efficiency).

Requirement for GDP, Sodium, and Magnesium

An important consideration in the [^{35}S]GTPγS assay is the reduction of basal [^{35}S]GTPγS binding (i.e., binding in the absence of agonist) so that agonist-stimulated [^{35}S]GTPγS binding is detectable. Several factors regarding G protein activity must be considered in the [^{35}S]GTPγS assay. First, since [^{35}S]GTPγS binds to G proteins with high affinity in the absence of receptor activation, excess GDP is included in the [^{35}S]GTPγS assay to minimize basal binding. This is a particular problem in brain where G proteins can represent up to 1% of total membrane protein (Sternweis and Robinshaw, 1984; Hepler and Gilman, 1992), and where basal [^{35}S]GTPγS binding is very high. Second, receptors and G proteins are not completely inactive in the absence of agonist, so that spontaneous activity must be minimized to detect agonist-stimulated [^{35}S]GTPγS binding. Strategies to minimize these potential problems are discussed in detail below.

The addition of micromolar concentrations of GDP to the [^{35}S]GTPγS membrane binding assay is necessary to detect agonist-stimulated [^{35}S]GTPγS binding. Although other nucleotides may also reduce basal binding, GDP is the most effective in decreasing basal binding while allowing receptor stimulation of [^{35}S]GTPγS binding (Hilf et al., 1989). The mechanism by which GDP decreases basal [^{35}S]GTPγS binding involves shifting the G protein into the inactive (GDP-bound) state. It is well established that receptor activation of the G protein decreases its affinity for GDP and increases its affinity for GTP (Cassel and Selinger, 1978). Thus, the addition of GDP would favor the inactivate state of the G protein, decreasing spontaneous G protein activity. GDP is an especially important component of the [^{35}S]GTPγS autoradiographic assay in tissue sections (Sim et al., 1995). In this case, millimolar concentrations of GDP are necessary to detect agonist-stimulated [^{35}S]GTPγS binding over basal [^{35}S]GTPγS binding (Fig. 3). The reason that such a high concentration of GDP is required in the tissue sections is not completely understood, but may be owing to the ability of GDP to penetrate intact sections effectively or be caused by the level of total protein in sections (*see* Troubleshooting).

Sodium is also required in the [^{35}S]GTPγS assay to decrease basal binding (Hilf et al., 1989; Tian et al., 1994). Although sodium decreases the affinity of agonist binding to G protein-coupled receptors (Pert and Snyder, 1974; Tsai and Lefkowitz, 1978) by acting at a site on the receptor (Horstman et al., 1990), this effect does not appear to interfere with agonist-stimulated [^{35}S]GTPγS binding when sufficient agonist concentrations are used. Rather, sodium is necessary to decrease basal [^{35}S]GTPγS binding so that agonist-stimulated [^{35}S]GTPγS binding is detectable. Sodium has also been shown to decrease basal G protein activity in the GTPase assay (Koski et al., 1982; Gierschik et al., 1989; Costa et al., 1990; Selley et al., 1993). The reason for this effect is, at least in part, that sodium decreases spontaneous G protein activation produced by unoccupied receptors (Costa et al., 1990; Tian et al., 1994). Thus, relatively high levels of G protein activity can occur in the absence of added agonist. This same effect is seen in the tissue sections with [^{35}S]GTPγS autoradiography, where elimination of sodium results in a high level of basal [^{35}S]GTPγS binding, above which agonist-simulated [^{35}S]GTPγS binding is undetectable (Sim et al., unpublished observations).

The third necessary component of the [^{35}S]GTPγS assay is magnesium (Hilf et al., 1989). Magnesium has a biphasic effect on G protein activity (Brandt and Ross, 1986; Gilman, 1987). Low (high-nanomolar to low-micromolar) concentrations of magnesium are required for hydrolysis of Gα-GTP and high (high-micromolar to low millimolar) concentrations promote agonist-stimulated G protein activation. However, very high (>10 m*M*) concentrations of magnesium may actually decrease [^{35}S]GTPγS binding to membranes (Gierschik et al.,

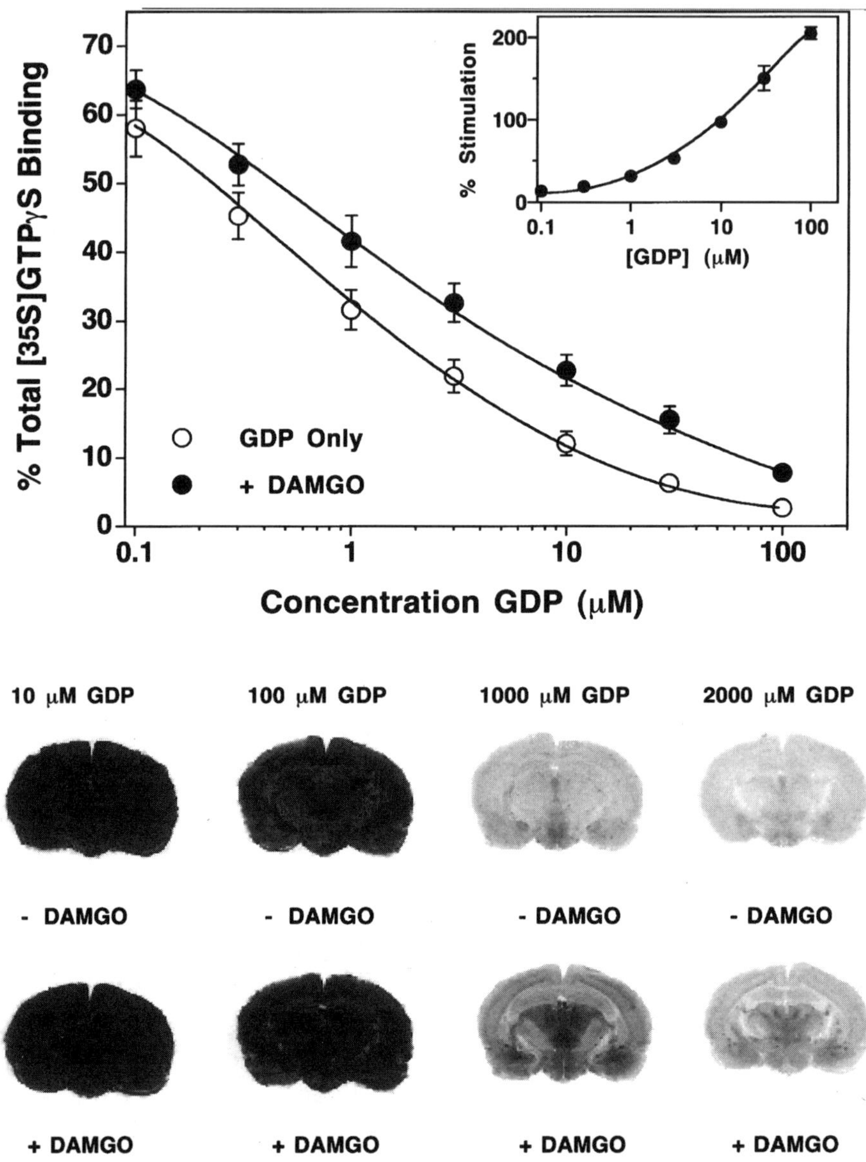

Fig. 3. Effect of GDP on basal and µ opioid-stimulated [35S]GTPγS binding in rat thalamic membranes and brain sections. (Top) Effect of GDP on basal and DAMGO-stimulated [35S]GTPγS binding in isolated thalamic membranes, expressed as percent total [35S]GTPγS bound in the absence of GDP or agonist (Inset) DAMGO-stimulated [35S]GTPγS binding, expressed as percent stimulation over basal [35S]GTPγS binding at each concentration of GDP. Basal and agonist-stimulated [35S]GTPγS binding are both decreased by increasing GDP concentrations, but the percent stimulation by agonist is increased. (Bottom) autoradiogram of brain sections at the level of the thalamus incubated with varying concentrations of GDP. Mu opioid-stimulated [35S]GTPγS binding is not detected until the GDP concentration reaches 1–2 m*M*.

1991) (Selley et al., unpublished observations). Thus, the magnesium concentration must be optimized to achieve maximal [^{35}S]GTPγS binding.

Methodology

Autoradiography

Tissue Preparation and Storage

The assay procedure described below has been used for brain tissue; however, agonist-stimulated [^{35}S]GTPγS autoradiography should also be applicable to other tissues. Animals are sacrificed by rapid decapitation, and the brains are removed. The brains are slowly immersed in isopentane maintained at –35°C on dry ice. After 3–5 min, the brains are removed from the isopentane with cold forceps and placed on dry ice for 3–5 min to evaporate remaining isopentane. The brains are then wrapped in cold foil, placed in sealed polypropylene containers, and stored at –80°C until sectioning. Brains are cut into 20-μm sections on a cryostat maintained at –20°C, and sections are thaw-mounted onto gelatin-coated slides. Slides are collected in racks in a humidified chamber on ice, desiccated under vacuum, and stored overnight at 4°C. The next day, slides are placed in slide boxes with desiccant pellets, and stored at –80°C until use. On the day of the assay, slides are removed from the freezer and brought to room temperature under a cool air-dryer for 30–60 min.

In Vitro [^{35}S]GTPγS Autoradiography

After slides have been brought to room temperature, they are transferred into slide racks and equilibrated in assay buffer (50 mM Tris-HCl, 3 mM MgCl$_2$, 0.2 mM EGTA, 100 mM NaCl, pH 7.4) for 10 min at 25°C. Incubations are performed in a water bath maintained at 25°C, so that the temperature remains constant. Slides are then transferred to mailers (5 slides) or polypropylene coplin jars (10 slides) and incubated for 15 min in assay buffer with 2 mM GDP. Slides are then transferred to incubation media that contain 0.04 nM

[^{35}S]GTPγS, 2 mM GDP, and desired agonist, and incubated for 2 h at 25°C. Basal binding is measured in the absence of agonist (i.e., with GDP and [^{35}S]GTPγS), and nonspecific binding is assessed in the presence of 10 μM GTPγS. For final rinses, slides are transferred into slide racks and rinsed in staining dishes. Slides are rinsed twice for 2 min each in Tris buffer (50 mM Tris-HCl, pH 7.0, at 25°C) on ice and once for 30 s in deionized H$_2$O on ice. For diagnostic purposes, sections can be wiped off of the slide with filter paper at this point and counted in a scintillation counter. For autoradiography, slides are dried well under a cool stream of air (30–60 min), then covered, and dried overnight at room temperature. Slides are exposed to autoradiography film for 48 h, although longer exposures (4–7 d) may be necessary for less efficacious agonists. Several types of autoradiographic film may be used, since [^{35}S] is a relatively high-energy radioisotope. Our laboratory uses Reflections® (New England Nuclear) film because of its low background, but other types of film appropriate for [^{35}S] have also been successfully used.

Analysis of Data

In our laboratory, films are digitized with a Sony XC-77 video camera and analyzed using the NIH IMAGE program for Macintosh computers. The regions measured are selected based on (1) interests of the particular study and (2) determination of regions with relatively high levels of agonist-stimulated [^{35}S]GTPγS binding. It is important to select regions with optical densities that fall within the limits of the standard, i.e., that are not too high or low to be measured reliably. The film exposure time may need to be modified so that the sections are not over- or under-exposed with respect to the standards. Images are captured by averaging 16 individual frames of the image to provide a more accurate density measurement.

Quantification is obtained by including [^{14}C] standards in the cassette. Since densitometric analysis is performed using [^{14}C] standards, a brain paste standard assay was performed to determine the correction factor necessary to calculate nCi/g [^{35}S] from nCi/g [^{14}C]. Briefly, known amounts of

[^{35}S]GTPγS were added to homogenized brain (brain paste), which was frozen into a block and sectioned at 20 μM. Sections were collected to measure:

1. Weight;
2. Radioactivity by scintillation counting; and
3. Densitometry by exposure to film.

From these values, a correction factor was obtained to transform the nCi/g based on [^{14}C] standard densitometry into nCi/g for [^{35}S]. This correction formula may vary based on the tissue and type of film used. For statistical analysis, sections are performed in duplicate (horizontal) or triplicate (coronal) of brains from at least five animals. Results are expressed as both net stimulated [^{35}S]GTPγS binding in nCi/g (agonist-stimulated [^{35}S]GTPγS binding minus basal binding) or percent stimulation (net binding divided by basal binding × 100).

[^{35}S]GTPγ S Membrane Binding Assay

Animals are sacrificed by decapitation, brains are removed, and desired regions are dissected on ice. Tissue is homogenized in 20 vol of cold Tris-Mg^{2+} buffer (50 mM Tris-HCl, 3 mM MgCl$_2$, 1 mM EGTA, pH 7.4) with a Polytron, centrifuged at 48,000g at 4°C for 10 min, resuspended in buffer, and centrifuged at 48,000 g at 4°C for 10 min. Tissue is resuspended in assay buffer, and protein levels are measured by the method of Bradford (Bradford, 1976). Tissue may be used immediately after preparation. However, since a very low concentration of protein is required for the assay, aliquots of extra tissue homogenate are stored at –80°C for future use. This is especially helpful when using treated animals or species in which tissue conservation is important. Membranes (4–15 μg protein depending upon region) are incubated in a 1-mL total volume with 0.05 nM [^{35}S]GTPγS, 20 μM GDP, and agonist in assay buffer at 30°C for 1 h. Basal binding is assessed in the absence of agonist (i.e., with GDP and [^{35}S]GTPγS alone), and nonspecific binding is measured in the presence of 10 μM GTPγS. The reaction is terminated by filtration under vacuum through Whatman GF/B glass fiber filters, followed by three washes with cold Tris buffer. Bound radioactivity is determined by liquid scintillation spectrophotometry at 95% efficiency after extraction overnight in 5 mL scintillation fluid.

Troubleshooting

Tissue Preparation

Once the tissue is sectioned for [35S]GTPγ S autoradiography, slides must be stored at –80°C until use. Storage at higher temperatures or freeze-thawing the sections decreases the [35S]GTPγ S signal, and may eliminate agonist stimulation of [35S]GTPγ S binding. The temperature is also a factor during the collection of sections from the cryostat. It is best to collect slides on ice, since decreased [35S]GTPγ S binding may result if slides are kept at room temperature for a prolonged time. Tissue storage is also an important consideration for tissue homogenate used in the [35S]GTPγ S membrane assay. Again, tissue may be prepared and stored in aliquots at –80°C. However, tissue should not be freeze-thawed and should be kept on ice during the assay procedure.

Materials

A high concentration (1–2 mM) of GDP is used in the autoradiographic assay, in both the 15-min preincubation period as well as the final incubation with agonist and [35S]GTPγ S. The final concentration of GDP to use in the assay must be determined empirically. GDP concentrations below 1 mM are not recommended, since basal levels of [35S]GTPγ S binding become too high to detect agonist-stimulated [35S]GTPγ S binding. High (2-mM) concentrations of GDP decrease both basal and stimulated labeling to produce a lighter image, and maximize the percent stimulation by agonist. Lower (1-mM) concentrations of GDP produce higher labeling and faster exposure times, but lower the percent stimulation by agonist. The GDP preincubation solution can be discarded, and a new incubation solution containing GDP, [35S]GTPγ S, and agonist can be added to the slides. However, to reduce the cost of GDP, it is possible to keep the preincubation solution and add [^{35}S]GTPγS and agonist. In this case, the solution must be mixed well, and this is best accomplished by removing the slides and quickly mixing the solution. It is also possible to perform subsequent runs of the assay, so that the same incubation mixture is used for multiple sets of slides. In our laboratory, the incubation solution is used a maximum of two times.

The relationship between protein concentration and GDP concentration is an important consideration in the [^{35}S]GTPγS binding assay. In assays of receptor-stimulated [^{35}S]GTPγS binding to isolated membranes, increasing protein concentrations requires concomitant increases in GDP concentrations owing to the increased level of basal [^{35}S]GTPγS binding. In brain membrane assays using 10–20 μM GDP, 2–15 μg protein/assay tube are generally used, depending on the region examined. In contrast, in autoradiographic assays, each 20-μm rat brain section (at the level of the caudate-putamen) contains approx 175 μg protein. This may explain why relatively high GDP concentrations are required in sections relative to membrane assays to reduce basal [^{35}S]GTPγS binding.

The storage of both [^{35}S]GTPγS and unlabeled GTPγS is an important consideration, since GTPγS is a somewhat unstable compound. For [^{35}S]GTPγS, the original solution is diluted 1:10 in water and stored at –80°C. Appropriate dilutions for assays are made from this stock on the day of the assays; the assay solutions are never stored for future use. Unlabeled GTPγS is stored desiccated at –20°C. Stock solutions are made and stored in aliquots at –80°C until use.

[^{35}S]GTPγS Binding Assay

During methods development for the [^{35}S]GTPγS assay, each experiment should contain a positive control, that is, a condition that duplicates a published result. Both μ opioid (using 1–10 μM DAMGO as an agonist) and GABA$_B$ (using 300 μM baclofen as an agonist) provide excellent signals and characteristic labeling patterns. The [^{35}S]GTPγS autoradiographic assay is also best performed in conjunction with

[^{35}S]GTPγS membrane binding assays to determine optimal conditions for the assay. This is especially important in determining the concentration of agonist to use for maximal stimulation of [^{35}S]GTPγS binding. As discussed in the Introduction, agonist-stimulated [^{35}S]GTPγS binding should be concentration-dependent and antagonist reversible. In the membrane [^{35}S]GTPγS assay, some agonists produce an artifactual overshoot of [^{35}S]GTPγS binding when used at high concentrations (i.e., over 30 μM). Thus, in determining the agonist concentration to use for maximal stimulation of [^{35}S]GTPγS binding, it is important to verify that the response is antagonist-reversible (i.e., receptor-mediated).

The assay buffer may be modified to optimize agonist-stimulated [^{35}S]GTPγS binding for some agonists. For instance, 0.5% bovine serum albumin is added to the incubation media when using the cannabinoid agonist WIN 55212-2. Protease inhibitors (10 μL/mL of a solution containing 0.2 mg/mL each of bestatin, leupeptin, pepstatin A, and aprotinin) may also be added to the preincubation buffer when using endogenous peptides. The pH of the buffers is especially important in the [^{35}S]GTPγS autoradiographic assay, since tissue may be damaged by acidic or basic pH. The pH of assay buffer is 7.4 at room temperature, and Tris buffer for rinses is 7.0 at room temperature.

The other assay parameters that may be adjusted for methods development are the concentration of [^{35}S]GTPγS, incubation time and temperature, concentrations of sodium and/or magnesium, protein levels (membranes), and concentration of GDP (generally 10–30 μM for membrane assay and 1–2 mM for sections).

General Applications of the Method

Advantages

[^{35}S]GTPγS autoradiography provides not only neuroanatomical localization, but also pharmacological data. For example, this technique clearly differentiates agonist and antagonist effects of drugs at specific receptor sites, which greatly increases the power of the technique. Moreover, this technique can help to address differences in efficiency of coupling of receptors to G proteins in specific brain nuclei, as discussed below. Although the basic methodology of [^{35}S]GTPγS autoradiography is similar to that of traditional receptor autoradiography, there are several important differences between these techniques. Perhaps most important is the concept that [^{35}S]GTPγS autoradiography demonstrates functional activity at the level of signal transduction, rather than binding at the level of the receptor.

From a technical perspective, the use of [^{35}S]GTPγS also provides several advantages. First, traditional receptor autoradiography generally uses a tritiated ligand, which requires film exposure on the order of weeks or months. Since [^{35}S] is a higher energy isotope, film exposure for [^{35}S]GTPγS autoradiography is generally 2–7 d. Furthermore, since [^{35}S]GTPγS is always the radioligand used, any available unlabeled agonist can be used to activate the receptor. Therefore, [^{35}S]GTPγS autoradiography is not restricted to systems in which specific radiolabeled ligands are available. This is particularly important in the case of newly isolated or synthesized ligands, for which radiolabeled ligands are not commercially available. For example, [^{35}S]GTPγS autoradiography has been used to map the distribution of ORL-1 receptor-activated G proteins in the rat brain (Sim et al., 1996) (Fig. 4). Once the sequence of the ORL-1 peptide was published (Reinscheid et al., 1995; Meunier et al., 1995), it was synthesized locally and used in the [^{35}S]GTPγS assay, so that the results of [^{35}S]GTPγS autoradiographic assays were obtained even though a radiolabeled peptide was not commercially available. Furthermore, ORL-1-stimulated [^{35}S]GTPγS binding in membrane assays was used to show (1) the dose-dependent nature of the response and (2) lack of reversibility by antagonists for other receptors.

Limitations

Since the [^{35}S]GTPγS assay examines function at the level of receptor–G protein coupling, many of the limitations of the

Basal

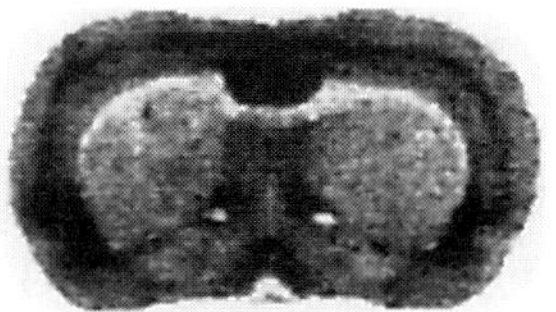

ORL-1

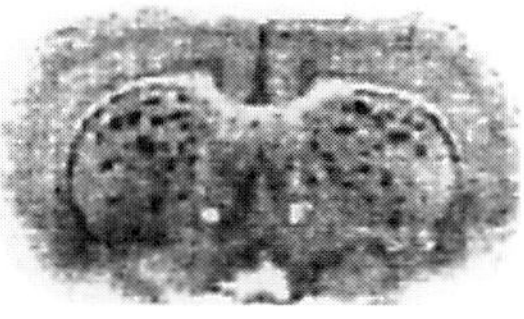

DAMGO

Fig. 4. Rat brain sections at the level of the caudate-putamen comparing ORL-1 peptide- and μ opioid-stimulated [^{35}S]GTPγS binding. μ Opioid-stimulated [^{35}S]GTPγS binding is seen in the patches of the caudate-putamen. In contrast, ORL-1 peptide-stimulated [^{35}S]GTPγS binding is absent in the caudate-putamen, but high in the cortex.

technique result from functional properties of the system. Perhaps the most obvious limitation of [^{35}S]GTPγS autoradiography is the difficulty in detecting certain classes of receptors owing to low receptor density and/or low efficiency of coupling (*see* Role of Coupling Efficiency).

Thus far, the assay has not been applicable to G_s-coupled systems in brain membranes or sections. This is because the relative abundance of G proteins in the brain is $G_o > G_i > G_s$ (Sternweis and Robishaw, 1984), so that the level of G_s is too low to produce detectable [^{35}S]GTPγS binding. However, [^{35}S]GTPγS binding stimulated by G_s-coupled receptors can be measured in other systems. Prostaglandin E_1-stimulated [^{35}S]GTPγS binding is measurable in NG108-15 cell membranes (Selley et al., submitted), and isoproterenol-stimulated [^{35}S]GTPγS binding is detectable in turkey erythrocyte membranes (Wieland and Jakobs, 1994).

Another limitation is the specificity of the technique for individual G proteins. Although the [^{35}S]GTPγS binding assay is highly specific for receptor type (by the use of selective agonists), there is currently no method to determine which type of G protein(s) the receptor activates in the tissue section. Thus, the [³⁵S]GTPγS binding signal results from the activation of the pool of G proteins coupled to the receptor of interest, which may be a heterogeneous G protein population.

Finally, several important factors must be carefully considered in determining catalytic amplification for G protein-coupled receptors. Several methods have been published for Scatchard analysis of [³⁵S]GTPγS binding in isolated membranes (Gierschik et al., 1991; Traynor and Nahorski, 1995; Sim et al., 1996b). Basically, this assay is performed using fixed concentrations of [³⁵S]GTPγS and GDP, with increasing concentrations of unlabeled GTPγS in the presence and absence of agonist. It would appear that these data can be analyzed like any other receptor binding data. However, the interpretation of [³⁵S]GTPγS binding is in actuality much more complex, since the [³⁵S]GTPγS binding assay violates certain assumptions of Scatchard analysis. First, the assay does not measure a direct bimolecular interaction, but rather the allosteric regulation (by agonist-bound receptor) of the competition of one ligand (GTPγS) for another (GDP). Also, the assay is not conducted under true equilibrium conditions, since at equilibrium the GTPγS will have displaced all of the GDP that it can in the basal condition, thus eliminating or greatly reducing the agonist-stimulated signal. Thus, the amplification factors calculated are relative based on the assay conditions, and can not provide an exact and accurate number of activated G proteins. However, if the assay results are precise, the GTPγS Scatchard assay is extremely useful for relative comparisons.

Role of Coupling Efficiency and Agonist Efficacy in the Sensitivity of Detection

One of the striking findings in comparing [³⁵S]GTPγS autoradiography with that of the traditional radioligand technique was the fact that the neuroanatomical distribution of agonist-stimulated [³⁵S]GTPγS binding closely paralleled that of receptor binding sites. From a qualitative point of view, the overall distribution for several receptor systems was virtually identical. However, from a quantitative perspective, interesting differences have emerged. One example is the cannabinoid receptor system, where the level of agonist-stimulated [^{35}S]GTPγS binding did not completely correlate with reported receptor densities (Sim et al., 1995). In this case, although there was a 10-fold excess of cannabinoid over opioid binding sites in the striatum, the level of opioid-stimulated [^{35}S]GTPγS binding was comparable to that of cannabinoid-stimulated [^{35}S]GTPγS binding. This similarity in agonist-stimulated [^{35}S]GTPγS binding between the two receptor systems was observed in both isolated membranes and by autoradiography of brain sections (Fig. 5). Thus, the 10-fold excess of cannabinoid compared to opioid binding sites was not reflected at the level of signal transduction. To explore the basis of this discrepancy, Scatchard analysis of [^{35}S]GTPγS binding was performed to determine the relative numbers of G proteins activated by cannabinoid and opioid receptors. Catalytic amplification factors were then calculated for each receptor based on receptor binding and G protein B_{max} values; these showed that each cannabinoid receptor activated three G proteins, whereas each μ and δ opioid receptor activated 17–20 G proteins. Therefore, opioid receptors in the striatum are more catalytically efficient than cannabinoid receptors. This finding may have important technical implications regarding the [^{35}S]GTPγS binding assay. Receptor density and catalytic activity of the system are important factors that determine the magnitude of the [^{35}S]GTPγS signal. Thus, it may be difficult or impossible to detect agonist-stimulated [^{35}S]GTPγS binding in a system with low receptor density and/or low catalytic activity.

Another important factor that affects [^{35}S]GTPγS labeling in both isolated membranes and brain sections is the efficacy of the agonist. To obtain high resolution localization of receptors using [^{35}S]GTPγS, it is necessary to ensure that the agonist utilized is a full agonist, since an agonist of partial efficacy will not provide the same maximal level of stimulation. This

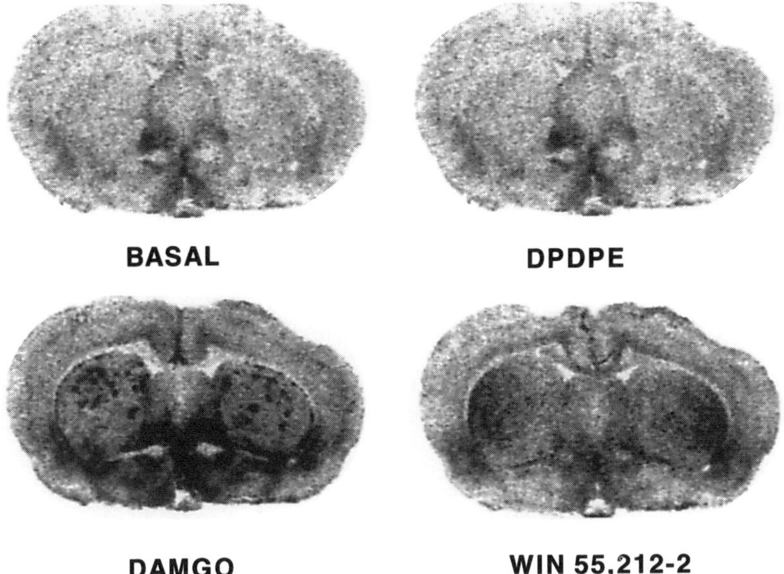

Fig. 5. Autoradiograms at the level of the caudate-putamen comparing opioid- and cannabinoid-stimulated [³⁵S]GTPγS binding. Although each agonist stimulated a unique distribution of [³⁵S]GTPγS binding, the overall levels of stimulation were similar when measured densitometrically, despite a tenfold excess of cannabinoid receptor binding sites.

is illustrated in Fig. 6, which compares the distribution of labeling in rat thalamus for several μ opioid agonists, including DAMGO, morphine, fentanyl, and buprenorphine. In this system, DAMGO is defined as a full agonist, whereas morphine and fentanyl produce approx 40% of the stimulation produced by DAMGO in the medial thalamus. Buprenorphine, in contrast, produced much less stimulation, approx 15% of the level produced by DAMGO. Nevertheless, it can be seen that the relative distribution of μ-stimulated binding in these sections is approximately the same whether a full or partial agonist is used.

Summary and Future Directions

[³⁵S]GTPγS autoradiography provides a rapid, versatile method of receptor localization, which provides qualitatively

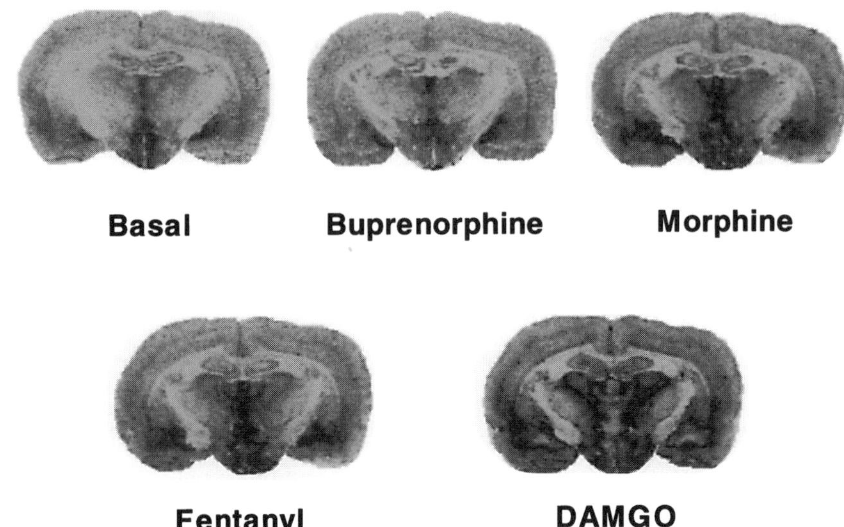

Fig. 6. Autoradiograms of brain sections at the level of the thalamus incubated with mu opioid agonists of different efficacies. DAMGO, the most efficacious agonist, stimulated robust [35S]GTPγS binding in the medial thalamus, hypothalamus and amygdala. Morphine- and fentanyl-stimulated [35S]GTPγS binding exhibited a similar anatomical distribution to that of DAMGO, but with a lower level of stimulation. Buprenorphine stimulated a very low level of [35S]GTPγS binding.

different information from traditional receptor autoradiography. [35S]GTPγS autoradiography has been successfully applied to several receptor systems in the brain: μ, δ and κ opioid, cannabinoid, ORL-1, and GABA$_B$. Further studies will extend this method to yet other G protein-coupled receptor systems. However, [35S]GTPγS binding stimulated by other receptors may be more difficult to detect owing to differences in receptor–G protein coupling. For instance, as discussed above, the efficiency of receptor–G protein coupling has a tremendous impact on the [35S]GTPγS signal, and the catalytic amplification of some receptors may be too low for significant detection.

Nevertheless, there are numerous anatomical applications of this procedure that are now being explored. For example, [³⁵S]GTPγS binding has been detected autoradiographically with film. It should also be possible to visualize the signal with emulsion, using protocols similar to *in situ* hybridization with [³⁵S] labeled probes. The application of emulsion autoradiography will provide detailed anatomical information at the cellular level. It will also be of interest to use [³⁵S]GTPγS autoradiography in conjunction with other anatomical techniques, including immunocytochemistry, receptor autoradiography, and *in situ* hybridization. Since the tissue is prepared for [³⁵S]GTPγS autoradiography using standard methods for receptor binding and *in situ* hybridization studies, alternate section analysis will be possible. This will be especially important in studies that involve chronic drug treatments or disease states, so that the neuronal system can be examined at the level of the receptor, mRNA, and G protein activity to provide a complete picture of the system. In this regard, it will also be interesting to expand [³⁵S]GTPγS autoradiography for use in human brain tissue. However, this may be difficult, since post mortem changes may be detrimental to the detection of receptor-activated G proteins. Finally, although this technique has been used successfully in brain tissue, it will be interesting to see whether it is applicable to other tissues, and to compare such factors as catalytic signal amplification by receptors among various tissues. All of these considerations suggest a widespread utility of this method for localizing G protein-coupled receptor-mediated signal transduction.

References

Asano, T., Pedersen, S. E., Scott, C. W., and Ross, E. M. (1984) Reconstitution of catecholamine-stimulated binding of guanosine 5′-*O*-(3-thiotriphosphate) to the stimulatory GTP-binding protein of adenylate cyclase. *Biochemistry* **23,** 5460–5467.

Birnbaumer, L., Abramowitz, J., and Brown, A. M. (1990) Receptor-effector coupling by G proteins. *Biochem. Biophys. Acta* **1031,** 163–224.

Bradford, M. M. (1976) A rapid and sensitive method for the quantification of microgram quantities of protein utilizing the principle of protein-dye binding. *Anal. Biochem.* **72,** 248–254.

Brandt, D. R. and Ross, E. M. (1986) Catecholamine-stimulated GTPase cycle: multiple sites of regulation by β-adrenergic receptor and Mg^{2+} studied in reconstituted receptor-Gs vesicles. *J. Biol. Chem.* **261,** 1656–1664.

Brown, A. M. and Birnbaumer, L. (1990) Ionic channels and their regulation by G protein subunits. *Ann. Rev. Physiol.* **52,** 197–213.

Cassel, D. and Selinger, Z. (1976) Catecholamine-stimulated GTPase activity in turkey erythrocyte membranes. *Biochim. Biophys. Acta* **452,** 538–551.

Cassel, D. and Selinger, Z. (1978) Mechanism of adenylate cyclase activation through the β-adrenergic receptor: catecholamine-induced displacement of bound GDP by GTP. *Proc. Natl. Acad. Sci. USA* **75,** 4155–4159.

Childers, S. R. (1991) Opioid receptor-coupled second messengers. *Life Sci.* **48,** 1991–2003.

Chu, D. C. M., Albin, R. L., Young, A. B., and Penney, J. B. (1990) Distribution and kinetics of GABA$_B$ binding sites in rat central nervous system: a quantitative autoradiographic study. *Neuroscience* **34,** 341–357.

Clapham, D. E. and Neer, E. J. (1993) New roles for G protein βγ-dimers in transmembrane signalling. *Nature (Lond.)* **365,** 403–406.

Costa, T., Lang, J., Gless, C., and Herz, A. (1990) Spontaneous association between opioid receptors and GTP-binding proteins in native membranes: specific regulation by antagonists and sodium ions. *Mol. Pharmacol.* **37,** 383–394.

Florio, V. A. and Sternweis, P. C. (1989) Mechanisms of muscarinic receptor action on G$_o$ in reconstituted phospholipid vesicles. *J. Biol. Chem.* **264,** 3909–3915.

Fung, B. K.-K. (1983) Characterization of transducin from bovine retinal rod outer segments: I. Separation and reconstitution of the subunits. *J. Biol. Chem.* **258,** 10,495–10,502.

Gierschik, P., Sidiropoulos, D., Steisslinger M., and Jakobs K. H. (1989) Na$^+$ regulation of formyl peptide receptor-mediated signal transduction in HL60 cells. Evidence that the cation prevents activation of the G protein by unoccupied receptors. *Eur. J. Pharmacol.* **172,** 481–492.

Gierschik, P., Moghtader, R., Straub, C., Dieterich, K., and Jakobs, K. H. (1991) Signal amplification in HL-60 granulocytes: evidence that the chemotactic peptide receptor catalytically activates guanine-nucleotide-binding regulatory proteins in native plasma membranes. *Eur. J. Biochem.* **197,** 725–732.

Gilman, A. G. (1987) G Protein: transducers of receptor-generated signals. *Ann. Rev. Biochem.* **56,** 615–649.

Goodman, R. R., Snyder, S. H., Kuhar, M. J., and Young, W. S. III. (1980) Differentiation of delta and mu opiate receptor localizations by light microscopic autoradiography. *Proc. Natl. Acad. Sci. USA* **77,** 6239–6243.

Hepler, J. R., and Gilman, A. G. (1992) G proteins. *Trends Biochem. Sci.* **17,** 383–387.

Herkenham, M. and Pert, C. B. (1980) *In vitro* autoradiography of opiate receptors in rat brain suggests loci of "opiatergic" pathways. *Proc. Natl. Acad. Sci. USA* **77,** 5532–5536.

Herkenham, M. and Pert, C. B. (1982) Light microscopic localization of brain opiate receptors: a general autoradiographic method which preserves tissue quality. *J. Neurosci.* **2,** 1129–1149.

Herkenham, M., Lynn, A. B., Johnson, M. R., Melvin, L. S., de Costa, B. R. and Rice, K. C. (1991) Characterization and localization of cannabinoid receptors in rat brain: a quantitative *in vitro* autoradiographic study. *J. Neurosci.* **11,** 563–583.

Hildebrandt, J. D., Sekura, R. D., Codina, J., Iyengar, R., Manclark, C. R. and Birnbaumer, L. (1983) Stimulation and inhibition of adenylyl cyclases mediated by distinct regulatory proteins. *Nature (Lond.)* **302,** 706–709.

Hilf, G., Gierschik, P., and Jakobs, K. H. (1989) Muscarinic acetylcholine receptor-stimulated binding of guanosine 5′-*O*-(3-thiotriphosphate) to guanine-nucleotide-binding proteins in cardiac membranes. *Eur. J. Biochem.* **186,** 725–731.

Horstman, D. A., Brandon, S., Wilson, A. L., Guyer, C. A., Cragoe, C. A., and Limbird, L. E. (1990) An aspartate conserved among G protein receptors confers allosteric regulation of alpha(2)-adrenergic receptors by sodium. *J. Biol. Chem.* **265,** 21,590–21,595.

Jansen, E. M., Haycock, D. A., Ward, S. J., and Seybold, V. S. (1992) Distribution of cannabinoid receptors in rat brain determined with aminoalkylindoles. *Brain Res.* **575,** 93–102.

Koski, G. and Klee, W. A. (1981) Opiates inhibit adenylate cyclase by stimulation of GTP hydrolysis. *Proc. Natl. Acad. Sci. USA* **78,** 4185–4189.

Koski, G., Streaty, R. A., and Klee, W. A. (1982) Modulation of sodium-sensitive GTPase by partial opiate agonists. *J. Biol. Chem.* **257,** 14,035–14,040.

Kuhar, M. J. and Yamamura, H. I. (1975) Light autoradiographic localisation of cholinergic muscarinic receptors in rat brain by specific binding of a potent antagonist. *Nature (Lond.)* **253,** 560–561.

Kurose, H., Katada, T., Haga, T., Haga, K., Ichiyama, A., and Uli, M. (1986) Functional interaction of purified muscarinic receptors with purified inhibitory guanine nucleotide regulatory proteins reconstituted in phospholipid vesicles. *J. Biol. Chem.* **261,** 6423–6428.

Lazareno, S., Farries, T., and Birdsall, N. J. M. (1993) Pharmacological characterization of guanine nucleotide exchange reactions in membranes from CHO cells stably transfected with human muscarinic receptors M1–M4. *Life Sci.* **52,** 449–456.

Lorenzen, A., Fuss, M., Vogt, H., and Schwabe, U. (1993) Measurement of guanine nucleotide-binding protein activation by A1 adenosine receptor agonists in bovine brain membranes: stimulation of guanosine-5′-O-(3-[^{35}S]thio)triphosphate binding. *Mol. Pharmacol.* **44,** 115–123.

Meunier, J.-C., Mollereau, C., Toll, L., Suaudeau, C., Moisand, C., Alvinerie, P., Butour, J.-L., Guillemot, J.-C., Ferrara, P., Monsarrat, B., et al. (1995) Isolation and structure of the endogenous agonist of opioid receptor-like ORL$_1$ receptor. *Nature (Lond.)* **377,** 532–535.

Pert, C. B. and Snyder, S. H. (1973) Opiate receptor: demonstration in nervous tissue. *Science* **179,** 1011–1014.

Pert, C. B. and Snyder, S. H. (1974) Opiate receptor binding of agonists and antagonists affected differentially by sodium. *Mol. Pharmacol.* **10,** 868–879.

Pert, C. B., Kuhar, M. J., and Snyder, S. H. (1975) Autoradiographic localization of the opiate receptor in rat brain. *Life Sci.* **16,** 1849–1854.

Reinscheid, R. K., Nothacker, H.-P., Bourson, A., Ardati, A., Henningsen, R. A., Bunzow, J. R., Grandy, D. K., Langen, H., Monsma, F. J. and Civelli, O. (1995) Orphanin, F. Q. A neuropeptide that activates an opioidlike G protein-coupled receptor. *Science* **270,** 792–794.

Selley, D. E. and Bidlack, J. M. (1992) Effects of β-endorphin on *Mu* and *Delta* opioid receptor-coupled G protein activity: low-*Km* GTPase studies. *J. Pharmacol. Exp. Ther.* **263,** 99–104.

Selley, D. E., Breivogel, C. S., and Childers, S. R. (1993) Modification of opioid receptor-G protein function by low pH pretreatment of membranes from NG108-15 cells: increase in opioid agonist efficacy by decreased inactivation of G Protein. *Mol. Pharmacol.* **44,** 731–741.

Sim, L. J., Selley, D. E., and Childers, S. R. (1995) *In vitro* autoradiography of receptor-activated G Protein in rat brain by agonist-stimulated guanylyl 5′-[γ-[^{35}S]thio]-triphosphate binding. *Proc. Natl. Acad. Sci. USA* **92,** 7242–7246.

Sim, L. J., Xiao R., and Childers, S. R. (1996a) Identification of opioid receptor-like (ORL1) peptide-stimulated [^{35}S]GTP$_γ$S binding in rat brain. *NeuroReport,* **7,** 729–733.

Sim, L. J., Xiao, R., and Childers, S. R. (1996b) Differences in G protein activation by mu and delta opioid, and cannabinoid, receptors in rat striation. *Eur. J. Pharmacol.* **307,** 97–105.

Sternweis, P. C. and Robishaw, J. D. (1984) Isolation of two proteins with high affinity for guanine nucleotides from membranes of bovine brain. *J. Biol. Chem.* **259,** 13,806–13,813.

Tang, W.-J. and Gilman, A. G. (1992) Adenylyl cyclases. *Cell* **70,** 869–872.

Taussig, R., Tang W.-J., Hepler, J. R., and Gilman, A. G. (1994) Distinct patterns of bidirectional regulation of mammalian adenylyl cyclases. *J. Biol. Chem.* **1994,** 6093–6100.

Tian, W.-N., Duzic, E., Lanier, S. M., and Deth, R. C. (1994) Determinants of α2-adrenergic receptor activation of G Proteins: evidence for a pre-coupled receptor/G protein state. *Mol. Pharmacol.* **45,** 524–531.

Traynor, J. R. and Nahorski, S. R. (1995) Modulation by μ-opioid agonists of guanosine-5′-O-(3-[^{35}S]thio)triphosphate binding to membranes from human neuroblastoma SH-SY5Y cells. *Mol. Pharmacol.* **47,** 848–854.

Tsai, B. S. and Lefkowitz, R. J. (1978) Agonist-specific effects of monovalent and divalent cations on adenylate cyclase-coupled alpha adrenergic receptors in rabbit platelets. *Mol. Pharmacol.* **14,** 540.

Wieland, T. and Jakobs, K. H. (1994) Measurement of receptor-stimulated guanosine 5′-O-(γ-thio)triphosphate binding by G Proteins. *Methods Enzymol.* **237,** 3–13.

Young, W. S. and Kuhar, M. J. (1979) A new method for receptor autoradiography: [^{3}H]Opioid receptors in rat brain. *Brain Res.* **179,** 255–270.

Small-Molecular-Weight G Proteins

Characterization and Significance

Rajinder Pal Bhullar

Introduction

Cells are capable of responding to extracellular stimuli because of their ability to detect external information and to transduce this signal to intracellular effectors that may result in the generation of second messengers and/or cause alteration in the cell's metabolism. The cell surface receptors are responsible for detecting the external information and a group of proteins, termed GTP-binding proteins, are responsible for receptor-effector coupling. In eukaryotic cells, there are two major families of GTP-binding proteins (G proteins) that participate in the various signal transduction pathways (Bourne et al., 1990, 1991). The first family consists of the heterotrimeric ($\alpha\beta\gamma$-subunit structure) GTP-binding proteins, which link cell surface receptors to effectors inside the cell (Birnbaumer et al., 1990; Neer, 1995; Rens-Domiano and Hamm, 1995). The second family includes GTP-binding proteins that consist of a single polypeptide chain and are commonly referred to as the small-mol-wt G proteins (Grand and Owen, 1991; Valencia et al., 1991; Takai et al., 1992; Bokoch and Der, 1993). This chapter will deal with the characterization and possible function(s) of small-mol-wt G proteins in mammalian cells.

Characterization of Small-Mol-Wt G Proteins

This group includes GTP-binding proteins that consist of a single polypeptide chain and have molecular weights be-

From: *Neuromethods, Vol. 31: G Protein Methods and Protocols*
Ed: R. K. Mishra, G. B. Baker, and A. A. Boulton Humana Press Inc.

tween 20 and 30 kDa. Initially, ras p21, the 21-kDa product of the *ras* oncogene (Barbacid, 1987), and the 20-kDa protein termed ADP-ribosylation factor (ARF) (Kahn and Gilman, 1984) were the best characterized members of this family. Now, six distinct genes coding for the ARF proteins have been cloned and sequenced (Kahn et al., 1991; Boman and Kahn, 1995). However, during the past 10–12 yr, oligonucleotide probes coding for the conserved regions of ras p21 have been used to screen cDNA libraries prepared from a variety of cells and/or tissues (Chardin, 1988). This has resulted in the cloning and sequencing of more than 50 additional genes that can potentially code for small-mol-wt G proteins (Bokoch and Der, 1993). The products of these genes are referred to as the small-mol-wt or ras-related GTP-binding proteins since they all share homology with the ras p21 protein.

Based on their homology with the ras p21 protein, the small-mol-wt G proteins have been divided into four distinct groups (Table 1). The group sharing the greatest degree of identity (~50%) with ras p21 includes genes coding for ral (Chardin and Tavitian, 1986, 1989; Ngsee et al., 1991), rap (Kawata et al., 1988; Pizon et al., 1988a,b), R-ras (Lowe et al., 1987), and TC21 (Drivas et al., 1990) proteins. Small-mol-wt G proteins included in rho (Yeramian et al., 1987; Chardin et al., 1988; Didsbury et al., 1989; Drivas et al., 1990; Munemitsu et al., 1990), rab (Haubruck et al., 1987; Vielh et al., 1989; Zahraoui et al., 1989; Chavrier et al., 1990, 1992; Nagata et al., 1990; Fischer von Mollard, 1994) and others (Kahn and Gilman, 1984; Drivas et al., 1990; Tsuchiya et al., 1991; Bischoff and Ponstingi, 1991; Clark et al., 1993; Nelson et al., 1994) share ~30% identity with the ras p21 protein. The group with the largest number of members is comprised of proteins coded for by the *rab*-related genes (Table 1). The screening of cDNA libraries prepared from a variety of cells and tissues have demonstrated the existence of ~30 *rab*-related genes (Chavrier et al., 1992). It is likely that additional, yet to be identified, genes that can code for small-mol-wt G proteins exist in eukaryotic cells.

Table 1
Mammalian Small-Mol-Wt GTP-Binding Proteins

Ras p21 related	Rho related	Rab related	Others
ralA and ralB	rhoA, -B, and -C	rab1A and -1B	Arf1
rap1A and -1B	rac1 and -2	rab2	Arf2
rap2A and -2B	G25K/CDC42Hs	rab3A–D	Arf3
R-ras	TC10	rab4	Arf4
TC21		rab5	Arf5
			Arf6
		rab30	cp20
		ram	Ran

The majority of the small-mol-wt G proteins are membrane-bound, although often cytosol also contains considerable amounts of these proteins (Bhullar and Haslam, 1987; Bhullar, 1992; Jilkina and Bhullar, 1996). The targeting of small-mol-wt G proteins to the membrane is facilitated by the attachment of a farnesyl or geranylgeranyl group to a cysteine at the fourth position from the C-terminal (Schafer and Rine, 1992; Omer and Gibbs, 1994; Casey, 1995). This step is catalyzed by isoprenyl transferases and the modification is commonly referred to as isoprenylation. Thus, ras p21 protein is farnesylated (Casey et al., 1989; Hancock et al., 1989) and some of the rab (Farnsworth et al., 1994; Wilson and Maltese, 1995), rap (Farrell et al., 1993), and ral (Kinsella et al., 1991; Jilkina and Bhullar, 1996) proteins undergo geranylgeranylation. The farnesyl and geranylgeranyl groups are intermediates in the cholesterol biosynthetic pathway and drugs that inhibit early steps (e.g., lovastatin inhibition of HMG CoA reductase) in this pathway cause the depletion of these intermediates in the cell (Alberts, 1988), resulting in the appearance in cytosol of a normally membrane-bound small-mol-wt G protein because it does not undergo the posttranslational modification necessary for association with the membrane (Hancock et al., 1989). In addition, some of the small-mol-wt G proteins are further modified at an additional C-terminal cysteine by the attachment of fatty acids and/or carboxymethylation, resulting in increased hydrophobicity of the protein and a tighter association with the membrane (Wright and Siegel, 1993). However, the C-terminal posttranslational modifications do not necessarily result in the association of the G protein with the membrane. For example, G25K/CDC42Hs (Maltese and Sheridan, 1990) and some other small-mol-wt G proteins (Bhullar, 1996) in human erythroleukemia cells remain in the cytosol after isoprenylation. In addition, isoprenylation may also promote interaction between small-mol-wt G proteins and other cellular proteins. For example, the isoprenylated form of rab3A preferentially associates with its guanine nucleotide dissociation inhibitor protein (Takahashi et al., 1996). Thus, isoprenylation may have functions in addition to its role in the targeting of proteins to the membrane.

Methods for the Detection of Small-Mol-Wt G Proteins

All GTP-binding proteins have the ability to bind GTP and thus, to detect these proteins in cells/tissues, GTP and/or GTPγS, an analog of GTP, are commonly used. A 25-kDa (G25K/CDC42Hs) protein was purified from human placenta (Evans et al., 1986) and bovine brain (Waldo et al., 1987) by following the binding of GTPγS to proteins. However, this method is time-consuming and requires purification of the protein of interest.

We have developed a method to quickly establish the presence of small-mol-wt G proteins in any type of cell or tissue. This method is termed the "nitrocellulose blot overlay assay" and is based on the ability of some of the small-mol-wt G proteins to renature after their separation using sodium dodecyl sulfate-polyacrylamide gel electrophoresis (SDS-PAGE) and transfer to nitrocellulose (Bhullar and Haslam, 1987; Bhullar, 1992; Jilkina and Bhullar, 1996). Samples containing protein (25 µg for a minigel or 70–100 µg for full-length gel) are precipitated with 10% (w/v) trichloroacetic acid and the protein pellets are solubilized in Laemmli's sample buffer (Laemmli, 1970). After neutralization of residual acid by the addition of 1N NaOH, the samples are heated for 3 min at 100°C and analyzed using SDS-PAGE with 13% (w/v) acrylamide in the separating gel. To transfer onto nitrocellulose, gels containing polypeptides separated using SDS-PAGE are incubated for 30 min in a buffer, pH 8.3, containing 25 mM Tris, 192 mM glycine, 20% methanol (v/v), and 0.05% SDS (w/v). The resolved polypeptides are transferred to nitrocellulose by electroblotting for 16 h at 30 V in the same buffer (Towbin et al., 1979; Bhullar and Haslam, 1987). To detect small-mol-wt G proteins, the nitrocellulose blot is incubated for 30 min at 20°C (room temperature) in a buffer containing 50 mM Tris-HCl, pH 7.5, 0.3% (w/v) Tween 20 (buffer B), and 5% milk powder (Bhullar, 1996). The blot is washed in buffer B (three times, 5 min/wash) and incubated for a further 30 min in buffer B containing 2 µM ATP, 4 µM MgCl$_2$, and 1.0 µCi/mL [α-^{32}P]GTP (final concentration 1 nM). After washing (three times, 5 min/wash) in the same buffer without GTP, the blot is air-dried and bound ^{32}P detected by autoradiography (12–24 h at –70°C). If required, the bound ^{32}P can be quantitated by counting Cerenkov radiation.

The above method was used to detect small-mol-wt G proteins in rabbit and bovine brain. The results presented in Fig. 1 demonstrate the presence of two major proteins of molecular weight 27 and 26 kDa, respectively, in the total rabbit and bovine brain homogenate (Fig. 1, lanes A and D). Further analysis of the cytosolic and particulate fractions demonstrated the presence of a major 26-kDa protein in the cytosol (Fig. 1, lanes B and E) and two major proteins of molecular weights 27 and 26 kDa in the particulate fraction (Fig. 1, lanes C and F). Some investigators have utilized [^{35}S]GTPγS as the probe in the nitrocellulose blot overlay assay (Comerford et al., 1989). However, there does not appear to be any major advantage to using [^{35}S]GTPγS. In fact, it has the disadvantage of being much more expensive than [α-^{32}P]GTP and the time required to detect a signal is also significantly longer. With [α-^{32}P]GTP as the probe, an overnight exposure (with

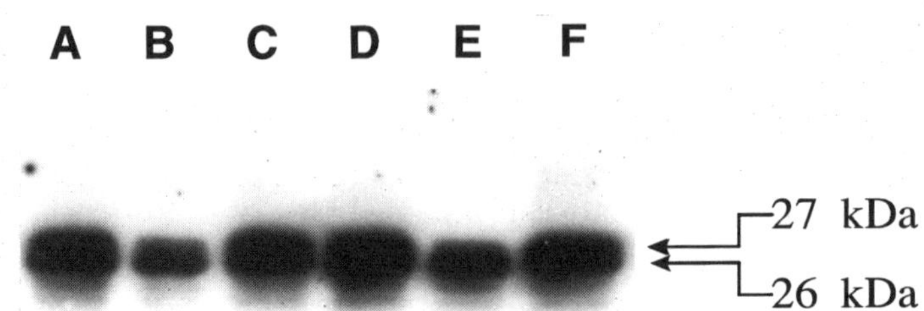

Fig. 1 Detection of small-mol-wt GTP-binding proteins in brain. Samples of protein (100 µg) from rabbit (lane A) and bovine (lane D) brain homogenate, rabbit (lane B) and bovine (lane E) brain 100,000 g_{av} cytosolic fraction, and rabbit (lane C) and bovine (lane F) brain 100,000 g_{av} particulate fraction were resolved using SDS-PAGE and electroblotted onto nitrocellulose. The blot was incubated with [α-^{32}P]GTP and an autoradiograph developed as shown above. Mobility of the major small-mol-wt GTP-binding proteins detected is indicated on the right. Reproduced from Bhullar (1992).

the use of an intensifying screen) is usually sufficient to detect the presence of small-mol-wt G proteins in a sample.

Analysis of Small-Mol Wt G Proteins by Isoelectric Focusing and Two Dimensional Polyacrylamide Gel Electrophoresis

Since up to 50 small-mol-wt G proteins could be present in a cell/tissue and some of these may have identical molecular weight, SDS-PAGE analysis alone is not sufficient to resolve these proteins. The ideal method for separating small-mol-wt G proteins in a cell or tissue sample requires isoelectric focusing in the first dimension followed by SDS-PAGE in the second dimension. The method described in O'Farrell (1975), with minor modifications (Bhullar, 1992), can be used for two-dimensional polyacrylamide gel electrophoresis. Briefly, samples containing protein (~100–200 µg) are precipitated by the addition of trichloroacetic acid to a final concentration of 10% (w/v) and pelleted by centrifugation. The protein pellet is resuspended in buffer (9.5M urea, 2% nonidet-P40, 5% β-mercaptoethanol, and 2% ampholines, pH 3.0–10.0) and loaded on

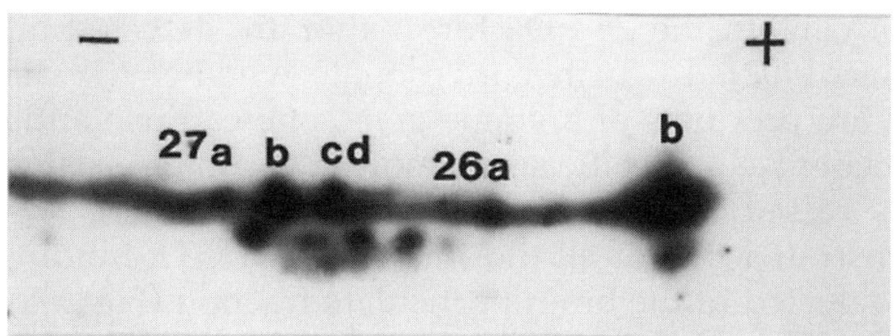

Fig. 2 Analysis of brain particulate fraction by two-dimensional poly-acrylamide gel electrophoresis. Two hundred micrograms of bovine brain particulate proteins were subjected to isoelectric focusing (400 V for 16 h and 800 V for 2 h, pH 3.0–10.0) in the first dimension with protein applied to the high pH (cathode) end of the gel. SDS-PAGE in the second dimension was carried out using 13% (w/v) acrylamide and the gel-blotted on to nitrocellulose. Small-mol-wt GTP-binding proteins were detected by incubating the blot with [α-^{32}P]GTP and an autoradiograph developed as shown above. The major forms of small-mol-wt GTP-binding proteins detected are indicated along with the orientation of the cathode (–) and anode (+) during the isoelectric focusing step. Reproduced from Bhullar (1992).

to a gel containing 4% acrylamide, 9.2M urea, 2% nonidet-P40, and 2% (w/v) pH 3.0–10.0 ampholines. The electrophoresis is carried out at room temperature for 16 h at 400 V followed by 2 h at 800 V using 0.01M H$_3$PO$_4$ in the lower reservoir (anode) and 0.1M NaOH in the upper reservoir (cathode). After iso-electric focusing is complete, gels are extruded from tubes and incubated with shaking for 1 h in Laemmli's sample buffer. At the conclusion of this incubation, gels are subjected to SDS-PAGE in the second dimension and the polypeptides are transferred to nitrocellulose and probed with [α-^{32}P]GTP as described above.

This approach was used to separate small-mol-wt G proteins from the bovine brain particulate fraction. As shown in Fig. 2, the 27- and 26-kDa GTP-binding proteins were separated into four (27a-d) and two (26a-b) forms, respectively. The pattern observed on two-dimensional analysis is usually tissue and/or cell-type specific. For example, human platelets do not contain the 26 kDa forms that are detected in brain (Bhullar, 1992). To further investigate whether the different forms are products of a single gene, one can use antibodies that recognize a specific small-mol-wt G protein. Using antibodies raised against recombinant ralA protein, we have demonstrated that of all the small-mol-wt GTP-binding proteins detected in the brain particulate fraction (Fig. 3A), only the 27a, 27b, and 27c forms are coded for by the *ral* gene(s) (Fig. 3B). The identity of the gene(s) coding for the remaining forms of small-mol-wt G proteins is not known. Antibodies to a variety of small-mol-wt G proteins are now commercially available and can help in the identification of these proteins. To help select the appropriate antibody, an approach that can be used requires determining pI values of the various forms of G proteins. The experimental values can be compared with the theoretical pI values for small-mol-wt G proteins (Huber et al., 1994), thus providing a clue to the potential identity of the unknown protein. To determine the pI of proteins, one gel from the isoelectric focusing step is sliced (0.5 cm/slice) and each gel piece is mixed with 0.5 mL of distilled water. After 1 h incubation, the pH is measured and a standard plot representing gel length vs pH is generated. The standard plot can be used to determine pI of various small-mol-wt G proteins detected using the nitrocellulose blot overlay assay. Using this approach, one can predict that the 26b protein in brain particulate fraction (Fig. 2) is likely to be coded for by the *rab3A* gene (Bhullar, 1992).

Limitations of the Nitrocellulose Blot Overlay Assay

An inherent drawback of the nitrocellulose blot overlay assay is that it will not detect all the small-mol-wt G proteins present in a cell or tissue and it cannot be used to establish the presence of heterotrimeric G proteins. The overlay method works best for small-mol-wt G proteins coded for by the *ral, rab,* and potentially *rap* gene families (Bhullar, 1992). Thus, ras p21 and the remaining ras-related G proteins will not be

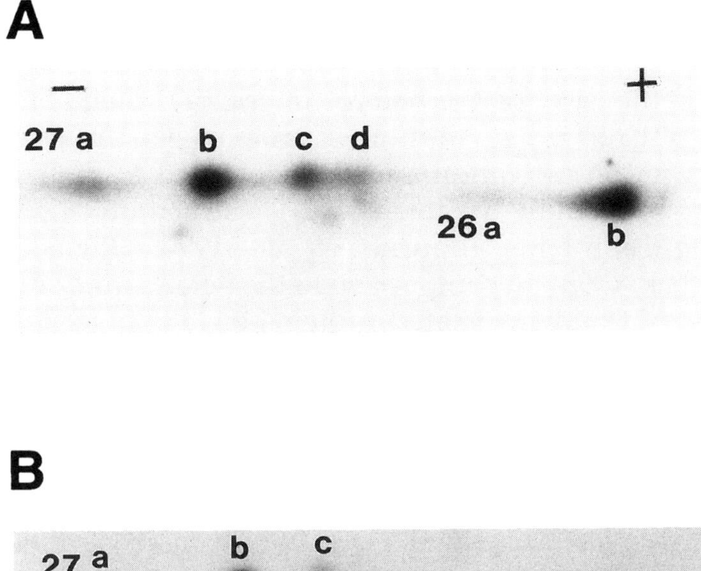

Fig. 3 Identification of ral GTP-binding protein in brain. Bovine brain particulate proteins (100 μg) were resolved using two-dimensional polyacrylamide gel electrophoresis as described in the legend to Fig. 2. One blot (**A**) was incubated with [α-^{32}P]GTP and the second blot (**B**) was probed with antibody against ralA protein followed by ^{125}I-labeled protein A. Autoradiographs are shown above. The major proteins detected with [α-^{32}P]GTP (A) are labeled 27a–d and 26a–b and those recognized by the ralA antibody (B) are labeled 27a–c. The orientation of the cathode (–) and anode (+) during the isoelectric focusing step is indicated. Reproduced from Bhullar (1992).

detected by this method even if they are present in the sample. However, the method is very valuable since products of *ral*, *rab*, and *rap* genes represent a majority of the members of the small-mol-wt G protein family (Table 1). Prolonged exposure of the autoradiograph may detect other additional small-mol-wt G proteins, but background may become a problem. However, incubation of blots in 5% milk powder prior to incubation with [α-^{32}P]GTP greatly reduces the background.

Regulation of the GTPase Activity of Small-Mol-Wt G Proteins

As is the case for the heterotrimeric G proteins, ras p21 and ras p21-related small-mol-wt proteins exist in the GTP-bound form on activation and are converted to the inactive GDP-bound form by the intrinsic GTPase activity associated with these proteins (Bourne et al., 1990; Boguski and McCormick, 1993). In comparison to the heterotrimeric G proteins, ras p21 and ras-related proteins have a lower intrinsic GTPase activity (Bourne et al., 1990). However, it has been shown that the GTPase activity of ras p21 can be stimulated by a cytoplasmic protein, termed the GTPase-activating protein (GAP) (Trahey and McCormick, 1987). The ras p21-specific GAP has been purified from bovine brain cytosol and was shown to consist of a single polypeptide chain of molecular mass 125 kDa (Gibbs et al., 1988). The gene coding for ras-GAP has been cloned and sequenced from a bovine brain cDNA library (Vogel et al., 1989) and the C-terminal region of GAP is believed to be necessary for its catalytic activity (Marshall et al., 1989). Insight into the structure of ras-GAP has led to the identification of another protein, termed neurofibromin, that acts as an activator of the intrinsic GTPase activity of ras p21 (Hattori et al., 1991; Basu et al., 1992). Studies on the regulation of GTPase activity of ras p21 suggested that the function of other small-mol-wt G proteins may be controlled in a similar fashion. Now, specific GAPs have been purified or identified for the rap1A (Nice et al., 1992), rho (Garrett et al., 1991), G25K/CDC42Hs (Barfod et al., 1993), and ral (Emkey et al., 1991; Bhullar and Seneviratne, 1996) proteins. In addition to the GAPs, other accessory proteins also exist that can either enhance (GDP dissociation stimulator [GDS] protein) or inhibit (GDP dissociation inhibitor [GDI] protein) the GTP-GDP exchange on small-mol-wt G proteins (Boguski and McCormick, 1993; Quilliam et al., 1995). In this chapter, methods for the detection of GTPase-activating protein will be discussed.

Methods for the Detection of GTPase-Activating Proteins

The main requirement for the detection of GAP for a specific small-mol-wt G protein is that the protein to be studied should be available in the pure form. This can be accomplished by purifying the protein of interest from a tissue or after its expression in a bacterial/eukaryotic system (Bhullar and Seneviratne, 1996). The method for measuring GAP activity is based on either quantitating the loss of [^{32}P] from the G protein that has been preloaded with [γ-^{32}P]GTP and/or by assessing the formation of [^{32}P]GDP when the protein is preloaded with [α-^{32}P]GTP. In the case of the assay that measures loss of ^{32}P from G protein that has been preloaded with [γ-^{32}P]GTP, the procedure is a modification of the method described in Trahey and McCormick (1987) and Bhullar and Seneviratne (1996). Briefly, in the first step of the assay, G protein (~2 μg) is incubated in a total volume of 100 μL for 15 min at room temperature in buffer containing 20 mM Tris-HCl, pH 7.5, 1 mM DTT, 0.1 mM AMPPNP, 2 mM EDTA, 40 mg/mL BSA, and 0.5 μM [γ-^{32}P]GTP (100–400 Ci/mmol). In the second step, an aliquot (15 ng) of [γ-^{32}P]GTP-loaded G protein is added to a mixture (total vol = 50 μL) containing 20 mM Tris-HCl, pH 7.5, 100 mM NaCl, and 10 mM MgCl$_2$ and incubated for 15 min at room temperature. The reaction is terminated by the addition of 1 mL of ice-cold stopping buffer (20 mM Tris-HCl, pH 7.5, 100 mM NaCl, and 5 mM MgCl$_2$) and the contents quickly filtered through nitrocellulose disks. After washing (five times, 2 mL/wash) with stopping buffer, radioactivity associated with filters is quantitated.

In the method that measures formation of [^{32}P]GDP during the GTPase assay, the G protein is loaded with [α-^{32}]GTP, as described above, and the GTPase assay products are analyzed by thin-layer chromatography (Bhullar and Seneviratne, 1996). To analyze products of the GTPase assay by thin layer chromatography, after filtration and washing, nitrocellulose filters are suspended in 500 μL of 0.1N HCl to extract the guanine nucleotides. The sample is adjusted to pH 7.2 by the addition of 130 μL 0.5*M* Tris. After making the extract 1 m*M* in GTP and GDP, an aliquot is spotted on poly(ethylene)imine cellulose sheets and chromatographed using 1*M* KH_2PO_4 (Kikuchi et al., 1989). The GTP and GDP spots are visualized using autoradiography (12–24 h using an intensifying screen) and radioactivity quantitated.

The first method described above was used to detect the presence in rat brain of GAP activity toward the ralA small-mol-wt G protein. As shown in Fig. 4, a GAP activity was detected in rat brain that stimulated the intrinsic GTPase activity of ralA. Further analysis demonstrated that the majority of the ralA-GAP activity was associated with the particulate fraction (Fig. 4). These results are comparable to those reported in Emkey et al. (1991).

An alternative *in situ* method for detecting GAP activity has also been developed (Manser et al., 1995). In this method, tissue/cellular proteins are separated by SDS-PAGE and the gel is overlaid with nitrocellulose-containing small-mol-wt G protein that has been preloaded with $[\gamma\text{-}^{32}P]GTP$. The advantage of this method is that one can quickly determine the molecular weight and the number of polypeptides that can stimulate the intrinsic GTPase activity of a particular G protein. Thus, multiple GTPase activating proteins for rho and rac proteins have been identified in brain and other tissues (Manser et al., 1995).

Significance of Small-Mol-Wt G Proteins

ras p21 protein has been known to regulate cell proliferation and differentiation (Barbacid, 1987). Mutations that cause a decrease in the intrinsic GTPase activity of ras p21 protein have been associated with the development of cancer (Barbacid, 1987). Recent studies have further elucidated the role of ras p21 in cell proliferation. It has been shown that on occupation of the growth factor receptor and its autophosphorylation, cytoplasmic proteins, termed Shc, Grb2, and Sos, are recruited to the phosphorylated receptor (Egan et al., 1993). The Shc-Grb2-Sos complex links with the ras p21 protein,

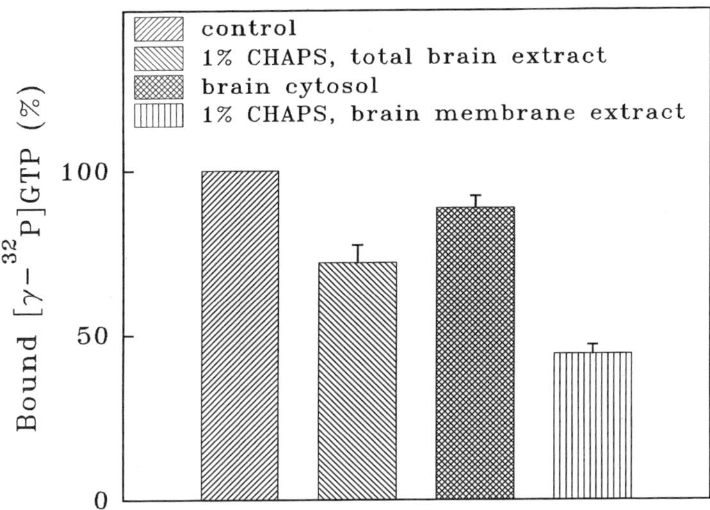

Fig. 4 Detection of ral-GAP activity in bovine brain. To measure GAP activity, proteins (50 µg) from detergent-solubilized total brain, and cytosolic- and detergent-solubilized particulate fraction were added to a mixture containing 15 ng of ralA protein purified after its expression in *E. coli* and loaded with [γ-^{32}P]GTP. After 15 min incubation, the mixture was passed through a nitrocellulose disk and radioactivity associated with the filter was quantitated using liquid scintillation counting. Data represent the average ± S.D. of at least three independent experiments.

causing a GTP for GDP exchange on ras p21 and converting it into the active form (Egan et al., 1993). The activated ras p21 is believed to be the signal for the recruitment of c-raf kinase to the plasma membrane and its activation (Leevers et al., 1994). This results in the initiation of the mitogen-activated protein (MAP) kinase cascade (Marshall, 1994). The MAP kinase can phosphorylate a variety of cellular proteins, including transcription factors, resulting in the increased expression of specific genes (Marshall, 1994).

In contrast, the precise function of other small-mol-wt G proteins in the cell is not well understood. However, several biochemical and genetic studies carried out in yeast have demonstrated a role for small-mol-wt G proteins, ypt1 and SEC4, in intracellular membrane trafficking processes and secretion (Rothman and Orci, 1992). Thus, it has been proposed that small-mol-wt G proteins may carry out similar functions in higher eukaryotes since the ypt1 and SEC4 proteins are not present in mammalian cells (Rothman and Orci, 1992). However, the mammalian homologs of ypt1 and SEC4 are coded for by the *rab* family of genes (Zahraoui et al., 1989). Thus, attention has focused on defining the role of rab proteins in mammalian cells. Initial studies demonstrated the association of these proteins with specific regions of the cell (Zerial and Stenmark, 1993). For example, rab4, rab5, rab7, and rab9 are associated with the endocytic apparatus of the cell (Zerial and Stenmark, 1993).

However, the rab3A protein has received the most attention in regard to regulation of the mammalian exocytic pathway (Lledo et al., 1994). This protein is found in high concentration in the brain and is associated exclusively with neuronal secretory vesicles (Fischer von Mollard et al., 1992; Bielinski et al., 1993) although a small proportion of neurons also express high levels of rab3C (Li et al., 1994). The association of rab3A with neuronal synaptic vesicles has led to the hypothesis that this protein may play an important role in regulating neurotransmitter release (Sudhof, 1995). Initial support for this hypothesis came from the observation that introduction of a peptide corresponding to the effector region of rab3A (a.a. 33–48) into permeabilized cells causes exocytotic fusion of secretory granules in the absence of cytosolic calcium (Oberhauser et al., 1992). Later it was shown that the rab3A protein dissociates from the vesicle membrane after Ca^{2+}-dependent exocytosis, suggesting an association-dissociation cycle during neurotransmitter release (Fischer von Mollard et al., 1991). Additional studies using antibodies have demonstrated that rab3A associated with synaptic vesicles exists in an integral membrane-bound and cytosolic form and that the cytosolic form of rab3 is complexed with GDI, which dissociates once rab3 is in the GTP-bound form (Ullrich et al., 1993). In addition, rab3A and rab3C have also been shown to interact with another protein, termed rabphilin-3A, that is also highly concentrated on synaptic vesicles (Shirataki et al., 1992, 1993; Mizoguchi et al., 1994). Rabphilin is a 80-kDa membrane-bound protein (Shirataki et al., 1992) that has a binding site for rab3A and a Ca^{2+}-phospholipid binding domain (Yamaguchi et al., 1993). It is believed that rab3A-GTP binds to rabphilin-3A, resulting in the docking of a secretory vesicle to the exocytotic site on the membrane. After the docking step, rab3A is released into the cytosol and is potentially ready to recruit another synaptic vesicle for exocytic cycle (Sudhof, 1995). However, recent results, obtained using antibodies highly specific for rab3A, have demonstrated that this protein does not dissociate from synaptic vesicles during calcium-dependent exocytosis (Bielinski et al., 1993). In addition, mice in which the rab3A expression had been abolished and the expression of rabphilin-3A was decreased by 70%, were viable (Geppert et al., 1994). However, in these mice synaptic depression was observed after repetitive stimuli, suggesting that rab3A may not be essential for synaptic vesicle exocytosis but may be important in recruiting synaptic vesicles (Geppert et al., 1994). Thus, the exact mechanism by which rab3A regulates exocytosis from synaptic vesicles is still unclear. In addition, other small-mol-wt G proteins (e.g., ral) are also associated with synaptic vesicles (Bielinski et al., 1993) and may also have an important role in regulating neurotransmitter release.

Conclusions

The small-mol-wt G protein field has seen exponential growth during the last ten years. This has resulted in the realization that a large number of these proteins are expressed in a cell-type and tissue-specific fashion. Furthermore, it has become clear that these proteins participate in a variety of intracellular processes. Although defining the precise function for each of these proteins in the cell is still at a preliminary stage, recently considerable progress has been made in this area and it has become apparent that small-mol-wt G proteins carry out their function in the cell by interacting with other intracellular proteins. However, many questions remain unanswered. For example, it is not clear what signals are required to activate a single species of small-mol-wt G protein since the majority of these proteins are not directly coupled to cell surface receptors. The identity of the effector proteins for each of the small-mol-wt G proteins is still not clear. In addition, a variety of small-mol-wt G proteins appear to regulate identical single effector molecules, whereas others regulate multiple effector molecules. For example, phospholipase D has been shown to be regulated by ARF (Brown et al., 1993), rho, and rac (Siddiqi et al., 1995). In addition, rho, rac, and G25K/CDC42Hs also interact with protein kinases (Manser et al., 1993, 1994; Leung et al., 1995). How does an activated small-mol-wt G protein discriminate between the multiple pathways that are under its regulation? Future research may provide answers to these questions and help define the precise function of each of the small-mol-wt G proteins in cell physiology.

References

Alberts, A. W. (1988) Discovery, biochemistry and biology of lovastatin. *Am. J. Physiol.* **62,** 10J–15J.

Barbacid, M. (1987) Ras genes. *Ann. Rev. Biochem.* **56,** 799–827.

Barfod, E. T., Zheng, Y., Kuang, W.-J., Hart, M. J., Evans, T., Cerione, R. A. and Ashkenazi, A. (1993) Cloning and expression of a human CDC42 GTPase-activating protein reveals a functional SH3-binding domain. *J. Biol. Chem.* **268,** 26,059–26,062.

Basu, T. N., Gutmann, D. H., Fletcher, J. A., Glover, T. W., Collins, F. S. and Downward, J. (1992) Aberrant regulation of ratproteins in malignant tumour cells from type 1 neurofibromatosis patients. *Nature (Lond.)* **356,** 713–715.

Bhullar, R. P. and Haslam, R. J. (1987) Detection of 23–27 kDa GTP-binding proteins in platelets and other cells. *Biochem. J.* **245,** 617–620.

Bhullar, R. P. (1992) Identification of some of the brain Gn27 as the ral gene product: comparison between the brain and platelet Gn proteins. *FEBS Lett.* **298,** 61–64.

Bhullar, R. P. (1996) Expression of a 24 kDa GTP-binding protein (Gn24) is increased in lovastatin treated human erythroleukemia cells. *Mol. Cell. Biochem.* **156,** 59–67.

Bhullar, R. P. and Seneviratne, H. D. (1996) Characterization of human platelet GTPase activating protein for the ral GTP-binding protein. *Biochim. Biophys. Acta* **1311,** 181–188.

Bielinski, D. F., Pyun, H.-Y., Linko-Stentz, K., Macara, I. G., and Fine, R. E. (1993) Ral and rab3A are major GTP-binding proteins of axonal rapid transport and synaptic vesicles and do not redistribute following depolarization stimulated synaptosomal exocytosis. *Biochim. Biophys. Acta* **1151**, 246–256.

Birnbaumer, L., Abramowitz, J., and Brown, A. M. (1990) Receptor-effector coupling by G proteins. *Biochim. Biophys. Acta* **1031**, 163–224.

Bischoff, F. R. and Ponstingi, H. (1991) Mitotic regulator protein RCC1 is complexed with a nuclear ras-related polypeptide. *Proc. Natl. Acad. Sci. USA* **88**, 10,830–10,834.

Boguski, M. S. and McCormick, F. (1993) Proteins regulating *ras* and its relatives. *Nature (Lond.)* **366**, 643–654.

Bokoch, G. M. and Der, C. J. (1993) Emerging concepts in the *ras* superfamily of GTP-binding proteins. *FASEB J.* **7**, 750–759.

Boman, A. L. and Kahn, R. A. (1995) Arf proteins: the membrane traffic police? *Trends Biochem. Sci.* **20**, 147–150.

Bourne, H. R., Sanders, D. A., and McCormick, F. (1990) The GTPase superfamily: conserved structure and molecular mechanism. *Nature (Lond.)* **348**, 125–132.

Bourne, H. R., Sanders, D. A., and McCormick, F. (1991) The GTPase superfamily: a conserved switch for diverse cell functions. *Nature (Lond.)* **349**, 117–127.

Brown, H. A., Gutowski, S., Moomaw, C. R., Slaughter, C., and Sternweis, P. C. (1993) ADP-ribosylation factor, a small GTP-dependent regulatory protein, stimulates phospholipase D activity. *Cell* **75**, 1137–1144.

Casey, P. J., Solski, P. A., Der, C. J., and Buss, J. E. (1989) p21ras is modified by a farnesyl isoprenoid. *Proc. Natl. Acad. Sci. USA* **86**, 8323–8327.

Casey, P. J. (1995) Protein lipidation in cell signaling. *Science* **268**, 221–225.

Chardin, P. (1988) The ras superfamily proteins. *Biochimie* **70**, 865–868.

Chardin, P. and Tavitian, A. (1986) The ral gene: a new ras related gene isolated by the use of a synthetic probe. *EMBO J.* **9**, 2203–2208.

Chardin, P., Madaule, P., and Tavitian, A. (1988) Coding sequence of human rho cDNAs clone 6 and clone 9. *Nucleic Acids Res.* **16**, 2717.

Chardin, P. and Tavitian, A. (1989) The coding sequence of human ralA and ralB. *Nucleic Acids Res.* **17**, 4380.

Chavrier, P., Vingron, M., Sander, C., Simons, K. and Zerial, M. (1990) Molecular cloning of YPT1/SEC4-related cDNAs from an epithelial cell line. *Mol. Cell. Biol.* **10**, 6578–6585.

Chavrier, P., Simons, K., and Zerial, M. (1992) The complexity of the Rab and Rho GTP-binding protein subfamilies revealed by a PCR cloning approach. *Gene* **112**, 261–264.

Clark, J., Moore, L., Krasinskas, A., Way, J., Battey, J., Tamkun, J., and Kahn, R. A. (1993) Selective amplification of additional members of the ADP-ribosylation factor (ARF) family: cloning of additional human and Drosophila ARF-like genes. *Proc. Natl. Acad. Sci. USA* **90,** 8952–8956.

Comerford, I. G., Gibson, J. R., Dawson, A. P., and Gibson, I. (1989) Ras p21 and other Gn proteins are detected in mammalian cell lines by [gamma-^{35}S]GTP gamma S binding. *Biochem. Biophys. Res. Comm.* **159,** 1269–1274.

Didsbury, J., Weber, R. F., Bokoch, G. M., Evans, T., and Snyderman, R. (1989) *rac* a novel ras-related family of proteins that are botulinum toxin substrates. *J. Biol. Chem.* **264,** 16,378–16,382.

Drivas, G. T., Shih, A., Coutavas, E., Rush, M. G., and D'Eustachio, P. (1990) Characterization of four novel ras-like genes expressed in a human teratocarcinoma cell line. *Mol. Cell. Biol.* **10,** 1793–1798.

Egan, S. E., Giddings, B. W., Brooks, M. W., Buday, L., Sizeland, A. M., and Weinberg, R. A. (1993) Association of Sos Ras exchange protein with Grb2 is implicated in tyrosine kinase signal transduction and transformation. *Nature (Lond.)* **363,** 45–51.

Emkey, R., Freedman, S., and Feig, L. (1991) Characterization of a GTPase-activating protein for the ras-related ral protein. *J. Biol. Chem.* **266,** 9703–9706.

Evans, T., Brown, M. L., Fraser, E. D., and Northup, J. K. (1986) Purification of the major GTP-binding proteins from human placental membranes. *J. Biol. Chem.* **261,** 7052–7059.

Farnsworth, C. C., Seabra, M. C., Ericsson, L. H., Gelb, M. H., and Glomset, J. A. (1994) Rab geranylgeranyl transferase catalyzes the geranylgeranylation of adjacent cysteines in the small GTPases Rab1A, Rab3A, and Rab5A. *Proc. Natl. Acad. Sci. USA* **91,** 11,963–11,967.

Farrell, F. X., Yamamoto, K., and Lapetina, E. G. (1993) Prenyl group identification of rap2 proteins: a *ras* superfamily member other than ras that is farnesylated. *Biochem. J.* **289,** 349–355.

Fischer von Mollard, G., Mignery, G. A., Baumert, M., Perin, M. S., Hanson, T. J., Burger, P. M., Jahn, R., and Sudhof, T. C. (1990) rab3 is a small GTP-binding protein exclusively localized to synaptic vesicles. *Proc. Natl. Acad. Sci. USA* **87,** 1988–1992.

Fischer von Mollard, G., Sudhof, T. C., and Jahn, R. (1991) A small GTP-binding protein dissociates from synaptic vesicles during exocytosis. *Nature (Lond.)* **349,** 79–81.

Fischer von Mollard, G., Stahl, B., Khokhlatchev, A., Sudhof, T. C., and Jahn, R. (1994) Rab3C is a synaptic vesicle protein that dissociates from synaptic vesicles after stimulation of exocytosis. *J. Biol. Chem.* **269,** 10,971–10,974.

Garrett, M. D., Major, G. N., Totty, N., and Hall, A. (1991) Purification and N-terminal sequence of the p21rho GTPase-activating protein, rho GAP. *Biochem J.* **276,** 833–836.

Geppert, M., Bolshakov, V. Y., Siegelbaum, S. A., Takei, K., De-Camilli, P., Hammer, R. E., and Sudhof, T. C. (1994) The role of Rab3A in neurotransmitter release. *Nature (Lond.)* **369,** 493–497.

Gibbs, J. B., Schaber, M. D., Allard, W. J., Sigal, I. S., and Scolnick, E. M. (1988) Purification of ras GTPase activating protein from bovine brain. *Proc. Natl. Acad. Sci. USA* **85,** 5026–5030.

Grand, R. J. and Owen, D. (1991) The biochemistry of ras p21. *Biochem. J.* **279,** 635–640.

Hancock, J. F., Magee, A. I., Childs, J. E., and Marshall, C. J. (1989) All ras proteins are polyisoprenylated but only some are palmitoylated. *Cell* **57,** 1167–1177.

Hattori, S., Ohmi, N., Maekawa, M., Hoshino, M., Kawakita, M., and Nakamura, S. (1991) Antibody against neurofibromatosis type 1 gene product reacts with a triton-insoluble GTPase activating protein toward ras p21. *Biochem. Biophys. Res. Commun.* **177,** 83–89.

Haubruck, H., Disela, C., Wagner, P. and Gallwitz, D. (1987) The ras related ypt protein is an ubiquitous eukaryotic protein: isolation and sequence analysis of mouse cDNA clones highly homologous to the yeast YPT1 gene. *EMBO J.* **6,** 4049–4053.

Huber, L. A., Ullrich, O., Takai, Y., Lutcke, A., Dupree, P., Olkkonen, V., Virta, H., Hoop, M. J., Alexander, K., Peter, M., Zerial, M., and Simons, K. (1994) Mapping of Ras-related GTP-binding proteins by GTP overlay following two-dimensional gel electrophoresis. *Proc. Natl. Acad. Sci. USA* **91,** 7874–7878.

Jilkina, O. and Bhullar, R. P. (1996) Generation of antibodies specific for the ralA and ralB GTP-binding proteins and determination of their concentration and distribution in human platelets. *Biochim. Biophys. Acta,* **1314,** 157–166.

Kahn, R. A. and Gilman, A. G. (1984) Purification of a protein cofactor required for ADP ribosylation of the stimulatory regulatory component of adenylate cyclase by cholera toxin. *J. Biol. Chem.* **259,** 6228–6234.

Kahn, R. A., Kern, F. G., Clark, J., Gelmann, E. P., and Rulka, C. (1991) Human ADP-ribosylation factors. A functionally conserved family of GTP-binding proteins. *J. Biol. Chem.* **266,** 2606–2614.

Kawata, M., Matsui, Y., Kondo, J., Hishida, T., Teranishi, Y., and Takai, Y. (1988) A novel small molecular weight GTP-binding protein with the same putative effector domain as the ras proteins in bovine brain membranes. Purification, determination of primary structure, and characterization. *J. Biol. Chem.* **263,** 18,965–18,971.

Kikuchi, A., Sasaki, T., Araki, S., Hata, Y., and Takai, Y. (1989) Purification and characterization from bovine brain cytosol of two GTPase-activating proteins specific for smg p21, a GTP-binding protein having the same effector domain as c-ras p21s. *J. Biol. Chem.* **264,** 9133–9136.

Kinsella, B. T., Erdman, R. A. and Maltese, W. A. (1991) Carboxyl-terminal isoprenylation of ras-related GTP-binding proteins encoded by rac1, rac2, and ralA. *J. Biol. Chem.* **266,** 9786–9794.

Laemmli, U. K. (1970) Cleavage of structural proteins during the assembly of the head of bacteriophage T4. *Nature (Lond.)* **227,** 680–685.

Leevers, S. J., Paterson, H. F., and Marshall, C. J. (1994) Requirement for Ras in Raf activation is overcome by targeting Raf to the plasma membrane. *Nature (Lond.)* **369,** 411–414.

Leung, T., Manser, E., Tan, L., and Lim, L. (1995) A novel serine/threonine kinase binding the ras-related rhoA GTPase which translocates the kinase to peripheral membranes. *J. Biol. Chem.* **270,** 29,051–29,054.

Li, C., Takei, K., Geppert, M., Daniell, L., Stenius, K., Chapman, E. R., Jahn, R., De-Camilli, P., and Sudhof, T. C. (1994) Synaptic targeting of rabphilin-3A, a synaptic vesicle Ca^{2+}/phospholipid-binding protein, depends on rab3A/3C. *Neuron* **13,** 885–898.

Lledo, P. M., Johannes, L., Vernier, P., Zorec, R., Darchen, F., Vincent, J. D., Henry, J. P. and Mason, W. T. (1994) Rab3 proteins: key players in the control of exocytosis. *Trends Neurosci.* **17,** 426–432.

Lowe, D. G., Capon, D. J., Delwart, E., Sakagutchi, A. Y., Naylor, S. L., and Goeddel, D. V. (1987) Structure of the human and murine R-ras genes, novel genes closely related to ras proto-oncogenes. *Cell* **48,** 137–146.

Maltese, W. A. and Sheridan, K. M. (1990) Isoprenoid modification of G25K (Gp), a low molecular mass GTP-binding protein distinct from p21ras. *J. Biol. Chem.* **265,** 17,883–17,890.

Manser, E., Leung, T., Salihuddin, H., Tan, L., and Lim, L. (1993) A non-receptor tyrosine kinase that inhibits the GTPase activity of p21cdc42. *Nature (Lond.)* **363,** 364–367.

Manser, E., Leung, T., Salihuddin, H., Zhao, Z.-S., and Lim, L. (1994) A brain serine/threonine protein kinase activated by Cdc42 and Rac1. *Nature (Lond.)* **367,** 40–46.

Manser, E., Leung, T., and Lim, L. (1995) Identification of GTPase-activating proteins by nitrocellulose overlay assay, in *Methods in Enzymology,* vol. 256 (Balch, W. E., Der, C. J., and Hall, A., eds.), Academic, San Diego, pp. 130–139.

Marshall, M. S., Hill, W. S., Ng, A. S., Vogel, U. S., Schaber, M. D., Scolnick, E. M., Dixon, R. A., Sigal, I. S., and Gibbs, J. B. (1989) A C-terminal domain of GAP is sufficient to stimulate *ras* p21 GTPase activity. *EMBO J.* **8,** 1105–1110.

Marshall, M. S. (1994) Ras target proteins in eukaryotic cells. *FASEB J.* **9,** 1311–1318.

Mizoguchi, A., Yano, Y., Hamaguchi, H., Yanagida, H., Ide, C., Zahraoui, A., Shirataki, H., Sasaki, T., and Takai, Y. (1994) Localization of Rabphilin-3A on the synaptic vesicle. *Biochem. Biophys. Res. Comm.* **202,** 1235–1243.

Munemitsu, S., Innis, M. A., Clark, R., McCormick, F., Ullrich, O., and Polakis, P. (1990) Molecular cloning and expression of a G25K cDNA, the human homolog of the yeast cell cycle gene CDC42. *Mol. Cell. Biol.* **10,** 5977–5982.

Nagata, K., Satoh, T., Itoh, H., Kozawa, T., Okano, Y., Doi, T., Kaziro, Y., and Nozawa, Y. (1990) The ram: a novel low molecular weight GTP-binding protein cDNA from a rat megakaryocyte library. *FEBS Lett.* **275,** 29–32.

Neer, E. J. (1995) Heterotrimeric G proteins: organizers of transmembrane signals. *Cell* **80,** 249–257.

Nelson, T. J., Yoshioka, T., Toyoshima, S., Han, Y. F., and Alkon, D. L. (1994) Characterization of a GTP-binding protein implicated in both memory storage and interorganelle vesicle transport. *Proc. Natl. Acad. Sci. USA* **91,** 9287–9291.

Ngsee, J. K., Elferink, L. A. and Scheller, R. H. (1991) A family of ras-like GTP-binding proteins expressed in electromotor neurons. *J. Biol. Chem.* **266,** 2675–2680.

Nice, E. C., Fabri, L., Hammacher, A., Holden, J., Simpson, R. J., and Burgess, A. W. (1992) The purification of a Rap1 GTPase-activating protein from bovine brain cytosol. *J. Biol. Chem.* **267,** 1546–1553.

Oberhauser, A. F., Monck, J. R., Balch, W. E., and Fernandez, J. M. (1992) Exocytotic fusion is activated by Rab3a peptides. *Nature (Lond.)* **360,** 270–273.

O'Farrell, P. H. (1975) High resolution two-dimensional electrophoresis of proteins. *J. Biol. Chem.* **250,** 4007–4021.

Omer, C. A. and Gibbs, J. B. (1994) Protein prenylation in eukaryotic microorganisms: genetics, biology and biochemistry. *Mol. Microbiol.* **11,** 219–225.

Pizon, V., Chardin, P., Lerosey, I., Oloffsson, B., and Tavitian, A. (1988a) Human cDNAs rapt and rap2 homologous to the Drosophila gene DMS3 encode proteins closely related to ras in the effecter region. *Oncogene* **3,** 201–204.

Pizon, V., Lerosey, I., Chardin, P., and Tavitian, A. (1988b) Nucleotide sequence of a human cDNA encoding a ras-related protein (rap1B). *Nucleic Acids Res.* **16,** 7719.

Quilliam, L. A., Khosravi-Far, R., Huff, S. Y. and Der, C. J. (1995) Guanine nucleotide exchange factors: activators of the ras superfamily of proteins. *Bioessays* **17,** 395–404.

Rens-Domiano, S. and Hamm, H. E. (1995) Structural and functional relationships of heterotrimeric G proteins. *FASEB J.* **9,** 1059–1066.

Rothman, J. E. and Orci, L. (1992) Molecular dissection of the secretory pathway. *Nature (Lond.)* **355,** 409–415.

Schafer, W. R. and Rine, J. (1992) Protein prenylation: genes, enzymes, targets, and functions. *Ann. Rev. Genet.* **30,** 209–237.

Shirataki, H., Kaibuchi, T., Yamaguchi, K., Wada, H., Horiuchi, H., and Takai, Y. (1992) A possible target protein for smg- 25A/rab3A small GTP-binding protein. *J. Biol. Chem.* **267,** 10,946–10,949.

Shirataki, H., Kaibuchi, K., Sakoda, T., Kishida, S., Yamaguchi, T., Wada, K., Miyazaki, M., and Takai, Y. (1993) Rabphilin-3A, a putative target protein for smg p25A/rab3A p25 small GTP-binding protein related to synaptotagmin. *Mol. Cell. Biol.* **13,** 2061–2068.

Siddiqi, A. R., Smith, J. L., Ross, A. H., Qiu, R. G., Symons, M., and Exton, J. H. (1995) Regulation of phospholipase D in HL60 cells. Evidence for a cytosolic phospholipase D. *J. Biol. Chem.* **270,** 8466–8473.

Sudhof, T. C. (1995) The synaptic vesicle cycle: a cascade of protein–protein interactions. *Nature (Lond.)* **375,** 645–653.

Takahashi, K., Sasaki, T., and Takai, Y. (1996) Heterodimer formation of prenylated rab3A with two rabphilin-3A molecules. *Biochem. Biophys. Res. Comm.* **217,** 979–986.

Takai, Y., Kaibuchi, K., Kikuchi, A., and Kawata, M. (1992) Small GTP-binding proteins. *Int. Rev. Cyt.* **133,** 187–230.

Towbin, H., Staehlin, T., and Gordon, J. (1979) Electrophoretic transfer of proteins from polyacrylamide gels to nitrocellulose sheets: procedure and some applications. *Proc. Natl. Acad. Sci. USA* **76,** 4350–4354.

Trahey, M. and McCormick, F. (1987) A cytoplasmic protein stimulates normal N-ras p21 GTPase, but does not affect oncogenic mutants. *Science* **238,** 542–545.

Tsuchiya, M., Price, S. R., Tsai, S. C., Moss, J., and Vaughan, M. (1991) Molecular identification of ADP-ribosylation factor mRNAs and their expression in mammalian cells. *J. Biol. Chem.* **266,** 2772–2777.

Ullrich, O., Stenmark, H., Alexander, K., Huber, L. A., Kaibuchi, K., Sasaki, T., Takai, Y., and Zerial, M. (1993) Rab GDP dissociation inhibitor as a general regulator for the membrane association of ras proteins. *J. Biol. Chem.* **268,** 18,143–18,150.

Valencia, A., Chardin, P., Wittinghofer, A., and Sander, C. (1991) The ras protein family: evolutionary tree and role of conserved amino acids. *Biochemistry* **30,** 4637–4648.

Vielh, E., Touchot, N., Zahraoui, A., and Tavitian, A. (1989) Nucleotide sequence of a rat cDNA: rab1B, encoding a rab1-YPT related protein. *Nucleic Acids Res.* **17,** 1770.

Vogel, U. S., Dixon, R. A., Schaber, M. D., Diehl, R. E., Marshall, M. S., Scolnick, E. M., Sigal, I. S. and Gibbs, J. B. (1989) Cloning of bovine GAP and its interaction with oncogenic ras p21. *Nature (Lond.)* **335,** 90–93.

Waldo, G. L., Evans, T., Fraser, E. D., Northup, J. K., Martin, M. W., and Harden, T. K. (1987) Identification and purification from bovine brain of a guanine-nucleotide-binding protein distinct from Gs, Gi and Go. *Biochem. J.* **246,** 431–439.

Wilson, A. L. and Maltese, W. A. (1995) Coupled translation/prenylation of Rab proteins in vitro, in *Methods in Enzymology,* vol. 250 (Casey, P. J. and Buss, J. E., eds.), Academic, San Diego, pp. 79–91.

Wright, L. S. and Siegel, F. L. (1993) Protein methylation in cerebellar synaptosomes. *J. Neurochem.* **60,** 1475–1482.

Yamaguchi, T., Shirataki, H., Kishida, S., Miyazaki, M., Nishikawa, J., Wada, K., Numata, S., Kaibuchi, K. and Takai, Y. (1993) Two functionally different domains of rabphilin-3A, Rab3A p25/smg p25A-binding and phospholipid- and Ca^{2+}-binding domains. *J. Biol. Chem.* **268,** 27,164–27,170.

Yeramian, P., Chardin, P., Madaule, P., and Tavitian, A. (1987) Nucleotide sequence of human rho cDNA clone 12. *Nucleic Acids Res.* **15,** 1869.

Zahraoui, A., Touchot, N., Chardin, P. and Tavitian, A. (1989) The human Rab genes encode a family of GTP-binding proteins related to yeast YPT1 and SEC4 products involved in secretion. *J. Biol. Chem.* **264,** 12,394–12,401.

Zerial, M. and Stenmark, H. (1993) Rab GTPases in vesicular transport. *Curr. Opin. Cell Biol.* **5,** 613–620.

G Proteins in the Etiology and Treatment of Mental Disorders

Jan O. Gordeladze and Per Wiik Johansen

Clinical Implications of G Proteins

Historical Remarks

Given their ubiquitous and crucial role in the integration and amplification of signal transduction pathways, it is not surprising that anomalies in the expression and/or function of various G proteins has been implicated in a multitude of pathophysiological states: pseudohypoparathyroidism (PsHP), heart failure, endocrine tumors, McCune-Albright syndrome, diabetes, alcoholism, schizophrenia, major valve prolapse, chronic cocaine or opiate ingestion, aging, hypo- or hyperthyroidism, and adrenalectomy or corticosteroid medication (Horn and Bilezkian, 1990; Hoffman and Tabakoff, 1990; Nestler et al., 1990; Kerwin and Beats, 1990; Dewas et al., 1990; Weinstein et al., 1991; Davies et al., 1991; Nomura et al., 1991; Paulssen et al., 1992).

G protein dysfunction appears to represent the primary pathology in Albright's Hereditary Osteodystrophy, McCune-Albright syndrome, and endocrine tumors; in several other conditions, G protein abnormalities are probably secondary, but are, notwithstanding, implicated in the pathophysiology of the condition. Among the examples given above, one may emphasize that in PsHP, distinct inactivating mutations have been identified in the gene encoding $G\alpha_s$ (Spiegel et al., 1992). Furthermore, certain endocrine tumors have been found to

From: *Neuromethods, Vol. 31: G Protein Methods and Protocols*
Ed: R. K. Mishra, G. B. Baker, and A. A. Boulton Humana Press Inc.

harbor activating point mutations of $G\alpha_s$ or $G\alpha_{i2}$ (Patten and Levine, 1990; Weinstein et al., 1990; Vallar et al., 1987; Lyons et al., 1990). More relevant to psychiatry, however, are the observations that: (1) markedly lower than normal lymphocyte $G\alpha_s$ levels were prevalent in subjects with congestive heart failure, and (2) successful treatment with Captopril was associated with an approx 60% restoration of the lymphocytic $G\alpha_s$ levels (Horn and Bilezkian, 1990).

Schizophrenia

Postmortem human brain tissue of schizophrenic patients exhibits a greater adenyl cyclase (AC) responsiveness to fluoride (replaces GTP on the α-subunit) stimulation in the nucleus caudatus and accumbens than in the corresponding brain areas of control subjects (Husseini and Manji, 1992; Memo et al., 1983). It appears that the dopamine D1 receptor might modulate ligand binding to the D2 receptor by modifying the levels of common G protein ($\beta\gamma$) subunits, and it was established that the D1–D2 receptor link was missing in more than 50% of tissue samples from patients with schizophrenia and Huntington's disease (Seeman et al., 1989). Furthermore, PTX-catalyzed ribosylation in the left putamen was some 40% lower in schizophrenic patients than in control subjects (Okada et al., 1990). These findings were compatible with enhanced dopaminergic AC stimulation (Husseini and Manji, 1992). In accordance with these lateralized abnormalities in postreceptor pathways in schizoprenic patients, greater than normal high-affinity forskolin binding was observed in the left parahippocampal gyrus and CA1 region in postmortem brains of schizophrenic patients (Kerwin and Beats, 1990). Since high-affinity forskolin binding reflects binding to the active form of the $G\alpha_s$–AC complex (Spiegel et al., 1992), higher than normal levels of free $G\alpha_s$ and/or AC should prevail. This may be owing to higher intrinsic $G\alpha_s$ and/or AC levels (Memo et al., 1983), a lower amount of $\beta\gamma$-subunits (Spiegel et al., 1992), a diminished PTX-catalyzed ribosylation of G_i/G_o (Memo et al., 1983), or a minimal or absent D1–D2 receptor link (Seeman et al., 1989) in the schizophrenic brain.

PTX-ribosylated G proteins ($G\alpha_i$ and/or $G\alpha_o$) are in fact reduced on the left hippocampal side (Okada et al., 1991), a finding that corresponds with abnormalities of the ventricular system. However, only $G\alpha_o$, not $G\alpha_{i2}$, measured on immunoblots was diminished in the hippocampus and caudate nucleus of the right hemisphere (Okada et al., 1994). Finally, in the left temporal cortex, immunodetected $G\alpha_i$ and $G\alpha_o$ were reduced, whereas concomitant cyclic AMP (cAMP) binding was enhanced (Nishino et al., 1993a). These findings implicate $G\alpha_{i1\ \&\ o}$ in the pathogenesis of schizophrenia.

Affective Disorders

Hyperactivity of several G proteins is observed in untreated manic patients. Antibipolar treatments (lithium, carbamazepine or electroconvulsive therapy [ECT]) attenuated both receptor-coupled mononuclear leukocyte G_s and non-G_s ($G_{i\ \&\ o}$) protein function; in contrast, only G_s activity was inhibited by antidepressant drugs (tricyclics or MAO inhibitors) (Avissar and Schreiber, 1992). An intact β-adrenoceptor system is necessary to demonstrate these effects, so it is probable that attenuation of β-adrenoceptor–G_s coupling may constitute the mechanism underlying their therapeutic antidepressant effect (Spiegel et al., 1992; Husseini and Manji, 1992; Avissar and Schreiber, 1992). In contrast, muscarinic receptor-coupled G proteins ($G_{i\ \&\ o}$) may be the site where the intervention in bipolar affective disorders is exerted (Avissar and Schreiber, 1992). A model taking G_s, G_{i2}, and G_{i1} into account, with protein kinase A (PKA) and protein kinase C (PKC) exerting negative reciprocal control on their respective G protein levels, shows that hyperactivity of the G proteins in question will lead to an unstable situation, with either supra- or subactivation of PKA, accounting for both depression and a hyperactive state. When litium is applied, G protein hyperfunction is reduced, and the PKA lability is obliterated (Avissar and Schreiber, 1992).

The levels of $G\alpha_s$ in postmortem brain tissues from manic depressive subjects have been reported to be significantly higher than those of control subjects (Young et al., 1991). Furthermore, hyperfunction of G proteins in manic subjects, as determined by isoproterenol-stimulated GTPγS binding, was demonstrated in leukocytes (Schreiber et al., 1991). However, abnormalities in lymphocyte β-adrenergic receptor and platelet α2-adrenergic receptor sensitivity in depressed patients (Pandey et al., 1990) are difficult to interpret in light of aberrations in circulating catecholamines and glucocorticoids in depression. For example, significantly lower lymphocyte isoproterenol-elicited cAMP accumulation was observed only in a subgroup of depressed patients with psychomotor agitation (Mann et al., 1985; Husseini and Manji, 1992). Similarly, a significant correlation between terminal insomnia and blunted lymphocytic cAMP accumulation in depressed patients has been observed (Ebstein et al., 1988). Finally, significant correlations between urinary measures of noradrenaline (NA) release and basal and postreceptor stimulated AC activities in both lymphocytes and platelets have been demonstrated (Husseini and Manji, 1992). It is presently uncertain whether these abnormalities in receptor and postreceptor susceptibilities are of primary origin or represent the sequelae of abnormalities in circulating levels of catecholamines and/or stress hormones.

In platelets, PTX-ribosylated G proteins appear unaltered in affective disorders (i.e., major depression) (Odagaki et al., 1994). However, both mononuclear leukocyte G_s and G_i are increased in bipolar patients (unaffected in major depression) (Young et al., 1994). Furthermore, in the frontal cortex of suicide victims with a history of depression, G_s and G_i were unaltered, but the AC activity (basal as well as stimulated) and the amount of the short form of G_s were increased (Cowburn et al., 1994). Finally, in the temporal and parietal cortex of depressive patients, $G_{i\ \&\ o}$ was increased and the $G_s/G_{i\ \&\ o}$

ratio was reduced (as evidenced by photoaffinity labeling) (Ozawa et al., 1993).

Aging: Parkinson's Disease and Alzheimer's Disease

With increased age, PTX and CTX ribosylation are enhanced in polymorphonuclear leukocytes (PMNLS), whereas in lymphocytes, PTX ribosylation is increased (however, G_o is selectively diminished), and CTX ribosylation is reduced. There is also a shift in the size of the preferred substrates of the ADP-ribosyl transferase (Fülöp et al., 1993). Also in other species, such as the fruit fly, senescence is associated with a reduction in G_o (Perez-Baun et al., 1994). In Parkinson patients, dopamine (DA)-sensitive AC appears to be increased in the putamen of treated individuals. Furthermore, PKA and PKC activities are normal in nondemented subjects. In contrast, steady-state AC, PKA, and PKC activities are reduced in the striatum of demented Parkinson patients. Surprisingly, in all subjects tested, both D1- and D2-receptor numbers in the putamen are normal (Nishino et al., 1993b). In Alzheimer's disease, $G\alpha_s$ mRNA levels are increased in hippocampal CA1, CA3, and CA4 regions as well as in the cortex. The long form of $G\alpha_s$ is increased, whereas the short form is reduced in the hippocampus, leading to a reduction of the G_s-elicited AC stimulation (Harrison et al., 1991).

As previously indicated, G_o and G_{i1} are the most abundant G proteins in brain tissue (Gierschik et al., 1986; Spiegel, 1990; Fülöp and Ildiko, 1994). Many G protein-coupled receptors and a variety of effectors, including AC, certain Ca^{2+} and K^+ channels, and phospholipase C (PLC) subtypes are regulated by one or more PTX-sensitive G proteins (Gilman, 1987; Milligan, 1993). As shown for several diseases, genetic changes in G protein expression and function can profoundly alter signal transduction (Spiegel et al., 1992). G proteins may also represent an important site for action of various drugs (e.g., glucocortioids) (Sitoh et al., 1989). There appears to be an age-related decrement in muscarinic receptor-stimulated low

K_m GTPase in the striatum and hippocampus of senescent subjects (Yamagami et al., 1992). These findings suggest that there may be an age-dependent decline in muscarinic receptor G protein coupling/uncoupling (Fülöp et al., 1993). Compared with young adults, adrenalin (A)-stimulated Ca^{2+} efflux and inositol triphosphate (IP_3) production was reduced by some 30 and 35%, respectively, in parotid cell membranes of old individuals. Moreover, A was much more effective in stimulating G protein GTPase in similar membrane fractions from young adults than old subjects. These data suggest that age-related impairments in α1-adrenergic responsiveness are mediated at least in part by alterations in the coupling of G proteins with α1-adrenergic receptors (Spiegel et al., 1993).

A disruption of the phosphoinositide second messenger system with age could be related to the impairment of neurological responsiveness and the behavioral deficits observed with aging. Carbachol-stimulated release of total inositol phosphates has been recently found to increase with age in hippocampal and cerebral cortex slices (Bartus et al., 1982; Mitchell et al., 1982; Nalepa et al., 1989; Mundy et al., 1991). In addition, NA-stimulated release of total inositol phosphates is also elevated with age in cerebral cortex slices (Nalepa et al., 1989). These effects occur in the absence of changes in the number of binding sites (Mundy et al., 1991).

A decrease in specific inositol phosphate release was reported after stimulation of cortical α1-adrenergic receptors, cortical muscarinic receptors and striatal muscarinic receptors (Mundy et al., 1991). In addition to age-related effects on agonist-stimulated IP_3 release, carbachol-stimulated inositol (1,2,4,5)-tetra phosphate (IP_4) formation was reduced in cerebral cortex slices of old individuals (Haba et al., 1988; Joseph et al., 1990; Pontzer et al., 1990). Concomitantly, there was a loss of muscarinic enhancement of K^+-evoked release of DA. This decreased agonist susceptibility was associated with lower IP_3 accumulation (Joseph et al., 1990). Age-related changes in receptor-stimulated inositol phosphate release may depend on receptor–effector coupling, or on a change in the stoichiometry of the receptor–G protein–PLC interaction. Several arguments support this latter contention, since the same phenomenon was also found in other cell systems, such as myocytes, lymphocytes, and granulocytes (Moscona Amir et al., 1989; Lipschitz et al., 1991; Fülöp et al., 1992).

An age-related uncoupling of the receptor and G protein PLC interaction could be caused by changes in membrane properties, such as membrane fluidity, and/or may be the result of two other plausible factors, e.g., reduction in receptor levels (e.g., M_2 or possibly M_1) and/or upstream deficits in the signal transduction processes (Springer et al., 1987; Joseph et al., 1988). Since both the AC and PLC holoenzymes are modulated by the ratio of G_s to G_{i2} (G_s/G_{i2}) and $G_{q/11}$ to G_o ($G_{q/11}/G_o$) (Gordeladze et al., 1994; Johansen et al., 1996), such deficits may in principle cause either enhanced hormonal activation or inhibition.

In general, the cholesterol content of brain membranes has been reported to be elevated with aging. Concomitantly, it seems that the phospholipid content in CNS membranes does not alter at any age (Calderini et al., 1983; Bothmer and Jolles, 1994). The ratio of CNS cholesterol to phospholipid is hence reduced (by some 30%), indicating that receptor–effector coupling may be perturbed (Roth et al., 1996). Both altered cholesterol content and its transbilayer distribution in nervous membranes of the aged may account for the general loss of α1-adrenergic receptors and receptor-PLC coupling with increasing age. Fortunately, it seems that these changes may be modulated by dietary interventions (Heron et al., 1981; Neelands and Clanidinin, 1983).

Psychotropic Drug Tolerance and Dependence

Enkephalin, β-endorphin, and dynorphin neurons are important in the development of drug tolerance and dependence, and their opiate receptors are all uncoupled from corresponding G proteins on overstimulation (Trujillo and Akil, 1991). In line with this concept, human μ-opoid receptors (which bind morphine, codeine, methadone, and fenantyl), when transfected into COS-7 cells, couple to G protein-activated K^+ channels (Mestek et al., 1995). The receptor–effector coupling was obliterated by homologous desensitization, a phenomenon that is potentiated by PKC or CaCaM kinase (Mestek et al., 1995).

A plausible model for acquisition of dependence, describing both latent and expressed dependence, is associated with alterations on three levels (Meunier, 1992), involving paragigantocellular (PGi) neurons, locus cerulus (LC), and the cerebral cortex/limbic system (CC/LS). Aberration in the integration of signals at the level of LC leads to enhanced NA secretion, chronically impinging on the CC/LS (Meunier, 1992). A variety (i.e., μ, δ, and κ types) of opiate receptors are coupled to their effectors (AC, K^+ and Ca^{2+} channels) by G proteins (Meunier, 1992). PTX has been shown to obliterate opiate-inhibited AC activation in cultured cells and LC neurons (Birnbaumer, 1990; Husseini and Manji, 1992). Moreover, it has been shown that alterations in G protein levels are temporally correlated with neuronal activity in the LC and with the behavioral manifestations (in the CC/LS) of the morphine withdrawal syndrome (Nestler et al., 1989; Rasmussen et al., 1990). These results suggest that alterations in the levels of G proteins may underlie the enhancement of neuronal excitability (and possibly withdrawal symptoms) observed with abrupt opiate discontinuation (Husseini and Manji, 1992). In vitro studies of cultured neurons from spinal cord dorsal root ganglia have demonstrated that exposure to κ-opiate agonists are accompanied by an approx 70% decrease in levels of $G\alpha_i$ and the absence of alterations in levels of $G\alpha_s$-, $G\alpha_o$-, and $G\beta$-subunits (Attali and Vogel, 1989). In contrast, rats subjected to chronic cocaine administration demonstrated low levels of $G\alpha_{i\,\&\,o}$ in the ventral tegmental area, nucleus accumbens, and LC (Nestler et al., 1990).

Alcoholism

The evidence for genetic factors in alcoholism is compelling (Shuckie, 1978). Therefore, it is not surprising that considerable effort has been made to reveal biological factors that may predispose individuals to alcoholism. A variety of putative eletrophysiologic, biochemical, and neuroendocrine markers have been studied, but research has now focused on G proteins, not only as the mediators of the biochemical effects of alcohol, but also as possible sites of underlying pathophysiology in individuals predisposed to the development of alcoholism (Hoffman and Tabakoff, 1990; Husseini and Manji, 1992). Subjects with a positive family history (FHPositive) of alcoholism exhibit increased steady-state levels of $G\alpha_s$ (as evidenced by CTX-ribosylation and immunoblot analyses) in both platelets, erythrocytes, and lymphocytes (Saito et al., 1994; Wand et al., 1994). Thus, $G\alpha_s$ may serve as a marker for the predisposal or development of future alcohol abuse.

Ethanol can profoundly affect the G protein AC signal transduction pathway (Hoffman and Tabakoff, 1990). Chronic ethanol ingestion or in vitro exposure results in a desensitization of cAMP production in a variety of tissues, including brain (Hoffman and Tabakoff, 1990; Nagy et al., 1989). Chronic exposure of cultured cell lines to ethanol results in a >30% decrease in both the $G\alpha_s$ protein and its mRNA levels (Mochly-Rosen et al., 1988; Charness et al., 1988). Incubation of neuroglial (NG) 108-15 cells with ethanol reduces steady-state levels of $G\alpha_s$ and $G\alpha_{i2}$, whereas $G\alpha_o$ is enhanced. As a consequence, both basal and agonost-elicited cAMP synthesis are enhanced. These effects are not mimicked by incubation with adenosine (Williams et al., 1993). In mice chronically fed ethanol, there is a reduction in the CTX-catalyzed ADP ribosylation of $G\alpha_s$ and a reduction of high-affinity forskolin binding (Hoffman and Tabakoff, 1990). One particular study showed that basal and adenosine-receptor-stimulated cAMP levels in fresh lymphocytes in alcoholic subjects were some 75% lower than those of normal subjects or of individuals with nonalcoholic liver disease (Diamond et al., 1987).

Cultured lymphocytes from alcoholics, grown in the absence of alcohol, also show some abnormalities in the cAMP signal transduction pathway (Nagy et al., 1989). In the same manner, platelet membranes of alcoholics reportedly abstinent from alcohol for 12–48 mo demonstrate lower receptor- and G_s-stimulated AC activity than do platelet membranes of age- and sex-matched control subjects (Tabakoff et al., 1988). These findings are supported by studies of CTX- and PTX-catalyzed ADP ribosylation of G_s and G_i, respectively, in platelet membranes from abstinent alcoholics, subjects with concomitant diagnosis of major depression and alcoholism (major depression as the primary diagnosis), and age- and sex-matched comparison subjects (Husseini and Manji, 1992). However, patients demonstrating abstinence symptomatology display enhanced lymphocyte $G\alpha_{i2}$ levels, whereas basal, PGE_1- and $GTP\gamma S$-elicited AC activities were all markedly decreased (Waltman et al., 1993).

Aluminum (Al) Intoxication of the CNS

Neurochemical responses to chronic Al administration have been studied (Jope and Johnson, 1992). Al was added to drinking water of adult rats for 4 wk or longer, and weanling rats were given Al for 8 wk. Selective cognitive impairment was demonstrated in the adult rats (Connor et al., 1988). Al inhibited IP_3 production and Ca^{2+} fluxes (Koenig and Jope, 1978). In weanling rats, Al decreased the in vivo concentration of IP_3 in the hippocampus (Jope and Johnson, 1992). An elevation of cAMP concentrations by some 50% in various brain regions in adult and weanling rats was also found. Al enhanced agonist-stimulated, but not basal, cAMP production in vitro (Koenig and Jope, 1978; Connor et al., 1988; Singer et al., 1990; Jope and Johnson, 1992). It was therefore postulated that Al inhibits the GTPase activity of $G\alpha_s$, leading to prolonged activation of G_s after receptor stimulation. Similar effects on both AC and PLC activities have also been demonstrated in bone cells (Gordeladze et al., 1993; Jablonski et al., 1996), but long-term, high-dose Al exposure led to attenuated hormonal stimulation of both signaling systems.

Rats treated with Al also showed increased phosphorylation and immunodetectable levels of microtubule-associated protein 2 (MAP-2) and the neurofilament protein NF-H (Johnson et al., 1990; Jope and Johnson, 1992). These studies indicate that chronic oral Al administration to rats has significant neurochemical consequences. Three sites of actions are implicated: altered Ca^{2+} homeostasis, enhanced cAMP production, and changes in cytoskeletal protein phosphorylation states and concentrations (Jope and Johnson, 1992).

Other Neuropsychiatric Conditions

Subnormal levels of platelet basal and postreceptor-stimulated AC activation have been reported in both panic and posttraumatic stress disorders (Charney et al., 1989; Lerer et al., 1990). However, it is highly probable that these platelet anomalies are secondary to aberrations of plasma catecholamines and/or stress hormones. Another illness with a constellation of symptoms with panic disorders is a form of hyperadrenergic disautonomia associated with mitral valve prolapse (Davies et al., 1987). Previous studies demonstrated that cyc⁻ lymphoma cell membranes reconstituted with erythrocyte G_s from patients with mitral valve prolapse are hyperresponsive with respect to both AC activation and receptor coupling (Davies et al., 1991). If a similar CNS-located G_s abnormality is present in panic disorder (i.e., desensitisation of β-adrenergic receptor-coupled AC), it would support the choice of either tricyclic antidepressants or monoamine oxidase inhibitors to treat the condition successfully (Husseini and Manji, 1992).

Psychotropic Drug Targeting

Lithium

Ample evidence suggests that lithium (Li) exerts substantial effects on second messenger-generating systems. Li has been shown to attenuate receptor- and postreceptor-stimulated AC (Mork and Geisler, 1989a, b; Manji et al., 1991; Risby et al., 1991). In contrast to many antidepressants, Li has been reported to induce few, if any, changes in the density of brain β-adrenergic receptors. Li also attenuates receptor-stimulated phosphoinositide turnover in brain and platelets (Risby et al., 1991). Since Li affects both phosphoinositide turnover and AC activity, recent attention has focused on mechanisms shared by these two signal transduction systems, namely G proteins. Li has been reported to attenuate markedly the agonist-induced increase in GTP binding in response to both isoproterenol and carbachol (Avissar et al., 1988). Furthermore, a greater isoproterenol-induced GTPγ_s binding was observed in leukocytes from untreated manic patients than in normal subjects or euthymic bipolar patients stabilized with Li (Schreiber et al., 1991). Subsequent to 2 wk of Li administration to normal volunteers, their platelets showed no alterations in the immunolabeling of G_s, G_i, or G_o, but there was an approx 35% increase in PTX-catalyzed ADP ribosylation in platelet membranes (Husseini and Manji, 1992). Since the undissociated inactive αβγ-heterotrimeric form of G_i is the substrate for PTX, the results suggest that Li inactivates G_i by stabilizing the inactive, undissociated conformation (Husseini and Manji, 1992).

A summary of more recent data show that in depressed patients receiving clomipramine (CMI) plus Li, forskolin-elicited AC activity inhibited by serotonin (5-hydroxytryptamine [5-HT]), via 5-HT1A receptors, is counteracted (Newman et al., 1992). Furthermore, Li administration to adult rats yielded a reduction in neostriatal D1 & 2-receptor G protein coupling, resulting in diminished D1-dopaminergic enhancement of AC activity (Carli et al., 1994). Moreover, Li administration also enhanced calpain (which is a Ca^{2+}-activated protease linked to neuronal plasticity) -mediated proteolysis of α_o ($\alpha_o\beta\gamma$) (Greenwood and Jope, 1994). Finally, in platelets and leukocytes of subjects with manic bipolar affective disorder, $G\alpha_s$ was enhanced, but $G\alpha_{i,o,q/11}$ remained unaffected. Li ingestion reduced both PTX-ribosylated G proteins and $G\alpha_{q/11}$ (Manji et al., 1995).

Since Li has been shown to inactivate the IP-phosphatases necessary for the resynthesis of PIP_2 from inositol

(Berridge and Irvine, 1989) and thereby reduces PLC substrate availability, Li probably obliterates the effect of PKC and Ca^{2+} on G protein activity (phosphorylation status) and gene expression in the CNS.

Neurotransmitter Receptor Intervention

The D2 receptors and associated G proteins ($G_{i2 \& 3}$ and G_o) (Corrette et al., 1995) have been implicated in the inhibition of neuroendocrine secretion. Of major clinical importance, however, is the observation that bromocriptine-resistant pituitary tumors demonstrate negligible $G\alpha_o$ expression (Bouvier et al., 1991). Recently, reduced $G\alpha_{i(2)}$ binding to D2 receptors was observed in the caudate nucleus of schizophrenic patients (Sumiyoshi et al., 1995). This indicates that $\alpha_{i2}\beta\gamma$ is frozen in the associated state, i.e., D2 receptors are dissociated, and free $\beta\gamma$ complex is scarce. Furthermore, it was demonstrated that measurable amounts of the D4 receptor in the caudate nucleus were more frequent in patients with schizophrenia than in normal controls or subjects with major depression (Sumiyoshi et al., 1995). As asserted earlier, it is highly probable that the single alteration in $\alpha_{i2}\beta\gamma$ levels most profoundly affects signaling pathways other than the AC system. This phenomenon underscores the need to locate the "primary" molecular lesion before instigating any kind of drug therapy.

Although there is some controversy over the functional importance of many of the 5-HT receptor subtypes, there is evidence that they fall into two major groups according to the nature of their coupling to secondary messengers or ion channels. Thus, the 5-HT1 and 5-HT2 receptors appear to occupy the G protein receptor subfamily, which may be coupled to either AC (5-HT1) or PLC (5-HT2). The central morphine (M) receptors (now known as 5-HT3 receptors) appear to occupy a ligand-gated ion channel superfamily (Leonard, 1994). The 5-HT receptor subtypes are implicated in thermoregulation, modulation of cardiovascular function, eating disorders, sleep, sexual activity, anxiety states, aggression, schizophrenia, and depression (Leonard, 1994).

Patients with a major depressive disorder exhibit increased basal cortisol secretion associated with decreased hypothermic and ACTH/cortisol responses. The attenuated neuroendocrine and thermoregulatory response to serotoninergic 5-HT1A receptor activation may reflect a glucocorticoid-dependent feedback inhibition of the hypothalamic–pituitary–adrenal (HPA) system and subsensitivity of the presynaptic 5-HT1A–G_i–AC complex activity (Lesch and Lerer, 1991). Differential regulation of 5-HT1A and 5-HT2 function leading to a relative 5-HT2–G_o–PLC complex supersensitivity may maintain HPA hyperactivity during the course of depression. These findings corroborate recent reports that guanyl cyclase (GC), via gene transcription, modifies (hamper) the expression of G_1 rather than G_o, rendering the G_o-mediated system hyperactive (Lesch and Lerer, 1991).

Most 5-HT1 receptor subtypes inhibit AC, whereas the 5-HT1C subclass activates PLC (like the 5-HT2 type). 5-HT4 receptors stimulate AC. Buspirone, a partial 5-HT1A agonist, is used in the treatment of generalized anxiety. Anxiolytic properties have also been ascribed to the 5-HT2 and 5-HT3 receptor antagonists. 5-HT3 antagonists also exert antipsychotic properties (Moulignier, 1994).

γ-Aminobutyric acid (GABA) receptors have long been associated with the use of benzodiazepines to treat anxiety states and insomnia (Costa, 1983), but their association with G proteins has not been extensively studied. Activation of $GABA_A$ receptors on melanotrophs is associated with an increase in membrane Cl^- conductance. Furthermore, GABA (through the $GABA_B$ receptors) inhibits nerve injury during anoxia, a process that is mediated via both PTX-sensitive G proteins and $G\alpha_{q/11}$ (i.e., activation of PLC and subsequently PKC) (Fern et al., 1995). Finally, concerted G protein activation through both $GABA_A$ and $GABA_B$ receptors is needed to activate K^+ channels and elicit electric discharges in the CNS (Destexhe and Sejnowski, 1995). Spillover of GABA may account for differences between inhibitory responses in the hippocampus and thalamus, providing insight into the functions of $GABA_B$ receptors in normal and epileptic discharges (Destexhe and Sejnowski, 1995).

Pharmacological and molecular studies have demonstrated two classes of excitatory amino acid or glutamate receptors in the mammalian CNS, ionotropic glutamate receptors (iGluRs) are ligand-gated integral ion channels that exist as heterotrimeric protein complexes. Metabotropic glutamate receptors (mGluRs) are coupled to G proteins. Characterization of mGluRs as a unique receptor class has involved PIP_2 hydrolysis, cAMP generation, Ca^{2+} mobilization, and arachidonic acid (AA) release (through subtypes 1α, 1β, and 5, respectively) (Schoepp and Conn, 1994). Contrastingly, mGluRs of subtypes 2–4 inhibit cAMP production. The physiological roles of these receptors have been examined in the hippocampus, a cortical structure involved in learning and memory and various neurological diseases like Alzheimer's disease and temporal lobe epilepsy (Schoepp and Conn, 1994).

Antidepressants

In spite of extensive research, the undisputed molecular mechanisms of antidepressant drugs have not yet been established. The most commonly observed effect is a reduction in the NA-stimulated production of cAMP (desensitization) accompanied by a reduction in the number of β-adrenergic receptors (downregulation) (Sulser, 1984). Desipramine appears to cause a functional uncoupling of the β-receptor from G_s in the cortex, and interferes with the high-affinity state breakdown and subsequent activation of AC, probably by decreasing the affinity of G_s for guanine nucleotides (Tsuchiya et al., 1988; Manji et al., 1991b). It has been reported that PTX treatment of rats overcomes desipramine-induced β-receptor desensitization without attenuating desipramine-induced β-receptor downregulation. These data suggest that G_i and/or

G_o is involved in the effect of desipramine. It is also possible that the removal of the inhibitory tone by PTX simply masks the desensitizing effects of desipramine (Husseini and Manji, 1992).

In the same manner as antidepressants, electroconvulsive shock (ECS) desensitized the β-adrenergic-mediated AC activity in rat cortex and hippocampus (Sulser, 1984). However, repeated ECS was shown to decrease both agonist-induced GTP and forskolin binding in the cortex (Avissar et al., 1990; Nishida et al., 1990). Multiple ECS therapy was also reported to decrease $GTP\gamma_s$ binding in the prefrontal cortex and hippocampus; it had no effect on the striatum, but it increased $GTP\gamma_s$ binding in the amygdala (Deckert et al., 1989).

Summary and Future Perspectives

G proteins serve as junctions between several signaling systems. Hence, extensive aberrations may occur in the CNS owing to alterations of steady-state levels and intrinsic activity, and receptor–effector coupling kinetics. To date, no psychiatric disease has been primarily linked to G protein anomalies. However, several dysfunctions have been associated with altered G protein activity.

Attempts to treat psychiatric disorders have been directed toward normalization of neurotransmitter activity through rectification at the receptor level. This approach is more often than not associated with extensive adverse effects.

In recent years, researchers have become aware of the complex intra- an intermolecular crosstalk between G protein-mediated signaling and steroid-like hormonal effects, as well as postranslational modifications (i.e., acylation and polyisoprenylation) of G protein subunits (Fukada and Kokame, 1994; Galbiata et al., 1994; Mumby et al., 1994; Shahinian and Silvius, 1995). In this context, it may be of particular interest to scrutinize the integrated input from hormones and metabolic factors converging at the level of hormone responsive elements (HREs). It has been shown that differentiation of preadipocytes

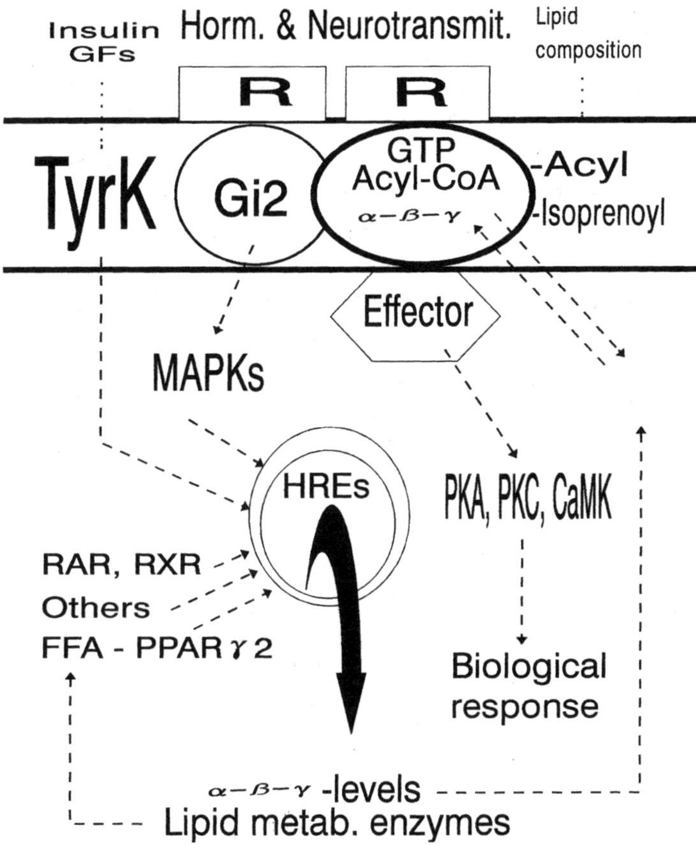

Fig. 1. Integrated influence of lipid metabolism and composition and posttranslational modifications on neurotransmitter-elicited signal transduction. Membrane fluidity, which is influenced by lipid composition, determines the coupling kinetics of the receptor–G protein–effector complex. Membrane attachment is dependent on the degree of acylation and isoprenylation of various G protein subunits. The intrinsic activity of the α-subunit depends on the presence of hydrolyzable GTP. However, its GTPase activity is somehow affected by the presence of acyl-CoA of various chain lengths. The gene expression of α-β-γ levels and lipid metabolizing enzymes depends grossly on the occupancy of retinoic acid (RAR, RXR) and other steroid-like receptors, as well as the occupancy of free fatty acid (FFA) metabolites on the peroxisome proliferator activated receptor (PPARγ2).

Conditions like those described above will eventually determine the differential activation of protein kinases (PKA, PKC, CaMK), which are all discriminants of the biological response, or activity, of the neuronal cell.

to adipocytes encompasses the integrated input of insulin, corticosteroids, G proteins ($G\alpha_s$ and $G\alpha_{i2}$), the peroxisome proliferator activated receptor PPAR ($PPAR\gamma_2$), and the retinoic acid receptor (RAR) subtypes (RAR and RXR), as well as the spectrum of fatty acids fed to the cells (Safoniva et al., 1994; Kamei et al., 1994; Tontonoz et al., 1995; Ailhaud et al., 1995; Gordeladze et al., 1996; Høvik et al., 1996). In this context, it is noteworthy that fatty acid acyl-CoA of various chain lengths allosterically, and in a bimodal fashion, affects intracellular GTP-activated Ca^{2+} translocation processes (Rys-Sikora et al., 1994).

Consequently, one should aim at drug design, not necessarily directed toward neurotransmitter receptors or G proteins *per se*. One approach may be to restore CNS lipid metabolism to ensure correct membrane lipid composition, posttranslational modification of G proteins, normalized G protein compartmentalization, and consequently, rectified ultimate G protein gene expression and function (Fig. 1).

References

Ailhaud, G., Amri, E. Z., and Grimaldi, P. A. (1995) Fatty acids and adipose cell differentiation. *Prostaglandins Leukot. Essent. Fatty Acids* **52,** 113–115.

Attali, B. and Vogel, Z. (1989) Long-term opiate exposure leads to reduction of the alpha i-1 subunit of GTP-binding proteins. *J. Neurochem.* **53,** 1636–1639.

Avissar, S. and Schreiber, G. (1992) The involvement of guanine nucleotide binding proteins in the pathogenesis and treatment of affective disorders. *Biol. Psychiatry,* **31,** 435–459.

Avissar, S., Schreiber, G., Danon, A., and Belmaker, R. H. (1988). Lithium inhibits adrenergic and cholinergic increases in GTP biding in rat cortex. *Nature (Lond.),* **331,** 440–442.

Avissar, S., Scheriber, G., Aulakh, C. S., Wozniak, K. M., and Murphy, D. L. (1990) Carbamazepine and electroconvulsive shock attenuate beta-adrenoceptor and muscarinic cholinoceptor coupling to G proteins in rat cortex. *Eur. J. Pharmacol.* **189,** 99–103.

Bartus, R., Dean, R. L., Beer, G., et al. (1982) The cholinergic hypothesis of geriatric memory dysfunction. *Science* **217,** 408–412.

Berridge, M. J. and Irvine, R. F. (1989) Inositol phosphates and cell signalling. *Nature (Lond.)* **341,** 197–205.

Birnbaumer, L. (1990) G Proteins in signal transduction. *Ann. Rev. Pharmacol. Toxicol.* **30,** 675–705.

Bothmer, J. and Jolles, J. (1994) Phosphoinositide metabolism, aging and Alzheimer's disease. *Biochim. Biophys. Acta* **1225,** 111–124.

Bouvier, C., Forge, H., Legacé, G., Drews, R., Sinnett, D., Labuda, D, et al. (1991) G proteins in normal rat pituitaries and in prolactin secreting rat pituitary tumors. *Mol. Cell. Endocrinol.* **78,** 33–44.

Calderini, G. A., Bonetti, A. C., Battistella, A., et al. (1983) Biochemical changes of rat brain membranes with aging. *Neurochem. Res.* **8,** 483–492.

Carli, M., Anand-Srivastava, M. B., Molina-Holgado, E., Dewar, K. M., and Reader, T. A. (1994) Effects of chronic lithium treatments on central dopaminergic receptor systems: G proteins as possible targets. *Neurochem. Int.* **24,** 13–22.

Charness, M. E., Querimit, L. A., and Henteleff, M. (1988) Ethanol differentially regulates G proteins in neural cells. *Biochem. Biophys. Res. Commun.* **155,** 138–143.

Charney, D. S., Innis, R. B., Duman, R. S., Woods, S. W., and Heninger, G. R. (1989) Platelet alpha-2-receptor binding and adenylate cyclase activity in panic disorder. *Psychopharmacology (Berl.)* **98,** 102–107.

Connor, D. J., Jope, R. S., and Harrell, L. E. (1988) Chronic oral aluminium administration to rats: cognition and cholinergic parameters. *Pharmacol. Biochem. Behav.* **157,** 684–694.

Corrette, B. J., Bauer, C. K., and Schwarz, J. R. (1995) Electrophysiology of anterior pituitary cells, in *The Electrophysiology of Neuroendocrine Cells* (Scherübl, H., Hescheler, J., eds.), CRC, Heidelberg, Germany, pp. 101–143.

Costa, E. (ed.) (1983) *The Benzodiazepines: From Molecular Biology to Clinical Practice.* Raven, New York.

Cowburn, R. F., Marcusson, J. O., Eriksson, A., Wiehager, B., and O'Neill, C. (1994) Adenylyl cyclase activity and G protein subunit levels in postmorten frontal cortex of suicide victims. *Brain Res.* **633(1–2),** 297–304.

Davies, A. O., Mares, A., Pool, J. L., and Taylor, A. A. (1987) Mitral valve prolapse with symptoms of beta-adrenergic hypersensitivity: beta 2-adrenergic receptor supercoupling with desensitization on isoproterenol exposure. *Am. J. Med.* **82,** 193–210.

Davies, A. O., Su, C. J., Balasubramanyam, C., Codina, J., and Birnbaumer, L. (1991) Abnormal guanine nucleotide regulatory protein in MVP dysautonomia: evidence from reconstitution of Gs. *J. Clin. Endocrinol. Metab.* **72,** 867–875.

Deckert, J., Nutt, D. J., and Marangos, P. J. (1989) Electroconvulsive shock (ECS) and the adenosine neuromedulatory system: effect of single and repeated ECS on the adenosine A1 and A2 receptors, adenylate cyclase, and the adenosine uptake site. *J. Neurochem.* **52,** 641–646.

Destexhe, A. and Sejnowski, T. J. (1995) G protein activation kinetics and spillover of gamma-aminobutyric acid may account for differences between inhibitory responses in the hippocampus and thalamus. *Proc. Natl. Acad. Sci. USA* **92(21)**, 9515–9519.

Dewas, D., Horsburgh, K., Graham, D. I., Brooks, D. N., and McCulloch, J. (1990) Selective alterations of high affinity [3H] forskolin binding sites in Alzheimer's disease: a quantitative autoradiographic study. *Brain Res.* **511**, 241–248.

Diamond, I., Wrubel, B., Estrin, W., and Gordon, A. (1987) Basal and adenosine receptor-stimulated levels of cAMP are reduced in lymphocytes from alcoholic patients. *Proc. Natl. Acad. Sci. USA* **84**, 1413–1416.

Ebstein, R. P., Lerer, B., Shapira, B., Schemesh, Z., Moscovich, D. G., and Kindler, S. (1988) Cyclic AMP second-messenger signal amplification in depression. *Br. J. Psychiatry* **152**, 665–669.

Fern, R., Waxman, S. G., and Ransom, B. R. (1995) Endogenous GABA attenuates CNS white matter dysfunction following anoxia. *J. Neurosci.* **15**, 699–708.

Fukada, Y. and Kokame, K. (1994) Covalent lipid modifications of heterotrimeric G proteins. Nippon Yakurigaku Zasshi, **103**, 263–272.

Fülöp, T. Jr. and Ildiko, S. (1994) Age-related changes in signal transduction. *Drugs & Aging* **5**, 366–390.

Fülöp, T. Jr., Barabas, G., Varga, Z., Jozsef, C., Cscabina, S., Szucs S., Seres, I., Szikszay, E., Jeney, Z., and Penyige, A. (1993) Age-dependent changes in transmembrane signalling: identification of G proteins in human lymphocytes and polymorphonuclear leukocytes. *Cell Signal* **5**, 593–603.

Fülöp, T. Jr., Barabas, G., and Varga, Zs. (1992) Transmembrane signalling changes with aging. *Ann. NY Acad. Sci.* **673**, 165–171.

Galbiati, G., Guzzi, F., Magee, A. I., Milligan, G., and Parenti, M. (1994) N-terminal fatty acylation of the alpha-subunit of the G protein Gi1: only the myristoylated protein is a substrate for palmitoylation. *Biochem. J.* **303**, 697–700.

Gierschik, P., Miligan, G., Pines, M. et al. (1986) Use of specific antibody to quantitate the guanine nucleotide binding protein G in brain. *Proc. Natl. Acad. Sci. USA* **83**, 2258–2262.

Gilman, A. (1987) G proteins: transducers of receptor generated signals. *Annu. Rev. Biochem.* **56**, 615–649.

Gordeladze, J. O., Jablonski, G., Paulssen, R. H., Paulssen, E. J., Mortensen, B. M., Gautvik, K. M. et al. (1993) G protein coupled signalling in osteosarcoma cell lines, in *Frontiers in Osteosarcoma Research* (Novak, J. F. and McMaster, J. H., eds.), Hogrefe & Huber, Bern, pp. 297–302.

Gordeladze, J. O., Johansen, P. W., Paulssen, R. H., Paulssen, E. P., and Gautvik, K. M. (1994) G proteins: implications for pathophysiology and disease. *Eur. J. Endocrinol.* **131**, 557–574.

Gordeladze, J. O., Merendino, J. J. Jr., Hermouet, S., Gutkind, J. S., and Acilli, D. (1997) The effect of activating and inactivating mutations of G_s- and G_{i2}-alpha protein subunits on growth and differentiation of 3T3-L1 preadipocytes. *J. Cell Biochem.*, **64**, 242–257.

Greenwood, A. F. and Jope, R. S. (1994) Brain G protein proteolysis by calpain: enhancement by lithium. *Brain Res.* **636**, 320–326.

Haba, K., Ogawa, N., Kawata, M., et al. (1988) A method for parallel determination and choline acetyltransferase and muscarinic cholinergic: application in aged-rat brain. *Neurochem. Res.* **13**, 951–955.

Harrison, P. J., Barton, A. J., McDonald, B., and Pearson, R. C. (1991) Alzheimer's disease: specific increases in a G protein subunit (Gs alpha) mRNA in hippocampal and cortical neurons. *Brain Res. Mol. Brain Res.* **10(1)**, 71–81.

Heron, D. S., Israeli, M., Hershkowitz, M., et al. (1981) Lipid induced modulation opiate receptor in mouse brain membrane. *Eur. J. Pharmacol.* **72**, 361–364.

Hoffman, P. L. and Tabakoff, B. (1990) Ethanol and guanine nucletode binding proteins: a selective interaction, *FASEB J.* **4**, 2612–2622.

Horn, E. M. and Bilezkian, J. P. (1990) Mechanism of transmembrane signaling of the beta-adrenergic receptor in congestive heart failure. *Circulation* **80**, 271–283.

Høvik, K. M., Wu, P.-F., Skrede, S., Bremer, J., and Gordeladze, J. O. (1996) The influence of activating and inactivating mutations of $G_{i2}\alpha$ proteins on fat metabolizing enzymes and lipid turnover in differentiated 3T3-L1 preadipocytes. *Biochim. Biophys. Acta*, in press.

Husseini, K. and Manji, M. D. (1992) G proteins: implications for psychiatry. *Am. J. Psychiatry* **149**, 746–760.

Jablonski, G., Klem, K. H., Danielsen, C. C., Mosekilde, L., and Gordeladze, J. O. (1996) Aluminium-induced bone disease in uremic rats: effect of deferoxamin. *Biosci. Rep.* **16**, 49–63.

Johansen, P. W., Paulssen, R. H., Bjøro, T., Gautvik, K. M., and Gordeladze, J. O. (1997) Dopaminergic signalling is coupled to distinct guanine nucleotide binding regulatory proteins (G proteins) in GH cell lines. *Biochim. Biophys. Acta*, in press.

Johnson, G. V. W., Cogdill, K. W., and Jope, R. S. (1990) Orally administered aluminium alters in vitro protein phosphorylation and protein kinase activities in rat brain. *Neurobiol. Aging* **11**, 209–216.

Jope, R. S. and Johnson, G. V. W. (1992) Neurotoxic effects of dietary aluminum, in *Aluminum in Biology and Medicine*, Wiley, Chichester (Ciba Foundation Symposium), pp. 254–267.

Joseph, J. A., Dalton, D. T. K., and Roth, G. S. (1988) Alteration in muscarinic control of striatal dopamine autoreceptor in senescence: a deficit at the ligand muscarinic receptor interface? *Brain Res.* **454**, 149–155.

Joseph, J. A., Kowatch, M. A., Maki, T., et al. (1990) Selective cross activation/inhibition of second messenger systems and the function of age-related deficits in the muscarinic control of dopamine release from periperfused rat striata. *Brain Res.* **537,** 40–48.

Kamei, Y., Kawada, T., Mizukami, J., and Sugimoto, E. (1994) The prevention of adipose differentiation of 3T3-L1 cells caused by retinoic acid is elicited through retinoic acid receptor alpha. *Life Sci.* **55,** 307–312.

Kerwin, R. W. and Beats, B. C. (1990) Increased forskolin binding in the left parahippocampal gyrus and CA1 region in post mortem schizophrenic brain determined by guanitative autoradiography. *Neurosci. Lett.* **119,** 164–168.

Koenig, M. L. and Jope, R. S. (1978) Aluminium inhibits the fast phase of voltage-dependent calcium influx into synaptosomes. *J. Neurochem.* **49,** 316–320.

Leonard, B. E. (1994) Serotonin receptors—where are they going? *Int. Clin. Psychopharmacol.* **9,** 7–17.

Lerer, B., Bleich, A., Bennett, E. R., Ebstein, R. P., and Balkin, J. (1990) Platelet adenylate cyclase and phospholipase C activity in posttraumatic stress disorder. *Biol. Psychiatry* **27,** 735–740.

Lesch, K. P. and Lerer, B. (1991) The 5-HT receptor-G protein-effector system complex in depression. I. Effect of glucocorticoids. *J. Neural. Transm. [Gen Sect],* **84,** 3–18.

Lipschitz, D. A., Upuda, K. D., Indelicato, S. R., et al. (1991) Effect of age on second messenger generation in neutrophils. *Blood* **78,** 1347–1354.

Lyons, J., Landis, C. A., Harsh, G., Vallar, L., Grunewald, K., Feichtinger, H., Duh, Q. Y., Clark, O. H., Kawasaki, E., Bourne, H. R., and McCormick, F. (1990) Two G protein oncogenes in human endocrine tumors. *Science* **249,** 655–659.

Manji, H. K., Hsiao, J. K., Risby, E. D., Oliver, J., Rudorfer, M. V., and Potter, W. Z. (1991a) The mechanisms of action of lithium, I: effects on serotoninergic and noradrenergic systems in normal subjects. *Arch. Gen. Psychiatry* **48,** 505–512.

Manji, H. K., Chen, G. A., Bitran, J. A., Gusovsky, F., and Potter, W. Z. (1991b) Chronic exposure of C6 glioma cells to desipramine desensitizes beta-adrenoceptors, but increase KL/KH ratio. *Eur. J. Pharmacol.* **206,** 159–162.

Manji, H. K., Chen, G., Shimon, H., Hsiao, J. K., Potter, W. Z., and Belmaker, R. H. (1995) Guanine nucleotide-binding proteins in bipolar affective disorder. Effects of long-term lithium treatment. *Arch. Gen. Psychiatry* **52,** 135–144.

Mann, J. J., Brown, R. P., Halper, J. P., Sweeney, J. A., Kocsis, J. H., Stokes, P. E., and Bilezikian, J. P. (1985) Reduced sensitivity of lymphocyte beta-adrenergic receptors in patients with endogenous depression and psychomotor agitation. *N. Engl. J. Med.* **313,** 715–720.

Memo, K., Kleinman, J. E., and Hanbauer, I. (1983) Coupling of dopamine D1 recognition sites with adenylate cyclase in nuclei accumbens and caudatus of schizophrenia. *Science* **221,** 1304–1307.

Mestek, A., Hurley, J. H., Bye, L. S., Campbell, A. D., Chen, Y., Tian, M., Liu, J., Schulman, H., and Yu, L. (1995) The human mu- opioid receptor: modulation of functional desensitization by calcium/calmodulin-dependent protein kinase and protein kinase C. *J. Neurosci.* **13**, 2396–2406.

Meunier, J.-C. (1992) Récepteurs des opoïdes, tolérance et dépendence. *Therapie* **47**, 495–502.

Milligan, G. (1993) Agonist regulation of cellular G protein levels and distribution: mechanisms and functional implications. *Trends Biochem. Sci.* **14**, 413–418.

Mitchell, S. J., Rawlins, J. N. P., and Steward, O. (1982) Medial septal area lesions disrupt theta rhythm and cholinergic staining in model entorhinal cortex and produce radial arm maze behavior in rats. *J. Neurosci.* **2**, 292–302.

Mochly-Rosen, D., Chang, F. H., Cheever, L., Kim, Diamond, I., and Gordon, A. S. (1988) Chronic ethanol causes heterologous desensitization of receptors by reducing alpha s messenger RNA. *Nature* **333**, 848–850.

Mork, A. and Geisler, A. (1989a) Effects of GTP on hormone-stimulated adenylate cyclase activity in cerebral cortex, striatum, and hippocampus from rats treated chronically with lithium. *Biol. Psychiatry* **26**, 279–288.

Mork, A. and Geisler, A. (1989b) The effects of lithium in vitro and ex vivo on adenylate cyclase in brain are exerted by distinct mechanisms. *Neuropharmacology,* **28**, 307–311.

Moscona Amir, E., Henis, Y. I., and Sokolovsky, M. (1989) Aging in rat heart myocytes disrupts muscarinic receptor coupling that leads to inhibition of cAMP accumulation and alters the pathway of muscarinic stimulated phosphoinositide hydrolysis. *Biochemistry* **28**, 7130–7137.

Moulignier, A. (1994) Récepteurs centraux de la sérotonine principaux aspects fondamentaux et fonctionelles applications thérapeutiques. *Rev. Neurol. (Paris)* **150**, 3–15.

Mumby, S. M., Kleuss, C., and Gilman, A. G. (1994) Receptor regulation of G protein palmitoylation. *Proc. Natl. Acad. Sci. USA* **91(7)**, 2800–2804.

Mundy, W., Tandon, P., Ali, S., et al. (1991) Age-related changes in receptor-mediated phosphoinsositide hydrolysis in various region of rat brain. *Life Sci.* **49**, 97–102.

Nagy, L. E., Diamond, I., Collier, K., Lopez, Ullman, B., and Gordon, A. S. (1989) Adenosine is required for ethanol-induced heterologous desensitization. *Mol. Pharmacol.* **357**, 744–748.

Nalepa, I., Pintor, A., and Fortuna, S. (1989) Increased responsiveness of the cerebral cortical phosphatidylinositol system to noradrenaline and carbachol in senescent rats. *Neurosci. Lett.* **107**, 195–199.

Neelands, P. J. and Clanidinin, M. T. (1983) Diet fat influences liver plasma-membrane lipid composition and glucagon-stimulated adenylate cyclase activity. *Biochem. J.* **215**, 573–583.

Nestler, E. J., Erdos, J. J., Terwilliger, R., Duman, R. S., and Tallman, J. F. (1989) Regulation of G proteins by chronic morphine in the rat locus coeruleus. *Brain Res.* **476,** 230–239.

Nestler, E. J., Terwilliger, R. Z., Walker, J. R., Sevarino, K. A., and Duman, R. S. (1990) Chronic cocanie treatment decreases levels of the G protein subunits Gi alpha and Go alpha in discrete regions of rat brain. *J. Neurochem.* **55,** 1079–1082.

Newman, M. E., Lerer, B., Lichtenberg, P., and Shapira, B. (1992) Platelet adenylate cyclase activity in depression and after clomipramine and lithium treatment: relation to serotonergic function. *Psychopharmacology (Berl.)* **109(1–2),** 231–234.

Nishida, A., Kaiya, H., Tohmatsu, T., Wakabyashi, S., and Nozawa, Y. (1990) Electroconvulsive treatment: effects on phospholipase C activity and GTP binding activity in rat brain. *J. Neural. Trans. (Gen. Sect.)* **81,** 121–130.

Nishino, N., Kitamura, N., Hashimoto, T., Kajimoto, Y., Shirai, Y., Murakami, N., Nakai, H. T., Komure, O., Shirakawa, O., Mita, T., et al. (1993) Increase in [^{3}H] cAMP binding sites and decrease in Gi alpha and Go alpha immunoreactivities in left temporal cortics from patients with schizophrenia. *Brain Res.* **615,** 41–49.

Nishino, N., Kitamura, N., Hashimoto, T., and Tanaka, C. (1993b) Transmembrane signalling systems in the brain of patients with Parkinson's disease. *Rev. Neurosci.* **4,** 213–222.

Nomura, Y., Kitamura, Y., Kawai, M. and Segawa, T. (1991) Alpha 2-adrenoceptor-GTP binding regulatory protein-adenylate cyclase system in cerebral cortical membranes of adult and senescent rats. *Brain Res.* **379,** 118–124.

Odagaki, Y., Koyama, T., and Yamashita, I. (1994) Platelet pertussis toxin-sensitive G proteins in affective disorders. *J. Affect Disord.* **31,** 173–177.

Okada, F., Crow, T. J., and Roberts, G. W. (1990) G proteins (Gi, Go) in the basal ganglia of control and schizophrenic brain. *J. Neural. Transm. (Gen. Sect.)* **118,** 164–168.

Okada, F., Crow, T. J., and Roberts, G. W. (1991) G proteins (Gi, Go) in the medial temporal lobe in schizophrenia: preliminary report of a neurochemical correlate of structural change. *J. Neural. Transm. (Gen. Sect.)* **84,** 147–153.

Okada, F., Tokumitsu, Y., Takahashi, N., Crow, T. J., and Roberts, G. W. (1994) Reduced concentration of the alpha-subunit of GTP-binding protein Go in schizophrenic brain. *J. Neural. Transm. (Gen. Sect.)* **95,** 95–104.

Ozawa, H., Gsell, W., Frolich, R., Pantucek, F., Beckmann, H., and Reiderer, P. (1993) Imbalance of the Gs and Gi/o function in post-mortem human brain of depressed patients. *J. Neural. Transm. (Gen. Sect.)* **92,** 63–69.

Pandey, G. N., Pandey, S. C., and Davis, J. M. (1990) Peripheral adrenergic receptors in affective illness and schizophrenia. *Pharmacol. Toxicol.* **3,** 13–36.

Patten, J. L. and Levine, M. A. (1990) Immunochemical analysis of the alpha-subunit of the stimulatory G protein of adenylyl cyclase in patients with Albright's hereditary osteodystryphy. *J. Clin. Endocrinol. Metab.* **71,** 1108–1214.

Paulssen R. H, Paulssen E. J., Gautvik K. M., and Gordeladze J. O. (1992) Modulation of G proteins and second messenger responsiveness by steroid hormones in GH_3 rat pituitary tumour cells. *Acta Physiol. Scand.* **146,** 511–518.

Perez-Baun, J. C., Galve, I., Ruiz-Verdu, A., Haro, A., and Guillen, A. (1994) Octopamine-sensitive adenylyl cyclase and G proteins in Ceratitis capitata brain during aging. *Neuropharmacology* **33,** 641–646.

Pontzer, J. G., Chandler, J., Stevens, B. R., et al. (1990) Receptors, phospho-inositol hydrolysis and plasticity of nerve cells. *Prog. Brain. Res.* **86,** 221–225.

Rasmussen, K., Beitner-Johnson, D. B., Krystal, J. H., Aghajanian, G. K., and Nestler, E. J. (1990) Opiate withdrawal and the rat locus coeruleus: behavioral, electrophysiological, and biochemical correlates. *J. Neurosci.* **10,** 2308–2317.

Risby, E. D., Hsiao, J. K. Manji, H. K., Bitran, J., Moses, F., Zhou, D. F., and Potter, W. Z. (1991) The mechanisms of action of lithium, II: effects on adenylate cyclase activity and beta-adrenergic receptor binding in normal subjects. *Arch. Gen. Psychiatry* **48,** 513–524.

Roth, G. S., Joseph, J. A., and Mason, R. P. (1996) Membrane alterations as causes of impaired signal transduction in Alzheimer's disease and aging. *Trends Neurosci.* **18,** 203–206.

Rys-Sikora, K. E., Ghosh, T. K., and Gill, D. L. (1994) Modification of GTP-activated calcium translocation by fatty acyl-CoA esters. Evidence for a GTP-induced perfusion event. *J. Biol. Chem.* **269,** 31607–31613.

Safoniva, I., Reichert, U., Schroot, B., Ailhaud, G., and Grimaldi, P. (1994) Fatty acids and retinoids act synergistically on adipose cell differentiation. *Biochem. Biophys. Res. Commun.* **204,** 498–504.

Saito, T., Katamura, Y., Ozawa, H., Hatta, S., and Takahata, N. (1994) Platelet GTP-binding protein in long-term abstinent alcoholics with an alcoholic first-degree relative. *Biol. Psychiatry* **36,** 495–497.

Schoepp, D. D. and Conn, P. J. (1994) Metabotropic glutamate receptors in brain function and pathology. *Trends Pharmacol. Sci.* **14,** 13–21.

Schreiber, G., Avissar, S., Danon, A., and Belmaker, R. H. (1991) Hyper-functional G proteins in mononuclear leukocytes of patients with mania. *Biol. Psychiatry* **29,** 273–280.

Seeman, P., Niznik, H. B., Guan, H. C., Booth, G., and Ulpian, C. (1989) Link between D1 and D2 dopamine receptors is reduced in schizophrenia and Huntington diseased brain. *Proc. Natl. Acad. Sci. USA* **86,** 10156–10160.

Shahinian, S. and Silvius, J. R. (1995) Doubly-lipid-modified protein sequence motifs exhibit long-lived anchorage to lipid bilayer membranes. *Biochemistry* **34,** 3812–3822.

Shuckie, M. A. (1978) Biology of risk for alcoholism, in *Psychopharmacology: The Third Generation of Progress.* (Meltzer, H. Y., ed.), Raven, New York.

Singer, H. S., Searles, C. D., Hahn, I.-H., March, J. L., and Tronocoso, J. C. (1990) The effect of aluminium on markers for synaptic neurotransmission, cyclic AMP and neurofilaments in a neuroblastoma × glioma hybridoma (NG108-15). *Brain Res.* **528,** 73–79.

Sitoh, N., Guitart, X., Hayward, M., et al. (1989) Corticosterone differentially regulates the expression of Gs and Gi messenger RNA and protein in rat cerebral cortex. *Proc. Natl. Acad. Sci. USA* **86,** 3096–3910.

Spiegel, A. M. (1990) Receptor-effector coupling by G proteins: implications for neuronal plasticity. *Prog. Brain Res.* **86,** 269–276.

Spiegel, A. M., Shenker, A., and Weinstein, L. S. (1992) Receptor-effector coupling by G proteins: implications for normal and abnormal signal transduction. *Endocrine Rev.* **13,** 536–565.

Spiegel, A. M., Weinstein, L. S., Shenker, H., et al. (1993) G proteins: from basic to clinical study, in: *Cell Signalling: Biology and Medicine of Signal Transcuction* (Brown, B. L., Dobson, P. R. N., eds.), Raven, New York, pp. 37–46.

Springer, J. Taynen, M., and Loy, R. (1987) Regional analysis of age-related change in the cholinergic system of the hippocampal formation and basal forebrain of the rat. *Brain Res.* **407,** 123–136.

Sulser, F. (1984) Antidepressant treatments and regulation of norepinephrine-receptor-coupled adenylate cyclase systems in brain. *Adv. Biochem. Psychopharmacol.* **39,** 249–261.

Sumiyoshi, T., Stockmeier, C. A., Overholser, J. C., Thompson, P. A., and Meltzer, H. Y. (1995) Dopamine D4 receptors and effects of guanine nucleotides on [^{3}H]-raclopride binding in postmortem caudate nucleus of subjects with schizophrenia or major depression. *Brain Res.* **681,** 109–116.

Tabakoff, B., Hoffman, P. L., Lee, J. M., Saito, T., Willard, B., and De-Leon Jones F. (1988) Differences in platelet enzyme activity between alcoholics and nonalcoholics. *N. Engl. J. Med.* **318,** 134–139.

Tontonoz, P., Hu, E., Devine, J., Beale, E. G., and Spiegelman, B. M. (1995) PPAR gamma 2 regulates adipose expression of the phosphoenolpyruvate carboxykinase Gene **15,** 351–357.

Trujillo, K. A. and Akil, H. (1991) Opiate tolerance and dependence: recent findings and synthesis. *New Biol.* **3,** 915–923.

Tsuchiya, F., Ikeda, H., Hatta, Y., and Saito, T. (1988) Effects of desipramine administration on receptor adenylate cyclase coupling in rat cerebral cortex. *Jpn. J. Psychiatry Neurol.* **42,** 858–860.

Vallar, L., Spada, A., and Giannattasio, G. (1987) Altered Gs and adenylate cyclase activity inhuman GH-secreting pituitary adenomas. *Nature (Lond.)* **330,** 566–568.

Waltman, C., Levine, M. A., McCaul, M. E., Svikis, D. S., and Wand, G. S. (1993) Enhanced expression of the inhibitory protein Gi2 alpha and decreased activity of adenylyl cyclase in lymphocytes of abstinent alcoholics. *Alcohol Clin. Exp. Res.* **17,** 315–320.

Wand, G. S., Waltman, C., Martin, C. S., McCaul, M. E., Levine, M. A., and Wolfgang, D. (1994) Differential expression of guanosine triphosphate binding proteins in men at high and low risk for the future development of alcoholism. *J. Clin. Invest.* **94,** 1004–1111.

Weinstein, L. S., Gejman, P. V., Friedman E., Kadowaki, T., Collins R. M., Gershon, W. S., and Spiegel A. M. (1990) Mutations of the Gs alpha-subunit gene in Albright hereditary osteodystrophy detected by denaturing gradient gel electrophoresis. *Proc. Natl. Acad. Sci. USA* **87,** 8287–8290.

Weinstein, L. S., Shenker, A., Gejman, P. V., Merino, M. J., Friedman, E., and Spiegel, A. M. (1991) Activating mutations of the stimulatory g protein in McCune-Albrights syndrome. *N. Engl. J. Med.* **325,** 1688–1695.

Williams, R. J., Veale, M. A., Horne, P., and Kelly, E. (1993) Ethanol differentially regulates guanine nucleotide-binding protein alpha subunit expression in NF108–15 cells independently of extracellular adenosine. *Mol. Pharmacol.* **43,** 158–166.

Yamagami, K., Joseph, J. A., and Roth, G. S. (1992) Decrement of muscarinic receptor stimulated low Km GTPase in striata and hippocampus from aged rat. *Brain Res.* **576,** 327–331.

Young, L. T., Li, P. P., Kish, S. J., Siu, K. P., and Warsh, J. J. (1991) Postmortem cerebral cortex G_s α-subunit levels are elevated in bipolar affective disorder. *Brain Res.* **553,** 323–326.

Young, L. T., Li, P. P., Kamble, A., Siu, K. P., and Warsh, J. J. (1994) Mononuclear leukocyte levels of G proteins in depressed patients with bipolar disorder or major depressive disorder. *Am. J. Psychiatry* **151,** 594–596.

G Protein α- and βγ-Subunits as Possible Mediators of Dopamine-D1:D2 Receptor Ligand Binding Interactions

Zdenek B. Pristupa, Roger K. Sunahara,
and Hyman B. Niznik

Introduction

Dopamine Receptors

Dopamine receptors are members of a large family of neurotransmitter/hormone receptors that exert their biological actions via signal transduction pathways that involve subtype-specific guanine nucleotide binding or G proteins (see Kaziro et al., 1991; Simon et al., 1991; Hille, 1992; Iyengar, 1993; Clapham and Neer, 1993; Gilman, 1995; Rodbell, 1995; Raymond, 1995). On the basis of biochemical, pharmacological and physiological criteria, receptors for dopamine within the central nervous system and periphery have been classified into two types, termed D1 and D2 (*see* reviews: Kebabian and Calne, 1979; Seeman, 1980; Niznik, 1987; Niznik and Jarvie 1989; Niznik and Van Tol, 1992; Kebabian, 1993; Hall, 1994; Lokhandwala and Chen, 1994). Operationally, native membrane-bound dopamine D1 receptors are defined by their ability to stimulate, via G_s or G_{olf} (Herve et al., 1993), adenylate cyclase activity which subsequently activates cAMP-dependent protein kinases (*see* Hemmings et al., 1987; Lovenberg et al., 1991; Niznik et al., 1992). D1 receptor stimulation in both

From: *Neuromethods, Vol. 31: G Protein Methods and Protocols*
Ed: R. K. Mishra, G. B. Baker, and A. A. Boulton Humana Press Inc.

brain and periphery has also been shown to result in the stimulation of phospholipase C (Felder et al., 1989; Mahan et al., 1990; Rodrigues and Dowling, 1990; Undie and Friedman, 1990, 1992; Dowling, 1991, 1994; Undie et al., 1994; Wang et al., 1995; Yu et al., 1996), the translocation of protein kinase C (McMillian et al., 1992), activation of facilitation Ca^{2+} channels (Artalejo et al., 1990; Lin et al., 1995), the inhibition of Na^+/H^+ exchange (Mailman et al., 1986; Felder et al., 1990, 1993), and Na^+/K^+-ATPase (Bertorello et al., 1990; Chen and Lokhandwala, 1993; Horiuchi et al., 1993; Shahedi et al., 1995), possibly through interaction with G_i-like proteins (Wang et al., 1995; Sidhu et al., 1991; Kimura et al., 1995) or indirectly via $G\beta\gamma$-subunits following the activation of cAMP-dependent protein kinases (Yu et al., 1996; Liu and Simon, 1996).

Dopamine D2 receptors are defined by their ability to inhibit the activity of adenylate cyclase, and also appear to have effects on numerous other effector systems, including stimulation of phosphatidylinositol turnover, potentiation of arachidonic acid release, regulation of K^+ and Ca^{2+} channels and regulation of Na^+-K^+-ATPase (Dal Toso et al., 1989; Neve et al., 1989; Albert et al., 1990; Senogles et al., 1990; Vallar et al., 1990; Kanterman et al., 1991; Elsholtz et al., 1991; Castellano et al., 1993; Lledo et al., 1994; Chio et al., 1994; Gardette et al., 1994; Schinelli et al., 1994; Greif et al., 1995; Shulman and Fox, 1996; Yamaguchi et al., 1996) via activation of subtype-specific G_i/G_o proteins (Missale et al., 1991; Burris et al., 1992; Lledo et al., 1992; Montmayeur et al., 1993; Di Marzo et al., 1993; Burris and Freeman, 1994; Liu and Lasater, 1994; Senoglies, 1994).

Many functional correlates of the stimulation of subtype-specific dopamine receptors have been documented (Cameron and Williams, 1993) and include: dopamine receptor regulation of neuron growth and differentiation (Lankford et al., 1988; McCobb and Kater, 1988; Swarzenski et al., 1994; Rodrigues and Dowling, 1990), activation of primary response gene transcription, in particular, immediate early genes, such as *c-fos* (Robertson et al., 1991; Young et al., 1991; Cole et al., 1992; Robertson and Robertson, 1994; Gerfen, 1995; Gerfen et al.,

1995), regulation of steroid hormone receptor translocation and gene expression (Power et al., 1991; Srivastava and Mishra, 1994; Lew and Elsholtz, 1994), induction of numerous behavioral responses and programs (e.g., Woolverton and Johnson, 1992; Bedard, 1995; Bardo et al., 1996; Sedvall, 1996; Starr, 1996), and the modulation and functional expression of dopamine receptor-mediated events (Clarke and White, 1987; Waddington and O'Boyle, 1989; Waddington, 1989; Seeman et al., 1989; Robertson et al., 1990; Robertson, 1992; Waddington et al., 1994; Apostolakis et al., 1996, and references below).

Dopamine Receptor Genes

The concept that two dopamine receptors were sufficient to account for all the effects mediated by dopamine was clearly an oversimplification. Recent molecular biological studies have identified five distinct genes, which can encode at least 23 functional dopamine receptors in humans. The members of the expanded human dopamine receptor family, however, can still be codified by way of the original D1 and D2 receptor dichotomy outlined above, and include two genes encoding dopamine-D1-like receptors, termed D1 (or D1A) and D5 (or D1B), and three genes encoding D2-like receptors with corresponding splice and polymorphic variants, termed $D2_{Long}$, $D2_{Short}$, D3 and its splice variants, and the numerous polymorphic forms of the D4 receptor. A review of the molecular biology of dopamine receptor family is beyond the scope of this work; the reader is referred to a number of excellent recent reviews on the subject (*see* Niznik and Van Tol, 1992; Sibley and Monsma, 1992; Civelli et al., 1993; Gingrich and Caron, 1993; Schwartz et al., 1993; Sibley et al., 1993; Grandy et al., 1994; Jackson and Westlind-Danielson, 1994; Jarvie et al., 1994; Sokoloff et al., 1994; Van Tol, 1994; Sokoloff and Schwartz, 1995; Jensen et al., 1996). Recently, two additional pharmacologically distinct members of the dopamine D1 receptor family have been cloned and characterized: D1C from *Xenopus laevis* (Sugamori et al., 1994), and D1D from *Gal-*

lus domesticus (Demchyshyn et al., 1995), although their presence as distinct receptor subtypes and their potential significance in humans or other mammals is currently unknown.

Dopamine Receptors in Schizophrenia

Of the many biological theories of schizophrenia (e.g., Jones and Murray, 1991; Lewine, 1992; Bloom, 1993; Joyce, 1993; Pilowsky et al., 1993, Rao and Moller, 1994; Ellison, 1994; Wolf and Weinberger, 1996) considerable evidence suggests an association between an abnormal dopaminergic neurotransmission/receptor system and schizophrenia (for reviews *see* Seeman, 1987, 1993; Seeman and Niznik, 1990; Cleghorn et al., 1991; Davis et al., 1991; Niznik and Van Tol, 1992; Goldstein and Deutch, 1992; Reynolds, 1992; Sunahara et al., 1993; Goldman-Rakic, 1995). Members of the dopamine-D2-like receptor family, in particular, appear to play a central role in the psychotic process (for review see Seeman, 1993). Thus, (1) D2-like receptor-selective neuroleptics block hallucinations and delusions in schizophrenia, in Huntington's or Alzheimer's disease psychosis, and in L-DOPA-induced psychosis, and (2) antipsychotic potencies of neuroleptics (as indexed by the therapeutic steady-state molarity in the plasma water of treated schizophrenics) correlate directly with their in vitro affinities for D2 receptors (Niznik and Van Tol, 1992; Seeman, 1987).

Direct biochemical evidence for the dopamine receptor hypothesis of schizophrenia is buttressed by the observations that higher densities or levels of D2-like receptors (as indexed by radioligand binding using [^{3}H]spiperone or [^{3}H]nemonapride) are seen in at least 50% of postmortem schizophrenic brain striata compared to other psychomotor diseased individuals, including those that have never received antipsychotic medication. Similar increases in D2-like receptors have also been observed using positron emission tomography (PET) in living patients (Wong et al., 1986; Seeman, 1987; Joyce et al., 1988; Seeman and Niznik, 1990; Cleghorn et al., 1991; Hietala et al., 1994).

The exact molecular form of the D2-like receptor that may give rise to the observed biochemical changes described above is currently unknown, since all molecular forms of the dopamine-D2 receptor family [$D2_{Long}$, $D2_{Short}$, $D3_{Long}$, $D3_{Short}$, and the numerous D4 receptor variants (Lichter et al., 1993)] appear to bind commonly used radioligands, such as spiperone and nemonapride (formerly YM-09151-2, or emonapride), with equal affinity. Moreover, radioligands that appear capable of selectively labeling only one set of D2-like receptors, such as 7-OH-DPAT for D3 receptors, have yet to be used to assess dopamine-D2-like receptor densities in the CNS of patients suffering from schizophrenia. Recent evidence, however, suggests that perturbations of the D3 receptor gene may be associated with schizophrenia [see Sokoloff et al., 1994; Morell, 1993 (although *see also* Yang et al., 1993; Nothen et al., 1993)] and that expression of a functional molecular form of the D3 receptor mRNA is missing or somewhat perturbed in some brain regions of schizophrenics (Schmauss et al., 1993).

Seeman et al. (1993) have recently proposed that the observed elevations of [³H]spiperone binding to post-mortem brain samples of schizophrenics is in fact due to an increase of the D4 rather than of D2 or D3 receptors. Exploiting the distinguishing sensitivity of the benzamide neuroleptic, raclopride for D2 and D3 (raclopride-sensitive) vs D4 (raclopride-insensitive) receptors, these authors putatively identified native striatal D4 receptors by subtracting the binding of [³H]raclopride (labeling only D2 and D3 receptors) from that of [³H]nemonapride (which labels D2, D3, and D4 receptors). Based on this subtractive methodology, a two- to sixfold induction of D4 receptor number was observed in schizophrenic postmortem striata compared to a variety of neuropsychiatric controls (Seeman et al., 1993; Murray et al.; 1995). The increase in estimated density of [³H]nemonapride and [³H]raclopride in postmortem brains of schizophrenics is taken to reflect the induction of D4 receptors, with accompanying increases in D4 mRNA synthesis. Conflicting reports have also been presented with regard to levels of both D4 receptors (Strange, 1994; Reynolds and Mason, 1994, 1995) and mRNA (Roberts et al., 1996). It should be also be noted that in some schizophrenic populations, the binding of [^{3}H]raclopride is substantially increased relative to controls (Ruiz et al., 1992) suggesting that it may be the D2 and/or D3 receptors that are be elevated in schizophrenia.

Dopamine-D1:D2 Receptor Interactions

Since the dopamine-D2 receptor family is central to antipsychotic action (Sunahara et al., 1993), it is important that factors be identified which modulate or influence the activity of these receptors. One of these factors appears to be the dopamine-D1/D5 receptor system.

Voluminous literature exists describing dopamine-D1:D2 receptor interactions at the behavioral, electrophysiological, pharmacological, and biochemical levels (reviewed in Clarke and White, 1987; Waddington and O'Boyle, 1989; Surmeier et al., 1993; Waddington, 1993; Waddington et al., 1994, 1995; White and Hu, 1995; Waddington and Deveney, 1996). Basically, the interactions can be either cooperative/synergistic, or antagonistic. Whereas some behaviors are predominantly influenced by the activity of D1-like receptors, such as grooming, self-mutilation, and repetitive jaw movements (Rosengarten et al., 1993), D2 receptors predominantly influence motor activity. In general, however, influences mediated by both receptors are noted with regard to actions on rapid jaw movements, vacuous chewing, rotation, adenylate cyclase activity, dopamine turnover, transmitter release (Cameron and Williams, 1993) and gene expression (Lahoste et al., 1993; Paul et al., 1992; Keefe and Gerfen, 1995). Whereas cooperative/synergistic D1:D2 interactions are important in the regulation of typical dopaminergic behaviors (e.g., locomotor activity, stereotypy, grooming, sniffing, yawning), antagonistic D1:D2 interactions appear to be involved in the regulation of atypical D2-stimulated behavior (e.g., myoclonic jerking) (*see* Waddington and O'Boyle, 1989; Code and Tang, 1991; Daly and Waddington, 1992; Waddington et al., 1994). Recent work suggests that prepulse inhibition of the startle reflex, which has found to be reduced in schizophrenics, appears to be regulated by D1:D2 synergistic interaction (Wan et al., 1996).

A number of important observations have been made regarding the functional and therapeutic significance of D1:D2 receptor interaction with regard to antipsychotic antiparkinsonian action, and in neuroleptic-induced extra-pyramidal syndromes. For example, it has been postulated that negative symptoms of schizophrenia may be associated with hypocortical D1/D5 dopaminergic activity (*see* Davis et al., 1991; de Beaurepaire et al., 1995; Den Boer et al., 1995; Karle et al., 1995), and that combination D1 agonist/D2 antagonist drug therapy can alleviate both the negative and positive symptoms of schizophrenia (Davidson et al., 1990). With regard to extrapyramidal syndromes as a result of chronic dopamine-D2-like receptor blockade, haloperidol-sensitized monkeys (*Cebus appela*) all develop extrapyramidal symptoms, which are antagonized with repeated daily administration of the selective D1-receptor antagonist SCH 39166 (McHugh and Coffin, 1991). D1 and D2 receptors have also been shown to interact in rat models of Parkinson's disease (Morelli, et al., 1991; Paul et al., 1992). Dyskinesias are probably among the most disabling complications of chronic L-DOPA therapy in Parkinson's disease. In MPTP-treated monkeys, (1) D2-like receptor stimulation by selective agonists can alleviate Parkinson-like syndrome without dyskinesia, provided D1 receptors are blocked, and (2) fairly selective D1 receptor stimulation in vivo by agonist will produce potent anti-Parkinson-like action with no dyskinesia, provided there is minimal stimulation of D2-like receptors by endogenous dopamine (Gomez-Mancilla and Bedard, 1991).

D1:D2 Receptor Crosstalk at the Ligand Binding Site

Most of the observed behavioral and biochemical consequences of dopamine D1:D2 receptor interactions cited above can be clearly attributed to processes and signal transduction systems "downstream" from the level of the dopamine receptor protein. However, using postmortem tissue, we discovered that D1- and D2-receptors can be found to link in vitro using a simple radioligand binding assay (Seeman et al., 1989). It was shown that endogenous dopamine, when added to striatal membrane homogenates, could lower the apparent density of D2 receptors labeled by [³H]raclopride. However, the prior addition of SCH 23390, a selective D1 receptor antagonist, prevented this apparent reduction in estimated receptor density, even though SCH 23390 alone had no effect on [³H]raclopride binding by D2 receptors (Fig. 1). Moreover, we ascertained that the dopamine-induced reduction of [³H]raclopride receptor density was strictly mediated by D2 receptor alone, since the same effects were seen in tissues not expressing the D1 receptor (e.g., in the pituitary). The reversal of this reduction is consistent with the pharmacological profile of dopamine-D1 receptors. At about the same time, Zhang and Segawa (1989) also demonstrated a slight, but significant, effect of SCH 23390 in reducing the potency of quinpirole (a potent D2 receptor agonist) for the high-affinity state of D2-like receptors labeled by [³H]spiperone. More significantly, however, we demonstrated that the D1:D2 receptor interaction, as indexed by [³H]raclopride binding, was absent or severely reduced in postmortem brain tissues of approximately half of the schizophrenics and Huntington's diseased population studied (Seeman et al., 1989). This was the first report suggesting that dopamine receptor crosstalk appeared to be mediated at the level of the ligand binding site, and that this interaction is perturbed in the brain tissues of many neuropsychiatrically diseased patients.

More recently, Seeman et al. (1994) have modified the basic [³H]raclopride saturation binding "link" method to one using competition binding assays in which the binding of [³H]raclopride is competed by increasing concentrations of unlabeled SCH 23390, either in the presence or absence of exogenous dopamine. In the presence of dopamine, SCH 23390, in a concentration-dependent manner, increases the binding of [³H]raclopride to D2 receptors, with a maximal effect corresponding to the dissociation constant of SCH

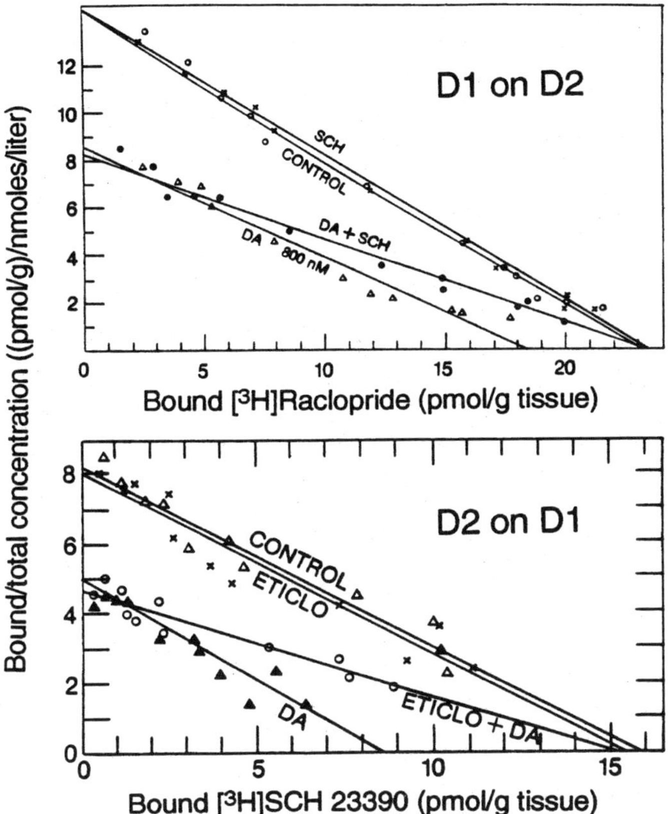

Fig. 1. Reciprocal dopamine-D1:D2 receptor interactions. Saturation analysis of [3H]raclopride **(top)** or [3H]SCH 23390 **(bottom)** binding to human caudate membranes. Addition of 800 nM dopamine reduces the estimated density of either D2 or D1 receptors when assayed with either [3H]raclopride or [3H]SCH 23390, respectively. Preincubation of membranes with the selective D1 antagonist SCH 23390 (top) or D2 antagonist, eticlopride (bottom), prevents the reduction of dopamine induced decreases in the B_{max} of these radioligands with a concomitant increase in the estimated *Kd*. (Figure adapted from Seeman et al., 1989.)

23390 for the D1 receptor (~500 p*M*) (*see* Fig. 2). In cells expressing only D2 receptors, SCH 23390 has no effect. An even more simplified assay using three points rather than curves is described in Figure 2. This assay of dopamine-D1:D2 receptor interactions is both simpler to conduct and much less

                                    Pristupa, Sunahara, and Niznik

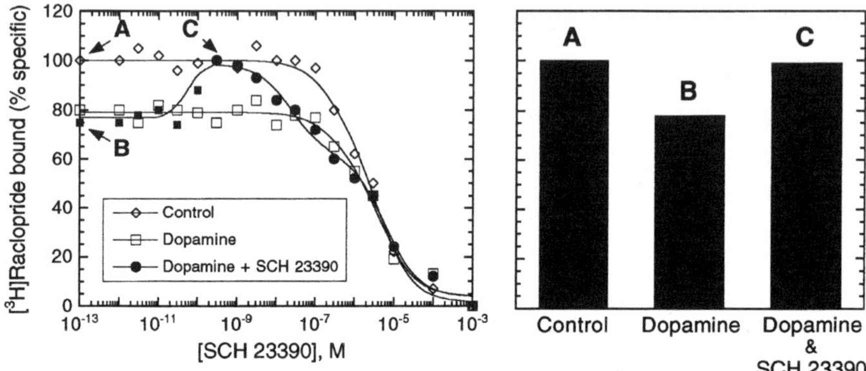

Fig. 2. [³H]Raclopride/SCH 23390 competition binding assay reveals D1:D2 receptor interactions. The binding of [³H]raclopride to membranes prepared from a stable Y1 cell line expressing human dopamine D2$_{Short}$ receptor. Figure at left depicts SCH 23390/[³H]raclopride competition curves: (**A**) Control; (**B**) in the presence of 500 n*M* dopamine; (**C**) 500 n*M* dopamine with Y1 cells expressing D1 as well as D2$_{Short}$ receptors. The increase in [³H]raclopride binding at (**C**) is interpreted as representing as interaction between D1 and D2 receptors. Note the concentration of SCH 23390 yielding maximal effects on [³H]raclopride binding corresponds to the estimated Kd of this compound at the D1 receptor (~ 500 p*M*). Bar graph at right depicts the main effects on [³H]raclopride binding. (Figure adapted from Niznik et al., 1995.)

costly than the one originally described. Although the mechanism causing the SCH 23390-induced increase in [³H]raclopride binding is unknown, it appears to reflect a cooperative mechanism, similar to that observed in other receptor systems (e.g., Wreggett and Wells, 1995).

Although various theories have been proposed (Agnati et al., 1993; Lemmon et al., 1992; Turgeon and Warring, 1992), the molecular mechanism(s) mediating the observed D1 and D2 receptor interaction, and its absence in some neuropsychiatrically diseased groups, is currently unknown. With the availability of the clones for dopamine-D1 and D2 receptors, it is now possible to assess at the molecular level the mechanism(s) responsible for the maintenance and expression of D1:D2 receptor interactions. In order to ascertain why some diseased individuals do not exhibit D1:D2 receptor interactions, we have embarked on a program examining D1:D2 receptor interactions in stable transfected cell lines coexpressing cloned D2 (Bunzow et al., 1988), and D1 (Sunahara et al., 1990) or D5 (Sunahara et al., 1991) receptors.

D1:D2 Receptor Coexpression Studies

In order to ascertain whether we could establish or reconstitute dopamine receptor interactions, initial experiments were conducted on GH4 cell lines stably expressing either the long or short form of the dopamine-D2 receptor. D1 or D5 receptors were obtained from genomic DNA of controls who display a D1:D2 link in vitro using radioligand binding techniques. As expected, cloned D1 receptors from control individuals that have been transiently transfected into GH4 cells expressing $D2_{Short}$ receptors were capable of reconstituting dopamine D1:D2 receptor crosstalk, as indexed by the dopamine-induced reduction of the maximal number of estimated [^{3}H]raclopride binding sites; this reduction was prevented by preincubation with SCH 23390. Of course, cells expressing only the dopamine-D2 receptor show no evidence of this link. Thus, although a dopamine-induced reduction in [^{3}H]raclopride binding is evident in cell lines expressing only the D2 receptor, SCH 23390 could not antagonize this reduction, an effect identical to that obtained in tissues of the pituitary gland (*see above*). Furthermore, when the D1 gene of a schizophrenic, which shows no D1:D2 receptor link in neural tissue in vitro, is transiently expressed in $GH4-D2_{Short}$ lines, D1:D2 receptor interactions are fully restored. These data, coupled with the observations that there appear to be no significant sequence mutations of the D1-like (O'Hara et al., 1993; Cichon et al., 1994; Liu et al., 1995; Sobell et al., 1995) or D2-like (Sarkar et al. 1991; Barr et al., 1993; Hallmayer et al., 1994; Macciardi, et al., 1994; Shaikh et al., 1994; Tanaka et al., 1995; Grassi et al., 1996; Rothschild et al., 1996, Tanaka et al., 1996) receptor genes in schizophrenics, suggest that the observed absence of D1:D2 receptor interactions in vitro in tissues from some schizophrenics is not due to a primary defect in the structure of either the dopamine D1 or D2 receptor proteins. In vitro measurement of the D1:D2 link is a phenomenon not specific to GH4 cell lines. Y1 cells, derived from mouse adrenocortical cells (Schimmer, 1979) co-expressing the human D1 and $D2_{Short}$ receptors also show the functional D1:D2 link.

Are $D1/D2_{Short}$ Receptor Interactions Uni- or Multicellular?

There is considerable debate regarding whether dopamine receptor interactions are uni- or multicellular in nature. There is evidence on one hand for dopamine receptor interactions at the level of the single cell, where D1 and D2 receptors and their mRNA are postulated to co-exist in up to ~40% of striatal neurons (Seeman et al., 1989; Weiner et al., 1991; Surmeier et al., 1992, 1993; Ariano et al., 1992; Lester et al., 1993; Rappaport et al., 1993; Curran and Watson, 1995; Hersch et al., 1995; Le Moine and Bloch, 1996; Shetreat et al., 1996). Coexpression of cloned D1 and D2 receptors enhances the release of arachidonic acid in CHO cells (Piomelli et al., 1991), and D1 and D2 receptors interact in single dissociated neurons in primary culture to modulate the activity of Na^+/K^+-ATPase activity (Bertorello et al., 1990). These systems may be responsible for the observed link between various dopamine receptor subtypes within the same cell, or even on adjacent neurons. Evidence from other laboratories, however, has suggested that if D1 and D2 receptors coexist on striatal interneurons, they do so on a very small percentage of selected cells (*see* Harrison et al., 1990; Gerfen et al., 1991; Le Moine et al., 1991; Gerfen, 1992a,b; Le Moine and Bloch, 1995).

In order to determine whether $D1:D2_{Short}$ receptor coexpression in the same cell is an obligatory requirement for the establishment of the link, we assayed $D1:D2_{Short}$ receptor interactions in cell lines separately expressing these receptors (Niznik et al., 1995; Sunahara et al., unpublished). The presence of the link in cells coexpressing D1 and $D2_{Short}$ was compared with a mixture of two populations of cells—one expressing only D1, the other expressing only $D2_{Short}$. Cells

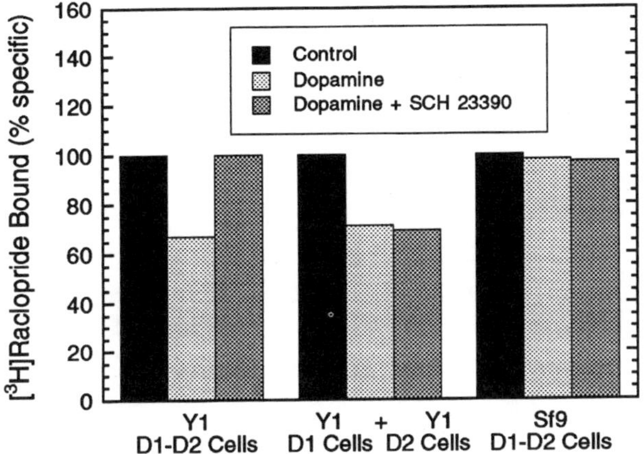

Fig. 3. D1:D2 Receptor interactions only occur when both receptors are expressed within the same cell. Y1 cell lines stably coexpressing D1 and D2$_{Short}$ receptors are shown to exhibit D1:D2 receptor interactions as determined by the "link" assay described in Fig. 2. Stable Y1 cell lines expressing each receptor individually, when pooled and homogenized, were found to be unable to reconstitute D1:D2 receptor interactions. In contrast, Sf9 cells coexpressing D1 and D2$_{Short}$ receptors did not exhibit the D1:D2 receptor link.

were mixed, homogenized, treated, and assayed for D1:D2 receptor crosstalk. Care was taken to ensure the equimolar equivalence of D1 and D2 receptors in these homogenates.

As depicted in Fig. 3, data obtained from these experiments clearly indicate, at least as indexed by our assay procedure, that D1 and D2 receptors need to be expressed in the same cell in order to establish and maintain receptor crosstalk. Moreover, D1:D2 receptor interactions are not solely restricted to these cloned receptor subtypes, since identical results were obtained when dopamine-D5 and D2 receptor interactions were assessed (data not shown).

The data clearly reveal that mixing or homogenizing cells that individually express equal amounts of either D1/D5- or D2-receptors is not sufficient to establish receptor crosstalk. Coexpression, and possibly close cellular colocalization, of D1 and D2 receptors is an absolute requirement for the maintenance of these dopamine-D1:D2 receptor interactions. These data provide strong support for the coexistence of subtype-specific dopamine receptors on select populations of striatal neurons. Indeed, there is mounting molecular, anatomical, and biochemical evidence for the existence of just such a set of striatal-nigral projection or local striatal interneurons in both rats and monkeys (*see* Surmeier et al., 1983; for review: Lester et al., 1993; Rappaport et al., 1993 Shetreat et al., 1996; Levey et al., 1993). Moreover, the data suggest the possibility that the observed absence of D1:D2 receptor crosstalk in some schizophrenics is restricted to a subpopulation of striatal cells co-expressing D1:D2 or D5:D2 receptors or receptor mRNAs. As such, these data also provide a testable hypothesis of the molecular mechanisms regulating the presence or absence of dopamine receptor crosstalk in schizophrenia. Is there selective cell death of striatal nigral projections neurons co-expressing D1/D5:D2 receptors in some schizophrenics? Is there a selective loss or perturbation of either dopamine-D1/D5:D2 receptor mRNA expression in cells that normally co-express both subtype-specific dopamine receptors? Is the cellular machinery (e.g., G protein-mediated pathways) downstream from these receptors perturbed, thereby inactivating normal dopamine receptor crosstalk? Although there is no evidence yet for specific cell death, or for selective aberrations of cell-specific and/or subtype-specific dopamine receptor mRNA expression, it does appear that specific G proteins may be involved in the molecular mechanisms driving dopamine-D1:D2 receptor interactions (*see also* Seeman et al., 1994). This idea is supported by the lack of observed D1:D2 receptor interaction with expression of D1 and either $D2_{Long}$ or $D2_{Short}$ receptors in Sf9 cells (Fig. 3; Lee, et al., unpublished), which display poor coupling to the complement of subtype-specific G proteins found in these cells (Boundy et al., 1996).

Role of G Proteins in the Maintenance of D1:D2 Receptor Interactions

We have attempted to assess whether signal transduction proteins, in particular, G protein subunits, are involved in the maintenance or expression of the observed D1:D2 link in these cells. We have previously hypothesized (Seeman et al., 1989) that guanine nucleotide binding proteins may be part of the cellular machinery required to modify or maintain dopamine D1:D2 receptor interactions in native membranes, and may even constitute the molecular basis for dopamine receptor subtype interactions. Cells coexpressing subtype-specific forms of dopamine D1 and D2 receptors treated with various agents known to modify G protein function and coupling to receptors (e.g., Gpp(NH)p, cholera toxin, pertussis toxin, G protein specific antisense RNA) may help to identify those G protein subunits responsible for dopamine receptor interactions.

G Protein α-Subunits

The involvement of subtype-specific G protein α-subunits in the generation of the D1:D2 link was investigated (Niznik et al., 1995; Sunahara et al., unpublished) by the use of bacterial toxins which cause ADP ribosylation and subunit dissociation of specific α-subunits of G proteins in these cell lines. Treatment of cells with pertussis toxin, which specifically activates G_i and G_o α-subunits and uncouples receptor-G protein complexes, prevents the reduction in [^{3}H]raclopride binding caused by dopamine. These data suggest and confirm original observations in native tissues that the reduction in [^{3}H]raclopride binding by dopamine is caused by a "locking in" of high-affinity D2 receptors, thereby occluding these sites to radioligand binding via a G_i-like mechanism (*see* Seeman et al., 1989, Seeman and Niznik, 1990 for discussion). Treatment with cholera toxin, which specifically activates $G\alpha_s$, has no effect on the ability of dopamine to reduce [^{3}H]raclopride binding to D2 receptors, but does prevent the ability of SCH 23390 to block/reverse the effect of dopamine and the establishment of D1:D2 receptor crosstalk. These results establish the involvement of the stimulatory G protein α-subunit, $G\alpha_s$, in the maintenance of the D1:D2 receptor interactions.

To evaluate further the involvement of specific G proteins in the D1:D2 link, cells coexpressing the human D1 and D2$_{Short}$ receptors were transfected with antisense oligonucleotides specific for mouse α-subunits of G$_s$, G$_{i3}$, and G$_q$ (Niznik et al., 1995; Sunahara et al., unpublished). Western blotting of cell extracts using antibodies to subtype-specific G proteins confirmed the reduction of protein expression after treatment. Anti-sense oligonucleotides to Gα_{i3} had the same effect as pertussis toxin, eliminating the inhibitory effect of dopamine on [^{3}H]raclopride binding. Antisense oligonucleotides to Gα_s have an effect similar to cholera toxin, allowing dopamine to reduce [^{3}H]raclopride binding, without allowing for the complete recovery of binding in the presence of the D1 antagonist SCH 23390. (It should be noted that, unlike the effects on Gα_{i3} or Gα_q, incomplete inhibition of Gα_s expression, as determined by Western blotting, was obtained using these oligonucleotides.) Moreover, no effect on the D1:D2 link was observed when cells were treated with antisense RNA to Gα_q, which mediates G protein receptor effects on turnover of phosphoinositides. These data are in accord with the preferential coupling of D2 receptors to the specific G protein α subunits α_{i2} and α_{i3} (O'Hara et al., 1996).

D1/D5 SYNTHETIC PEPTIDES AND ACTIVATION OF Gα_s-PROTEINS

One additional approach to determine the role of G proteins in the maintenance of dopamine receptor interactions has been to attempt to block D1:D2$_{Short}$ receptor interactions in recombinant cells by the addition of D1/D5 receptor-specific peptides encoding amino acid residues within the carboxyl portion of the third cytoplasmic loop responsible for D1 activation of adenylate cyclase (Niznik et al., 1995, Sunahara et al., unpublished). Site-specific synthetic peptides to defined sequences in the second and third cytoplasmic loop regions of G protein-linked receptors have been used to identify G protein-activating or inhibitory functions (Okamoto et al., 1990, 1991; Okamoto and Nishimoto, 1992; Mukai et al., 1992; Dalman and Neubig, 1991; Cheung et al.,

1991, 1992; Deutsch and Sun, 1992; Cheng et al., 1992; Taylor et al., 1994, 1996; Wade et al., 1994). First, these studies establish that functional activation of G proteins and adenylate cyclase does not require the entire receptor—short peptides containing critical amino acid sequences are also effective. Second, the effects of modifications of the peptide by phosphorylation or site-directed mutagenesis may be used to assess receptor–G protein coupling. Thus, synthetic β2 adrenergic peptides, which activate $G\alpha_s$ and adenylate cyclase, when phosphorylated by PKA, lose their ability to interact with G_s, but display increased affinity for G_i (Okamoto et al., 1991).

Dopamine-D1 and D5 receptors display significant sequence divergence within the second and third cytoplasmic loops. Site-specific synthetic peptides to these regions may define critical amino acid residues required for G protein activation or inhibition and maintenance of dopamine receptor interactions. Two peptides were synthesized to correspond to the region of the D1 and D5 receptor that display strong sequence homology to a β2-adrenergic receptor sequence that has been shown to activate G_s, the C-terminal portion of the third cytoplasmic loop. The peptides described above all contain classical consensus sequences for PKA or pseudo-PKC phosphorylation (*see* Kemp and Pearson, 1990).

Since activation of G_s results in the stimulation of adenylate cyclase, these peptides were tested for their effects on the accumulation of cAMP in cells coexpressing D1- and $D2_{Short}$ receptors (Niznik et al., 1995, Sunahara et al., unpublished). A D2 receptor peptide (identical to the 29 amino acid residues encoded by exon 6, which is present in $D2_{Long}$, but absent in $D2_{Short}$) had no effect on levels of cAMP, which is consistent with recent observations (Malek et al., 1993) whereas the D1- and D5-receptor peptides stimulated the accumulation of cAMP by four- to fivefold. The approx 10-fold difference in the estimated EC_{50} of D1 and D5 peptides on cAMP accumulation parallels the observed differences in binding affinity displayed by these receptors for dopamine.

Similar results have been obtained on adenylate cyclase activity in these cells (Jarvie et al., 1993; Tiberi and Caron, 1994).

INHIBITION OF D1/D5:D2$_{Short}$ INTERACTIONS
VIA SITE-SPECIFIC PEPTIDES

Interaction between D1 and D2$_{Short}$ was measured in the presence of chemically synthesized site-specific peptide from the D2 receptor, as well as peptides derived from analogous regions of the D1 and D5 receptors. The D2-derived peptide has no inhibitory effect on the D1:D2 link, suggesting that D2$_{Short}$ and D2$_{Long}$ may not have differential involvement in the mechanism of the receptor interaction in this cell line, as suspected from previous results obtained in other cells (*see below*). The D1- and D5-specific peptides, which are capable of activation of Gα_s and subsequent formation of cAMP, interfere with the ability of SCH 23390 to inhibit the effect of dopamine on [^{3}H]raclopride binding, indicating that the D1:D2 receptor link is mediated via Gα_s.

Although we did not test for peptide stability and/or interaction with membrane components, the data suggest that activation of Gα_s abolishes dopamine receptor interactions. Although indirect, the data would support the finding that SCH 23390 can display some partial agonist activity (*see* Tiberi and Caron, 1994). We next undertook to assess the role of other components of the G protein system, such as $\beta\gamma$ subunits, for their influence on the expression of dopamine D1:D2 receptor interactions. $\beta\gamma$-subunits are known to activate adenylate cyclase and other effector systems on their own (Tang and Gilman, 1991; Federman et al., 1992; Coso et al., 1996; Herlitze et al., 1996; Nakajima et al., 1996; Suh and Hamm, 1996; Sunahara et al., 1996), influence the ability of receptor coupling to Gα_s-subunits (Schmidt et al., 1994), and translocate receptor-specific protein kinases (Kim et al., 1993; Haske et al., 1996, and see Clapham and Neer, 1993; Gudermann et al., 1996; Hamm and Gilchrist, 1996 for reviews).

G Protein $\beta\gamma$-Subunits

We have recently obtained some data (Niznik et al., 1995; Sunahara et al., unpublished) that strongly suggest that G$_s$-

protein βγ-subunits, which are associated with the D1 receptor, are the mediators of the dopamine-D1:D2 link in vitro. Thus, membrane pretreatment with dopamine and SCH 23390 appears to favor the formation of $G\alpha_s$-βγ complexes, whereas stimulation by dopamine alone allows for their uncoupling as assessed by Western blotting with antibodies to βγ-subunits. Further work will of course be necessary to prove that βγ-subunits are the regulators of D1:D2 interactions, but the available data already suggest a mechanism whereby we may explain dopamine receptor interactions. In essence, we propose that the binding of SCH 23390 to D1 receptors, at a time when dopamine is allowed to stimulate D2 receptors, sequesters free βγ-subunits via the formation of a strong D1 receptor–G protein complex (possibly owing to partial agonist activity of SCH 23390; e.g., *see* Tiberi and Caron, 1994). The altered stoichiometry of receptor-G protein subunit distribution and availability of free βγ-subunits favor the dissociation of $G\alpha_i$-like subunits from the D2 receptor and the formation of a low-affinity or uncoupled form of the receptor. This low-affinity form of the D2 receptor is easily recognized and successfully labeled by [^{3}H]raclopride. Whether a similar mechanism is responsible for these receptor interactions in native human cells is currently unknown. The data suggest, however, that at least in schizophrenic tissues not expressing the D1:D2 receptor link, specific aberrations of particular molecular forms of βγ-subunits, or inappropriate expression of these proteins in cell-specific striatal-nigral populations, might be responsible for the lack of receptor crosstalk. Interestingly, Seeman et al. (1993) have noted perturbation of D2 receptor coupling to G protein α-subunits; similar observations have been made with the D1 receptor (Mamelak et al., 1993).

D2 Receptor Subtype (D2$_{Long}$/D2$_{Short}$) Link with D1/D5

We have obtained evidence that subtype-specific G protein α-subunits, and possibly βγ-subunits, are required to maintain dopamine-D1:D2 receptor interactions. Indeed, although $D2_{Long}$ and $D2_{Short}$ are pharmacologically indistinguishable, some differences are evident in terms of their interactions with specific G proteins (Dal Toso et al., 1989; Senogles et al., 1990; Missale et al., 1991; Lledo et al., 1992; Montmayeur et al., 1993; Boundy et al., 1996; Falardeau, 1994), responsiveness to posttranslational modifications (Liu et al., 1992), glycosylation and intracellular trafficking (Fishburn et al., 1995), and dopamine-induced sequestration (Itokawa et al., 1996). Regions within the third cytoplasmic loop of G proteins have been shown to be required for selectivity of G protein coupling (Cheung et al., 1992; Cotecchia et al., 1992; and *see* Dohlman et al., 1991; Bockaert, 1991; Kobilka, 1992; Ostrowski et al., 1992, for reviews).

The presence or absence of D1:D2-receptor interactions in cell lines expressing these receptors may be solely dependent upon which of the two molecular forms of D2 receptor are present in these cells. We have established through the Y1 stable cell lines that either D1 or D5 receptors are capable of establishing linkage to the $D2_{Short}$ receptor. As such, there is no functional difference between these receptors in regulation of D2 receptor activity. Preliminary data suggest that although $D2_{Short}$ may always link to D1, $D2_{Long}$ appears incapable of doing so, at least in GH4 cells in which D1 was transiently cotransfected (Niznik et al., 1995; Sunahara et al., unpublished). Since, as outlined above, the maintenance of D1:D2 receptor interactions appears to be dependent on the appropriate complement of G proteins, it is not surprising that different cell lines, each with its own population of G protein subunits, will differentially allow for the expression of D1:D2 receptor interactions. This issue, of course, complicates the picture in human native membranes, since we certainly do not know which of the multiple G protein α- and/or $\beta\gamma$-subunits are specifically expressed in striatal nigral cells coexpressing D1 and D2 receptors. Future work, using *in situ* hybridization and immunohistochemical detection of subtype-specific G proteins and receptors via double-labeling techniques may resolve this issue. We are currently attempting such a study, since we have recently developed specific human dopamine D1/D5 and D2 receptor antibodies (Levey et al., 1993) and have visualized these proteins via EM.

Significance

Determination of the mechanism of the normal D1:D2 receptor interactions (e.g., involvement of altered G protein structure, expression or function)

1. May assist in explaining the observed lack of these events in about half of psychotic individuals;
2. Will allow for the elucidation of genetic roles in the maintenance of the diseased state, moreover, this work may point toward novel screening method for the evaluation of G protein involvement in disease; and
3. May allow us come to a better understanding of the actions of dopamine receptors, which may point the way to a more beneficial use of either selective dopamine D1 and/or D2 receptor drugs in the treatment of psychomotor disease.

References

Agnati, L. F., Fuxe, K., Benfenati, F., von Euler, G., and Fredholm, B. (1993) Intramembrane receptor-receptor interactions: integration of signal transduction pathways in the nervous system. *Neurochem. Int.* **22,** 213–222.

Albert, P. R., Neve, K. A., Bunzow, J. R., and Civelli, O. (1990) Coupling of a cloned rat dopamine D2 receptor to inhibition of adenylate cyclase and prolactin secretion. *J. Biol. Chem.* **265,** 2098–2104.

Apostolakis, E. M., Garai, J., Fox, C., Smith, C. L., Watson, S. J., Clark, J. H., and O'Malley, B. W. (1996) Dopaminergic regulation of progesterone receptors: brain D5 dopamine receptors mediate induction of lordosis by D1-like agonists in rats. *J. Neurosci.* **16,** 4823–4834.

Ariano, M. A., Stromski, C. J., Smyk-Randall, E. M., and Sibley, D. R. (1992) D2 dopamine receptor localization on striatonigral neurons. *Neurosci. Lett.* **144,** 215–220.

Artalejo, C. R., Ariano, M. A., Perlman, R. L., and Fox, A. P. (1990) Activation of facilitation calcium channels in chromaffin cells by D1 dopamine receptors through a cAMP/protein kinase A-dependent mechanism. *Nature* **348,** 239–242.

Bardo, M. T., Donohew, R. L., and Harrington, N. G. (1996) Psychobiology of novelty seeking and drug seeking behavior. *Behav. Brain Res.* **77,** 23–43.

Barr, C. L., Kennedy, J. L., Lichter, J. B., Van Tol, H. H. M., Wetterberg, L., Livak, K. J., and Kidd, K. K. (1993) Alleles at the dopamine D4 receptor locus do not contribute to the genetic susceptibility to schizophrenia in a large swedish kindred. *Am. J. Med. Genet.* **48,** 218–222.

Bedard, P. J. (1995) D-1 receptors in neurology and psychiatry—an overview. *Prog. Neuro.-Psychopharmacol. Biol. Psychiatr.* **19,** 713–726.

Bertorello, A. M., Hopfield, J. F., Aperia, A., and Greengard, P. (1990) Inhibition by dopamine of Na$^+$-K$^+$-ATPase activity in neostriatal neurons through D1 and D2 dopamine receptor synergism. *Nature* **347,** 386–388.

Bloom, F. E. (1993) Advancing a neurodevelopmental origin for schizophrenia. *Arch. Gen. Psychiatry* **50,** 224–227.

Bockaert, J. (1991) G proteins and G protein-coupled receptors: structure, function and interactions. *Curr. Opinion Neurobiol.* **1,** 32–42.

Boundy, V. A., Lu, L. H., and Molinoff, P. B. (1996) Differential coupling of rat D2 dopamine receptor isoforms expressed in *Spodoptera frugiperda* insect cells. *J. Pharmacol. Exp. Ther.* **276,** 784–794.

Bunzow, J. R., Van Tol, H., Grandy, D. K., Albert, P., Salon, J., Christie, M., Machida, C. A., Neve, K. A., and Civelli, O. (1988) Cloning and expression of a rat D2 dopamine receptor cDNA. *Nature* **336,** 783–787.

Burris, T. P. and Freeman, M. E. (1994) Comparison of the forms of the dopamine D2 receptor expressed in GH(4)C(1) cells. *Proc. Soc. Exp. Biol. Med.* **205,** 226–235.

Burris, T. P., Nguyen, D. N., Smith, S. G., and Freeman, M. E. (1992) The stimulatory and inhibitory effects of dopamine on prolactin secretion involve different G proteins. *Endocrinology* **130,** 926–932.

Cameron, D. L. and Williams, J. T. (1993) Dopamine D1 receptors facilitate transmitter release. *Nature* **366,** 344–347.

Castellano, M. A., Liu, L. X., Monsma, F. J., Sibley, D. R., Kapatos, G., and Chlodo, L. A. (1993) Transfected D2 short dopamine receptors inhibit voltage-dependent potassium current in neuroblastoma x glioma hybrid (NG108–15) cells. *Mol. Pharmacol.* **44,** 649–656.

Chen, C. and Lokhandwala, M. F. (1993) Inhibition of Na$^+$-K$^+$-ATPase in rat renal proximal tubules by dopamine involved DA-1 receptor activation. *Naunyn-Schmiedebergs Arch. Pharmacol.* **347,** 289–295.

Cheng, H. C., Nishio, H., Hatase, O., Ralph, S., and Wang, J. H. (1992) A synthetic peptide derived from p34cdc2 is a specific and efficient substrate of src-family tyrosine kinases. *J. Biol. Chem.* **267,** 9248–9256.

Cheung, A. H., Huang, R. R., Graziano, M. P., and Strader, C. D. (1991) Specific activation of Gs by synthetic peptides corresponding to an intracellular loop of the beta-adrenergic receptor. *FEBS Lett.* **279,** 277–280.

Cheung, A. H., Huang, R. R., and Strader, C. D. (1992) Involvement of specific hydrophobic, but not hydrophilic, amino acids in the third intracellular loop of the beta-adrenergic receptor in the activation of Gs. *Mol. Pharmacol.* **41,** 1061–1065.

Chio, C. L., Lajiness, M. E., and Huff, R. M. (1994) Activation of heterologously expressed D3 dopamine receptors—comparison with D2 dopamine receptors. *Mol. Pharmacol.* **45,** 51–60.

Cichon, S., Nothen, M. M., Rietschel, M., Korner, J., and Propping, P. (1994) Single-strand conformation analysis (SSCA) of the dopamine D1 receptor gene (DRD1) reveals no significant mutation in patients with schizophrenia and manic depression. *Biol. Psychiatry* **36,** 850–863.

Civelli, O., Bunzow, J. R., and Grandy, D. K. (1993) Molecular diversity of the dopamine receptors. *Ann. Rev. Pharmacol. Toxicol.* **32,** 281–307.

Clapham, D. E. and Neer, E. J. (1993) New roles for G protein βγ-dimers in transmembrane signalling. *Nature* **365,** 403–406.

Clarke, D. and White, F. J. (1987) Review: D1 dopamine receptor—the search for a function: a critical evaluation of the D1/D2 dopamine receptor classification and its functional implications. *Synapse* **1,** 347–388.

Cleghorn, J. M., Zipursky, R. B., and List, S. J. (1991) Structural and functional brain imaging in schizophrenia. *J. Psychiatr. Neurosci.* **16,** 53–74.

Code, R. and Tang, A. (1991) Yawning produced by dopamine agonists in rhesus monkeys. *Eur. J. Pharmacol.* **201,** 235–238.

Cole, A. J., Bhat, R. B., Patt, C., Worley, P. F., and Baraban, J. M. (1992) D1 dopamine receptor activation of multiple transcription factor genes in rat striatum. *J. Neurochem.* **58,** 1420–1426.

Coso, O. A., Teramoto, H., Simonds, W. F., and Gutkind, J. S. (1996) Signaling from G protein-coupled receptors to c-Jun kinase involves beta gamma subunits of heterotrimeric G proteins acting on a ras and rac1-dependent pathway. *J. Biol. Chem.* **271,** 3963–3966.

Cotecchia, S., Ostrowski, J., Kjelsberg, M. A., Caron, M. G., and Lefkowitz, R. J. (1992) Discrete amino acid sequences of the α1-adrenergic receptor determine the selectivity of coupling to phosphatidylinositol hydrolysis. *J. Biol. Chem.* **267,** 1633–1639.

Curran, E. J. and Watson, S. J., Jr. (1995) Dopamine receptor mRNA expression patterns by opioid peptide cells in the nucleus accumbens of the rat: a double in situ hybridization study. *J. Comp. Neurol.* **361,** 57–76.

Dalman, H. M. and Neubig, R. R. (1991) Two peptides from the alpha 2A-adrenergic receptor alter receptor G protein coupling by distinct mechanisms. *J. Biol. Chem.* **266,** 11,025–11,029.

Dal Toso, R., Sommer, B., Ewert, M., Herb, A., Pritchett, D., Bach, A., Shivers, B., and Seeberg, P. (1989) The dopamine D2 receptor: two molecular forms generated by alternative splicing. *EMBO J.* **8,** 4025–4034.

Daly, S. A. and Waddington, J. L. (1992) Two directions of dopamine D1/D2 receptor interaction in studies of behavioural regulation: a finding generic to four new, selective dopamine D1 receptor antagonists. *Eur. J. Pharmacol.* **213,** 251–258.

Davidson, M., Harvey, P. D., Bergman, R. L., Powchik, P., Kaminsky, R., Losonczy, M. F., and Davis, K. L. (1990) Effects of the D1 agonist SKF-38393 combined with haloperidol in schizophrenic patients. *Arch. Gen. Psychiatry* **47,** 190–191.

Davis, K. L., Kahn, R. S., Ko, G., and Davidson, M. (1991) Dopamine in schizophrenia: a review and reconceptualization. *Am. J. Psychiatry* **148,** 1474–1486.

de Beaurepaire, R., Labelle, A., Naber, D., Jones, B. D., and Barnes, T. R. (1995) An open trial of the D1 antagonist SCH 39166 in six cases of acute psychotic states. *Psychopharmacology* **121,** 323–327.

Demchyshyn, L. L., Sugamori, K. S., Lee, F. J. S., Hamadanizadeh, S. A., and Niznik, H. B. (1995) The dopamine D1D receptor. Cloning and characterization of three pharmacologically distinct D1-like receptors from *Gallus domesticus. J. Biol. Chem.* **270,** 4005–4012.

Den Boer, J. A., van Megen, H. J., Fleischhacker, W. W., Louwerens, J. W., Slaap, B. R., Westenberg, H. G., Burrows, G. D., and Srivastava, O. N. (1995) Differential effects of the D1-DA receptor antagonist SCH39166 on positive and negative symptoms of schizophrenia. *Psychopharmacology* **121,** 317–322.

Deutsch, P. J. and Sun, Y. (1992) The 38-amino acid form of pituitary adenylate cyclase-activating polypeptide stimulates dual signaling cascades in PC12 cells and promotes neurite outgrowth. *J. Biol. Chem.* **267,** 5108–5113.

DiMarzo, V., Vial, D., Sokoloff, P., Schwartz, J. C., and Piomelli, D. (1993) Selection of alternative Gi-mediated signaling pathways at the dopamine D2 receptor by protein kinase-C. *J. Neurosci.* **13,** 4846–4853.

Dohlman, H. G., Thorner, J., Caron, M. G., and Lefkowitz, R. J. (1991) Model systems for the study of seven transmembrane segment receptors. *Ann. Rev. Biochem.* **60,** 653–688.

Dowling, J. E. (1991) Retinal neuromodulation: the role of dopamine. *Vis. Neurosci.* **7,** 87–97.

Dowling, J. E. (1994) The neuromodulatory role of dopamine in the teleost retina, in *Dopamine Receptors and Transporters. Pharmacology, Structure, and Function* (Niznik, H. B., ed.), Marcel Dekker, New York, pp. 37–57.

Ellison, G. (1994) Stimulant-induced psychosis, the dopamine theory of schizophrenia, and the habenula. *Brain Res. Rev.* **19,** 223–239.

Elsholtz, H. P., Lew, A. M., Albert, P. R., and Sundmark, V. C. (1991) Inhibitory control of prolactin and Pit1 gene promotors by dopamine: dual signaling pathways required for D2 receptor regulated expression of the prolactin gene. *J. Biol. Chem.* **266,** 22,915–22,925.

Falardeau, P. (1994) Functional distinctions of D2Long and D2Short receptors., in *Dopamine Receptors and Transporters. Pharmacology, Structure, and Function* (Niznik, H. B., ed.), Marcel Dekker, New York, pp. 323–342.

Federman, A. D., Conklin, B. R., Schrader, K. A., Reed, R. R., and Bourne, H. R. (1992) Hormonal stimulation of adenylyl cyclase through Gi-protein beta gamma subunits. *Nature* **356,** 159–161.

Felder, C. C., Blecher, M. and Jose, P. A. (1989) Dopamine D1 mediated stimulation of phospholipase C activity in rat cortical membranes. *J. Biol. Chem.* **264,** 8739–8745.

Felder, C. C., Campbell, T., Albrecht, F., and Jose, P. A. (1990) Dopamine inhibits Na^+-H^+ exchanger activity in renal BBMV by stimulation of adenylate cyclase. *Am. J. Physiol.* **259,** F297–303.

Felder, C. C., Albrecht, F. E., Campbell, T., Eisner, G. M., and Jose, P. A. (1993) cAMP-independent, G protein-linked inhibition of Na^+/H^+ exchange in renal brush border by D1 dopamine agonists. *Am. J. Physiol.* **264,** F1032–F1037.

Fishburn, C. S., Elazar, Z., and Fuchs, S. (1995) Differential glycosylation and intracellular trafficking for the long and short isoforms of the D2 dopamine receptor. *J. Biol. Chem.* **270,** 29,819–29,824.

Gardette, R., Rasolonjanahary, R., Kordon, C., and Enjalbert, A. (1994) Epidermal growth factor treatment induces D2 dopamine receptors functionally coupled to delayed outward potassium current (I-K) in GH4C1 clonal anterior pituitary cells. *Neuroendocrinol,* **59,** 10–19.

Gerfen, C. R. (1992a) The neostriatal mosaic—multiple levels of compartmental organization. *J. Neural Transm.—Gen. Sect.* 43–59.

Gerfen, C. R. (1992b) The neostriatal mosaic: multiple levels of compartmental organization in the basal ganglia. *Ann. Rev. Neurosci.* **15,** 285–320.

Gerfen, C. R. (1995) Dopamine-receptor function in the basal ganglia. *Clin. Neuropharmacol.* **18,** S162–S177.

Gerfen, C. R., McGinty, J. F., and Young, W. S. I. (1991) Dopamine differentially regulates dynorphin, substance P, and enkephalin expression in striatal neurons: in situ hybridization histochemical analysis. *J. Neurosci.* **11,** 1016–1031.

Gerfen, C. R., Keefe, K. A., and Gauda, E. B. (1995) D1 and D2 dopamine receptor function in the striatum: coactivation of D1- and D2-dopamine receptors on separate populations of neurons results in potentiated immediate early gene response in D1-containing neurons. *J. Neurosci.* **15,** 8167–8176.

Gilman, A. G. (1995) G proteins and regulation of adenylate cyclase (Nobel lecture). *Angew. Chem. Int. Ed.* **34,** 1406–1419.

Gingrich, J. A. and Caron, M. G. (1993) Recent advances in the molecular biology of dopamine receptors. *Ann. Rev. Neurosci.* **16,** 299–321.

Goldman-Rakic, P. S. (1995) Cellular basis of working memory. *Neuron* **14,** 477–485.

Goldstein, M. and Deutch, A. Y. (1992) Dopaminergic mechanisms in the pathogenesis of schizophrenia. *FASEB J.* **6,** 2413–2421.

Gomez-Mancilla, B. and Bedard, P. J. (1991) Effect of D1 and D2 agonists and antagonists on dyskinesia produced by L-DOPA in 1-methyl-4-phenyl-1,2,3,6-tetrahydropyridine-treated monkeys. *J. Pharmacol. Exp. Ther.* **259,** 409–413.

Grandy, D. K., Bunzow, J. R., and Civelli, O. (1994) The dopamine-D2 receptor, in *Dopamine Receptors and Transporters. Pharmacology, Structure, and Function* (Niznik, H. B., ed.), Marcel Dekker, Inc, New York, pp. 151–164.

Grassi, E., Mortilla, M., Amaducci, L., Pallanti, S., Pazzagli, A., Galassi, F., Guarnieri, B. M., Petruzzi, C., Bolino, F., Ortenzi, L., Nistico, R., Decataldo, S., Rossi, A., and Sorbi, S. (1996) No evidence of linkage between schizophrenia and D2 dopamine receptor gene locus in italian pedigrees. *Neurosci. Lett.* **206,** 196–198.

Greif, G. J., Lin, Y. J., Liu, J. C., and Freedman, J. E. (1995) Dopamine-modulated potassium channels on rat striatal neurons: specific activation and cellular expression. *J. Neurosci.* **15,** 4533–4544.

Gudermann, T., Kalkbrenner, F., and Schultz, G. (1996) Diversity and selectivity of receptor-G protein interaction. *Ann. Rev. Pharmacol. Toxicol.* **36,** 429–459.

Hall, H. (1994) Dopamine receptors: radioligands for pharmacological and biochemical characterization, in *Dopamine Receptors and Transporters. Pharmacology, Structure, and Function* (Niznik, H. B., ed.), Marcel Dekker, New York, pp. 3–35.

Hallmayer, J., Maier, W., Schwab, S., Ertl, M. A., Minges, J., Ackenheil, M., Lichtermann, D., and Wildenauer, D. B. (1994) No evidence of linkage between the dopamine D2 receptor gene and schizophrenia. *Psychiatry Res.* **53,** 203–215.

Hamm, H. E. and Gilchrist, A. (1996) Heterotrimeric G proteins. *Curr. Opinion Cell. Biol.* **8,** 189–196.

Harrison, M. B., Wiley, R. G., and Wooten, G. F. (1990) Selective localization of striatal D1 receptors to striatonigral neurons. *Brain Res.* **528,** 317–322.

Haske, T. N., Deblasi, A., and Levine, H. (1996) An intact N terminus of the gamma subunit is required for the G beta gamma stimulation of rhodopsin phosphorylation by human beta-adrenergic receptor kinase-1 but not for kinase binding. *J. Biol. Chem.* **271,** 2941–2948.

Hemmings, H. C., Walaas, S., Oiumet, C., and Greengard, P. (1987) Dopaminergic regulation of protein phosphorylation in the striatum: DARP-32. *Trends Neurosci.* **10,** 77–82.

Herlitze, S., Garcia, D. E., Mackie, K., Hille, B., Scheuer, T., and Catterall, W. A. (1996) Modulation of Ca^{2+} channels by G protein beta gamma subunits. *Nature* **380,** 258–262.

Hersch, S. M., Ciliax, B. J., Gutekunst, C. A., Rees, H. D., Heilman, C. J., Yung, K. K. L., Bolam, J. P., Ince, E., Yi, H., and Levey, A. I. (1995) Electron microscopic analysis of D1 and D2 dopamine receptor proteins in the dorsal striatum and their synaptic relationships with motor corticostriatal afferents. *J. Neurosci.* **15,** 5222–5237.

Herve, D., Levi-Strauss, M., Marey-Semper, I., Verney, C., Tassin, J. P., Glowinski, J., and Girault, J. A. (1993) Golf and Gs in rat basal ganglia—possible involvement of Golf in the coupling of dopamine D1 receptor with adenylyl cyclase. *J. Neurosci.* **13,** 2237–2248.

Hietala, J., Syvalahti, E., Vuorio, K., Nagren, K., Lehikoinen, P., Ruotsalainen, U., Rakkolainen, V., Lehtinen, V., and Wegelius, U. (1994) Striatal D2 dopamine receptor characteristics in neuroleptic-naive schizophrenic patients studied with positon emission tomography. *Arch. Gen. Psychiatry* **51,** 116–123.

Hille, B. (1992) G protein-coupled mechanisms and nervous signaling. *Neuron* **9,** 187–195.

Horiuchi, A., Takeyasu, K., Mouradian, M. M., Jose, P. A., and Felder, R. A. (1993) D1A dopamine receptor stimulation inhibits Na^+/K^+-ATPase activity through protein kinase A. *Mol. Pharmacol.* **43,** 281–285.

Itokawa, M., Toru, M., Ito, K., Tsuga, H., Kameyama, K., Haga, T., Arinami, T., and Hamaguchi, H. (1996) Sequestration of the short and long isoforms of dopamine D2 receptors expressed in chinese hamster ovary cells. *Mol. Pharmacol.* **49,** 560–566.

Iyengar, R. (1993) Molecular and functional diversity of mammalian G(s)-stimulated adenylyl cyclases. *FASEB J.* **7,** 768–775.

Jackson, D. M. and Westlind-Danielson, A. (1994) Dopamine receptors: molecular biology, biochemistry and behavioural aspects. *Pharmacol. Ther.* **64,** 291–370.

Jarvie, K. R., Tiberi, M., Silvia, C., Gingrich, J. A., and Caron, M. G. (1993) Molecular cloning, stable expression and desensitization of the human dopamine D1b/D5 receptor. *J. Receptor Res.* **13,** 573–590.

Jarvie, K. R., Tiberi, M., and Caron, M. G. (1994) Dopamine D_{1a} and D_{1b} receptors, in *Dopamine Receptors and Transporters. Pharmacology, Structure, and Function* (Niznik, H. B., ed.), Marcel Dekker, Inc, New York, pp. 133–150.

Jensen, A. A., Pedersen, U. B., Din, N., and Andersen, P. H. (1996) The dopamine D1 receptor family: structural and functional aspects. *Biochem. Soc. Trans.* **24,** 163–169.

Jones, P. and Murray, R. M. (1991) The genetics of schizophrenia is the genetics of neurodevelopment. *Brit. J. Psychiatry* **158,** 615–623.

Joyce, J. N., Lexow, N., Bird, E., and Winokur, A. (1988) Organization of D1 and D2 receptors in human striatum: receptor autoradiographic studies in Huntington's disease and schizophrenia. *Synapse* **2,** 546– 557.

Joyce, J. N. (1993) The dopamine hypothesis of schizophrenia: limbic interactions with 5HT and NE. *Psychopharmacol.* **112,** 16–34.

Kanterman, R. Y., Mahan, L. C., Briley, E. M., Monsma, F. J., Jr., Sibley, D. R., Axelrod, J., and Felder, C. C. (1991) Transfected D2 dopamine receptors mediate the potentiation of arachadonic acid release in CHO cells. *Mol. Pharmacol.* **39,** 364–369.

Karle, J., Clemmesen, L., Hansen, L., Andersen, M., Andersen, J., Fensbo, C., Sloth-Nielsen, M., Skrumsager, B. K., Lublin, H., and Gerlach, J. (1995) NNC 01-0687, a selective dopamine D1 receptor antagonist, in the treatment of schizophrenia. *Psychopharmacology* **121,** 328, 329.

Kaziro, Y., Itoh, H., Kozasa, T., Nakafuka, M., and Satoh, T. (1991) Structure and function of signal transducing GTP binding proteins. *Ann. Rev. Biochem.* **60,** 349–400.

Kebabian, J. W. (1993) Brain dopamine receptors—20 years of progress. *Neurochem. Res.* **18,** 101–104.

Kebabian, J. W. and Calne, D. B. (1979) Multiple receptors for dopamine. *Nature* **277,** 93–96.

Keefe, K. A. and Gerfen, C. R. (1995) D1–D2 dopamine receptor synergy in striatum: effects of intrastriatal infusions of dopamine agonists and antagonists on immediate early gene expression. *Neurosci.* **66,** 903–913.

Kemp, B. E. and Pearson, R. B. (1990) Protein kinase recognition sequence motifs. *Trends Biochem. Sci.* **15,** 342–346.

Kim, C. M., Dion, S. B., and Benovic, J. L. (1993) Mechanism of b-adrenergic receptor kinase activation by G proteins. *J. Biol. Chem.* **268,** 15,412–15,418.

Kimura, K., White, B. H., and Sidhu, A. (1995) Coupling of human D1 dopamine receptors to different guanine nucleotide binding proteins—evidence that D1 dopamine receptors can couple to both Gs and Go. *J. Biol. Chem.* **270,** 14,672–14,678.

Kobilka, B. K. (1992) Adrenergic receptors as models for G protein coupled receptors. *Ann. Rev. Neurosci.* **15,** 87–114.

Lahoste, G. J., Yu, J., and Marshall, J. F. (1993) Striatal fos expression is indicative of dopamine D1/D2 synergism and receptor supersensitivity. *Proc. Natl. Acad. Sci. USA* **90,** 7451–7455.

Lankford, K. L., DeMello, F. G., and Klein, W. L. (1988) D1-type dopamine receptors inhibit growth cone motility in cultured retina neurons: evidence that neurotransmitters act as morphogenic growth regulators in the developing central nervous system. *Proc. Natl. Acad. Sci. USA* **85,** 2839–2843.

Lemmon, M. A., Flanagan, J. M., Treutlein, H. R., Zhang, J., and Engleman, D. M. (1992) Sequence specificity in the dimerization of transmembrane helices. *Biochem.* **31,** 12,719–12,725.

Le Moine, C. and Bloch, B. (1995) D1 and D2 dopamine receptor gene expression in the rat striatum: sensitive cRNA probes demonstrate prominent segregation of D1 and D2 mRNAs in distinct neuronal populations of the dorsal and ventral striatum. *J. Comp. Neurol.* **355,** 418–426.

Le Moine, C. and Bloch, B. (1996) Expression of the D3 dopamine receptor in peptidergic neurons of the nucleus accumbens: comparison with the D1 and D2 dopamine receptors. *Neuroscience* **73,** 131–143.

Le Moine, C., Normand, E., and Bloch, B. (1991) Phenotypical characterization of rat striatal neurons expressing the D1 dopamine receptor gene. *Proc. Natl. Acad. Sci. USA* **88,** 4205–4209.

Lester, J., Fink, S., Aronin, N., and Difiglia, M. (1993) Colocalization of D1 and D2 dopamine receptor messenger RNAs in striatal neurons. *Brain Res.* **621,** 106–110.

Levey, A. I., Hersch, S. M., Rye, D. B., Sunahara, R. K., Niznik, H. B., Kitt, C. A., Price, D. L., Maggio, R., Brann, M. R., and Ciliax, B. J. (1993) Localization of D1 and D2 dopamine receptors in brain with subtype-specific antibodies. *Proc. Natl. Acad. Sci. (USA)* **90,** 8861–8865.

Lew, A. M. and Elsholtz, H. P. (1994) Dopaminergic signalling and regulation of pituitary hormone genes, in *Dopamine Receptors and Transporters. Pharmacology, Structure, and Function* (Niznik, H. B., ed.), Marcel Dekker, Inc, New York, pp. 473–491.

Lewine, R. R. J. (1992) Brain morphology in schizophrenia. *Curr. Opin. Psychiatry* **5,** 92–97.

Lichter, J. B., Barr, C. L., Kennedy, J. L., Van Tol, H. H. M., Kidd, K. K., and Livak, K. J. (1993) A hypervariable segment in the human dopamine receptor D4 (DRD4) gene. *Hum. Mol. Genet.* **2,** 767–773.

Lin, C. W., Miller, T. R., Witte, D. G., Bianchi, B. R., Stashko, M., Manelli, A. M., and Frail, D. E. (1995) Characterization of cloned human dopamine D1 receptor-mediated calcium release in 293 cells. *Mol. Pharmacol.* **47,** 131–139.

Liu, M. Y. and Simon, M. I. (1996) Regulation by cAMP-dependent protein kinase of a G protein-mediated phospholipase C. *Nature* **382,** 83–87.

Liu, Q., Sobell, J. L., Heston, L. L., and Sommer, S. S. (1995) Screening the dopamine D1 receptor gene in 131 schizophrenics and eight alcoholics: identification of polymorphisms but lack of functionally significant sequence changes. *Am. J. Med. Genet.* **60,** 165–171.

Liu, Y. and Lasater, E. M. (1994) Calcium currents in turtle retinal ganglion cells. 2. Dopamine modulation via a cyclic AMP-dependent mechanism. *J. Neurophysiol.* **71,** 743–752.

Liu, Y. F., Civelli, O., Grandy, D. K., and Albert, P. R. (1992) Differential sensitivity of the short and long human dopamine-D2 receptor subtypes to protein kinase-C. *J. Neurochem.* **59,** 2311–2317.

Lledo, P.-M., Homburger, V., Bockaert, J., and Vincent, J. D. (1992) Differential G protein mediated coupling of D2 dopamine receptors to K^+ and Ca^{2+} currents in rat anterior pituitary cells. *Neuron* **8,** 455–463.

Lledo, P.-M., Vernier, P., Kukstas, L. A., Vincent, J.-D., Homburger, V., and Bockaert, J. (1994) Coupling of dopamine receptors to ionic channels in excitable tissues, in *Dopamine Receptors and Transporters. Pharmacology, Structure, and Function* (Niznik, H. B., ed.), Marcel Dekker, New York, pp. 59–88.

Lokhandwala, M. F. and Chen, C. (1994) Peripheral dopamine receptors, in *Dopamine Receptors and Transporters. Pharmacology, Structure, and Function* (Niznik, H. B., ed.), Marcel Dekker, New York, pp. 89–102.

Lovenberg, T. W., Nichols, D. E., Nestler, E. J., Roth, R. H., and Mailman, R. B. (1991) Guanine nucleotide binding proteins and the regulation of cAMP synthesis in NS20Y neuroblastoma cells: role of D1 dopamine and muscarinic receptors. *Brain Res.* **556,** 101–107.

Macciardi, F., Verga, M., Kennedy, J. L., Petronis, A., Bersani, G., Pancheri, P., and Smeraldi, E. (1994) An association study between schizophrenia and the dopamine receptor genes DRD3 and DRD4 using haplotype relative risk. *Hum. Hered.* **44,** 328–336.

Mahan, L., Burch, R., Monsma, F., and Sibley, D. (1990) Expression of striatal D1 dopamine receptors coupled to inositol phosphate production and calcium mobilization in *Xenopus* oocytes. *Proc. Natl. Acad. Sci. USA* **87,** 2196–2200.

Mailman, R. B., Schulz, D. W., Kilts, C. D., Lewis, M. H., Rollema, H., and Wyrick, S. (1986) Multiple forms of the D1 receptor: its linkage to adenylate cyclase and psychopharmacological effects. *Psychopharmacol. Bull.* **22,** 593–598.

Malek, D., Munch, G., and Palm, D. (1993) 2 sites in the 3rd inner loop of the dopamine-D2 receptor are involved in functional G protein-mediated coupling to adenylate cyclase. *FEBS Lett.* **325,** 215–219.

Mamelak, M., Chiu, S., and Mishra, R. K. (1993) High- and low-affinity states of dopamine D1 receptors in schizophrenia. *Eur. J. Pharmacol.* **233,** 175, 176.

McCobb, D. P. and Kater, S. B. (1988) Membrane voltage and neurotransmitter regulation of neuronal growth cone motility. *Devel. Biol.* **130,** 599–609.

McHugh, D. and Coffin, V. (1991) The reversal of extrapyramidal side effects with SCH 39166, a dopamine D1 receptor antagonist. *Eur. J. Pharmacol.* **202,** 133, 134.

McMillian, M. K., He, X. P., Hong, J. S., and Pennypacker, K. R. (1992) Dopamine stimulates [³H] phorbol 12,13-dibutyrate binding in cultured striatal cells. *J. Neurochem.* **58,** 1308–1312.

Missale, C., Boroni, F., Castelletti, L., Dal Toso, R., Gabellini, N., Sigala, S., and Spano, P. F. (1991) Lack of coupling of D2 receptors to adenylate cyclase in GH3 cells exposed to epidermal growth factor: possible role of a differential expression of Gi protein subtypes. *J. Biol. Chem.* **266,** 23,392–23,398.

Montmayeur, J. P., Guiramand, J., and Borrelli, E. (1993) Preferential coupling between dopamine-D2 receptors and G proteins. *Mol. Endocrinol.* **7,** 161–170.

Morell, R. (1993) Association between schizophrenia and homozygosity at the dopamine-D3 receptor gene. *J. Med. Genet.* **30,** 708.

Morelli, M., Fenu, S., Cozzolino, A., and Di Chiara, G. (1991) Positive and negative interactions in the behavioural expression of D1 and D2 receptor stimulation in a model of Parkinsonism: role of priming. *Neuroscience* **42,** 41–48.

Mukai, H., Munekata, E., and Higashijima, T. (1992) G protein antagonists. A novel hydrophobic peptide competes with receptor for G protein binding. *J. Biol. Chem.* **267,** 16,237–16,243.

Murray, A. M., Hyde, T. M., Knable, M. B., Herman, M. M., Bigelow, L. B., Carter, J. M., Weinberger, D. R., and Kleinman, J. E. (1995) Distribution of putative D4 dopamine receptors in postmortem striatum from patients with schizophrenia. *J. Neurosci.* **15,** 2186–2191.

Nakajima, Y., Nakajima, S., and Kozasa, T. (1996) Activation of G protein-coupled inward rectifier K^+ channels in brain neurons requires association of G protein beta gamma subunits with cell membrane. *FEBS Lett.* **390,** 217–220.

Neve, K. A., Henningsen, R. A., Bunzow, J. R., and Civelli, O. (1989) Functional characterization of a rat dopamine D2 receptor cDNA expressed in a mammalian cell line. *Mol. Pharmacol.* **36,** 446–451.

Niznik, H. B. (1987) Dopamine receptors; molecular structure and function. *Mol. Cell. Endocrinol.* **54,** 1–22.

Niznik, H. B. and Jarvie, K. R. (1989) Dopamine receptors, in *Receptor Pharmacology and Function* (Williams, N., Glennon, R. A., and Timmermans, P., eds.), Marcel Dekker, Inc., New York, pp. 717–768.

Niznik, H. B. and Van Tol, H. H. M. (1992) Dopamine receptor genes: new tools for molecular psychiatry. *J. Psychiatr. Neurosci.* **17,** 158–180.

Niznik, H., O'Dowd, B., Sunahara, R., Van Tol, H., Seeman, P., Weiner, D., and Brann, M. (1992) The dopamine D1 receptors., in *Molecular Biology of Receptors that Couple to G Proteins* (Brann, M., ed.), Birkhauser Boston, pp. 142–159.

Niznik, H. B., Sunahara, R. K., Pristupa, Z. B., and Jarvie, K. R. (1995) Molekulare grundlagen der interaktion zwischen dopamin-(D1-/D2-) rezeptoren, in *Schizophrenie. Dopaminrezeptoren und Neuroleptika* (Gerlach, J., ed.), Springer, Berlin, pp. 1–29.

Nothen, M. M., Cichon, S., Propping, P., Fimmers, R., Schwab, S. G. and Wildenauer, D. B. (1993) Excess of homozygosity at the dopamine

D3 receptor gene in schizophrenia not confirmed. *J. Med. Genet.* **30,** 708.

Ohara, K., Ulpian, C., Seeman, P., Sunahara, R. K., Van Tol, H. H. M., and Niznik, B. (1993) Schizophrenia: dopamine D1 receptor sequence is normal, but has DNA polymorphisms. *Neuropsychopharmacology* **8,** 131–135.

O'Hara, C. M., Tang, L., Taussig, R., Todd, R. D., and O'Malley, K. L. (1996) Dopamine D2L receptor couples to G alpha(i2) and G alpha(i3) but not G alpha(i1), leading to the inhibition of adenylate cyclase in transfected cell lines. *J. Pharmacol. Exp. Ther.* **278,** 354–360.

Okamoto, T. and Nishimoto, I. (1992) Detection of G protein-activator regions in M4 subtype muscarinic, cholinergic, and alpha 2-adrenergic receptors based upon characteristics in primary structure. *J. Biol. Chem.* **267,** 8342–8346.

Okamoto, T., Katada, T., Murayama, Y., Ui, M., Ogata, E., and Nishimoto, I. (1990) A simple structure encodes G protein-activating function of the IGF-II/mannose 6-phosphate receptor. *Cell* **62,** 709–717.

Okamoto, T., Murayama, Y., Hayashi, Y., Inagaki, M., Ogata, E., and Nishimoto, I. (1991) Identification of a Gs activator region of the beta 2-adrenergic receptor that is autoregulated via protein kinase A-dependent phosphorylation. *Cell* **67,** 723–730.

Ostrowski, J., Kjelsberg, M. A., Caron, M. G., and Lefkowitz, R. J. (1992) Mutagenesis of the β_2 adrenergic receptor: how structure elucidates function. *Ann. Rev. Pharmacol. Toxicol.* **32,** 167–183.

Paul, M. L., Graybiel, A. M., David, J. C., and Robertson, H. A. (1992) D1-like and D2-like dopamine receptors synergistically activate rotation and c-*fos* expression in the dopamine-depleted striatum in a rat model of Parkinson's disease. *J. Neurosci.* **12,** 3729–3742.

Pilowsky, L. S., Kerwin, R. W., and Murray, R. M. (1993) Schizophrenia—a neurodevelopmental perspective. *Neuropsychopharmacology* **9,** 83–91.

Piomelli, D., Pilon, C., Giros, B., Sokoloff, P., Martres, M. P., and Schwartz, J. C. (1991) Dopamine activation of the arachidonic acid cascade as a basis for D1/D2 receptor synergism. *Nature* **353,** 164–167.

Power, R. F., Mani, S. K., Codina, J., Conneely, O. M., and O'Malley, B. W. (1991) Dopaminergic and ligand-independent activation of steroid hormone receptors. *Science* **254,** 1636–1639.

Rao, M. L. and Moller, H. J. (1994) Biochemical findings of negative symptoms in schizophrenia and their putative relevance to pharmacologic treatment. *Neuropsychobiology* **30,** 160–172.

Rappaport, M. S., Sealfon, S. C., Prikhozhan, A., Huntley, G. W., and Morrison, J. H. (1993) Heterogeneous distribution of D1-receptor, D2-receptor and D5-receptor messenger RNAs in monkey striatum. *Brain Res.* **616,** 242–250.

Raymond, J. R. (1995) Multiple mechanisms of receptor-G protein signaling specificity. *J. Physiol.-Renal Fluid Electrolyte Physiol.* **38,** F141–F158.

Reynolds, G. P. (1992) Developments in the drug treatment of schizophrenia. *Trends Pharmacol. Sci.* **13,** 116–121.

Reynolds, G. P. and Mason, S. L. (1994) Are striatal dopamine D4 receptors increased in schizophrenia? *J. Neurochem.* **63,** 1576–1577.

Reynolds, G. P. and Mason, S. L. (1995) Absence of detectable striatal dopamine D4 receptors in drug-treated schizophrenia. *Eur. J. Pharmacol.* **281,** R5, R6.

Roberts, D. A., Balderson, D., Pickeringbrown, S. M., Deakin, J. F. W., and Owen, F. (1996) The relative abundance of dopamine D4 receptor mRNA in post mortem brains of schizophrenics. *Schizophr. Res.* **20,** 171–174.

Robertson, G. S. and Robertson, H. A. (1994) Dopamine receptor regulation of immediate-early genes in the basal ganglia, in *Dopamine Receptors and Transporters. Pharmacology, Structure, and Function* (Niznik, H. B., ed.), Marcel Dekker, Inc, New York, pp. 419–436.

Robertson, H. A. (1992) Dopamine receptor interactions: some implications for the treatment of Parkinson's disease. *Trends Neurosci.* **15,** 201–206.

Robertson, H. A., Paul, M. L., and Robertson, G. S. (1990) Interactions between D1 and D2 dopamine receptors: the immediate early response gene c-fos in long term changes in the striatum, in *The Basal Ganglia* (Bernardi, G., ed.), Plenum Press, New York, pp. 417–423.

Robertson, H. A., Paul, M. L., Moratalla, R. and Graybiel, A. M. (1991) Expression of the immediate early gene c-fos in basal ganglia: induction by dopaminergic drugs. *Can. J. Neurol. Sci.* **18,** 380–383.

Rodbell, M. (1995) Signal transduction: evolution of an idea (Nobel lecture). *Angew. Chem. Int. Ed.* **34,** 1420–1428.

Rodrigues, P. and Dowling, J. E. (1990) Dopamine induces neurite retraction in retinal horizontal cells via diacylglycerol and protein kinase C. *Proc. Natl. Acad. Sci. USA* **87,** 9693–9697.

Rosengarten, H., Schweitzer, J. W., and Friedhoff, A. J. (1993) A subpopulation of dopamine-D1 receptors mediate repetitive jaw movements in rats. *Pharmacol. Biochem. Behav.* **45,** 921–924.

Rothschild, L. G., Badner, J., Cravchik, A., Gershon, E. S., and Gejman, P. V. (1996) No association detected between a D3 receptor gene-expressed variant and schizophrenia. *Amer. J. Med. Gen.* **67,** 232–234.

Ruiz, J., Gabilondo, A. M., Meana, J. J., and Garcia-Sevilla, J. A. (1992) Increased [^{3}H] raclopride binding sites in postmortem brains from schizophrenic violent suicide victims. *Psychopharmacol.* **109,** 410–414.

Sarkar, G., Kapelner, S., Grandy, D. K., Marchionni, M., Civelli, O., Sobell, J., Heston, L., and Sommer, S. S. (1991) Direct sequencing of the dopamine D2 receptor (DRD2) in schizophrenics reveals three polymorphisms but no structural change in the receptor. *Genomics* **11,** 8–14.

Schimmer, B. (1979) Adrenocortical Y1 cells. *Meth. Enzymol.* **52,** 570–574.

Schinelli, S., Paolillo, M., and Corona, G. L. (1994) Opposing actions of D1- and D2-dopamine receptors on arachidonic acid release and cyclic AMP production in striatal neurons. *J. Neurochem.* **62,** 944–949.

Schmauss, C., Haroutunian, V., Davis, K. L., and Davidson, M. (1993) Selective loss of dopamine-D3-type receptor messenger RNA expression in parietal and motor cortices of patients with chronic schizophrenia. *Proc. Natl. Acad. Sci. (U.S.A.)* **90,** 8942–8946.

Schmidt, C. J., Thomas, T. C., Levine, M. A., and Neer, E. J. (1992) Specificity of G protein beta and gamma subunit interactions. *J. Biol. Chem.* **267,** 13,807–13,810.

Schwartz, J. C., Giros, B., Martres, M. P., and Sokoloff, P. (1993) Multiple dopamine receptors as molecular targets for antipsychotics, in *New Generation of Antipsychotic Drugs: Novel Mechanisms of Action* (Brunello, N., Mendlewicz, J. and Racagni, M., ed.), Karger, Basel, pp. 1–14.

Sedvall, G. C. (1996) Neurobiological correlates of acute neuroleptic treatment. *Int. Clin. Psychopharmacol.* **11,** 41–46.

Seeman, P. (1980) Dopamine receptors. *Pharmacol. Rev.* **32,** 229–313.

Seeman, P. (1987) Dopamine receptors and the dopamine hypothesis of schizophrenia. *Synapse* **1,** 133–152.

Seeman, P. (1993) Schizophrenia as a brain disease. The dopamine receptor story. *Arch. Neurol.* **50,** 1093–1095.

Seeman, P. and Niznik, H. B. (1990) Dopamine receptors and transporters in Parkinson's disease and schizophrenia. *FASEB J.* **4,** 2737–2744.

Seeman, P., Niznik, H. B., Guan, H. C., Booth, G., and Ulpian, C. (1989) Link between D1 and D2 dopamine receptors is reduced in chizophrenia and Huntington diseased brain. *Proc. Natl. Acad. Sci. USA* **86,** 10156–10160.

Seeman, P., Guan, H. C. and Van Tol, H. H. M. (1993) Dopamine D4 receptors elevated in schizophrenia. *Nature* **365,** 441–445.

Seeman, P., Sunahara, R. K., and Niznik, H. B. (1994) Receptor-receptor link in membranes revealed by ligand competition—example for dopamine D1 and D2 receptors. *Synapse* **17,** 62–64.

Senogles, S. E. (1994) The D2 dopamine receptor isoforms signal through distinct G(i alpha) adenylate cyclase—A study with site-directed mutant G(i alpha) proteins. *J. Biol. Chem.* **269,** 23,120–23,127.

Senogles, S. E., Spiegel, A. M., Padress, W., Iyengar, R., and Caron, M. G. (1990) Specificity of receptor-G protein interactions: discrimination of Gi subtypes by the D2 dopamine receptor in a reconstituted system. *J. Biol. Chem.* **265,** 4507–4514.

Shahedi, M., Laborde, K., Azimi, S., Hamdani, S., and Sachs, C. (1995) Mechanisms of dopamine effects on Na^+-K^+-ATPase activity in Madin-Darby canine kidney (MDCK) epithelial cells. *Pflugers Archiv.—Eur. J. Physiol.* **429,** 832–840.

Shaikh, S., Gill, M., Owen, M., Asherson, P., McGuffin, P., Nanko, S., Murray, R. M., and Collier, D. A. (1994) Failure to find linkage between a functional polymorphism in the dopamine D4 receptor gene and schizophrenia. *Am. J. Med. Genet.* **54,** 8–11.

Shetreat, M. E., Lin, L., Wong, A. C., and Rayport, S. (1996) Visualization of D1 dopamine receptors on living nucleus accumbens neurons and their colocalization with D2 receptors. *J. Neurochem.* **66,** 1475–1482.

Shulman, L. M. and Fox, D. A. (1996) Dopamine inhibits mammalian photoreceptor Na^+,K^+-ATPase activity via a selective effect on the alpha 3 isozyme. *Proc. Natl. Acad. Sci. USA* **93,** 8034–8039.

Sibley, D. R. and Monsma, F. J. (1992) Molecular biology of dopamine receptors. *Trends Pharmacol. Sci.* **13,** 61–69.

Sibley, D. R., Monsma, F. J., Jr., and Shen, Y. (1993) Molecular neurobiology of dopaminergic receptors. *Int. Rev. Neurobiol.* **35,** 391–415.

Sidhu, A., Sullivan, M., Kohout, T., Balen, P., and Fishman, P. (1991) D_1 dopamine receptors can interact with both stimulatory and inhibitory guanine nucleotide binding proteins. *J. Neurochem.* **57,** 1445– 1451.

Simon, M. I., Strathmann, M. P., and Narasimhan, G. (1991) Diversity of G proteins in signal transduction. *Science* **252,** 802–808.

Sobell, J. L., Lind, T. J., Sigurdson, D. C., Zald, D. H., Snitz, B. E., Grove, W. M., Heston, L. L., and Sommer, S. S. (1995) The D5 dopamine receptor gene in schizophrenia: identification of a nonsense change and multiple missense changes but lack of association with disease. *Hum. Mol. Genet.* **4,** 507–514.

Sokoloff, P. and Schwartz, J. C. (1995) Novel dopamine receptors half a decade later. *Trends Pharmacol Sci.* **16,** 270–275.

Sokoloff, P., Martres, M.-P., Giros, B., Levesque, D., Diaz, J., Pilon, C. Griffon, N., and Schwartz, J.-C. (1994) The dopamine-D_3 receptor, in *Dopamine Receptors and Transporters. Pharmacology, Structure, and Function* (Niznik, H. B., ed.), Marcel Dekker, New York, pp. 165–188.

Srivastava, L. K. and Mishra, R. K. (1994) Dopamine receptor gene expression: effects of neuroleptics, denervation, and development, in *Dopamine Receptors and Transporters. Pharmacology, Structure, and Function* (Niznik, H. B., ed.), Marcel Dekker, New York, pp. 437–457.

Sugamori, K. S., Demchyshyn, L. L., Chung, M., and Niznik, H. B. (1994) D1A, D1B, and D1C dopamine receptors from *Xenopus laevis. Proc. Natl. Acad. Sci. USA* **91,** 10,536–10,540.

Starr, M. S. (1996) The role of dopamine in epilepsy. *Synapse* **22,** 159–194.

Strange, P. G. (1994) Dopamine D4 receptors: curiouser and curiouser. *Trends Pharmacol. Sci.* **15,** 317–319.

Suh, K. H. and Hamm, H. E. (1996) Cyclic AMP-dependent phosphoprotein components I and II interact with beta gamma subunits of transducin in frog rod outer segments. *Biochemistry* **35,** 290–298.

Sunahara, R., Niznik, H., Weiner, D., Stormann, T., Brann, M., Kennedy, J., Gelerntner, J., Rozmahel, R., Yang, Y., Israel, Y., Seeman, P., and O'Dowd, B. (1990) Human dopamine D1 receptor encoded by an intronless gene on chromosome 5. *Nature* **347,** 80–83.

Sunahara, R., Guan, H., O'Dowd, B., Seeman, P., Laurier, L., Ng, G., George, S., Van Tol, H., and Niznik, H. (1991) Cloning of the gene for a human dopamine D5 receptor with higher affinity for dopamine than D1. *Nature* **350,** 614–619.

Sunahara, R. K., Seeman, P., Van Tol, H. H. M., and Niznik, H. B. (1993) Dopamine receptors and antipsychotic drug response. *Br. J. Psychiatry* **163,** 31–38.

Sunahara, R. K., Dessauer, C. W., and Gilman, A. G. (1996) Complexity and diversity of mammalian adenylyl cyclases. *Ann. Rev. Pharmacol. Toxicol.* **36,** 461–480.

Surmeier, D. J., Eberwine, J., Wilson, C. J., Cao, Y., Stefani, A., and Kitai, S. T. (1992) Dopamine receptor subtypes colocalize in rat striatonigral neurons. *Proc. Natl. Acad. Sci. (USA)* **89,** 10,178–10,182.

Surmeier, D. J., Reiner, A., Levine, M. S., and Ariano, M. A. (1993) Are neostriatal dopamine receptors co-localized? *Trends Neurosci.* **16,** 299–305.

Swarzenski, B. C., Tang, L., Oh, Y. J., O'Malley, K. L., and Todd, R. D. (1994) Morphogenic potentials of D2, D3, and D4 dopamine receptors revealed in transfected neuronal cell lines. *Proc. Natl. Acad. Sci. (U.S.A.)* **91,** 649–653.

Tanaka, T., Igarashi, S., Onodera, O., Tanaka, H., Kameda, K., Takahashi, K., Tsuji, S., and Ihda, S. (1995) Lack of association between dopamine D4 receptor gene and schizophrenia. *Am. J. Med. Genet.* **60,** 580–582.

Tanaka, T., Igarashi, S., Onodera, O., Tanaka, H., Takahashi, M., Maeda, M., Kameda, K., Tsuji, S., and Ihda, S. (1996) Association study between schizophrenia and dopamine D3 receptor gene polymorphism. *Am. J. Med. Genet.* **67,** 366–368.

Tang, W. J., and Gilman, A. G. (1991) Type-specific regulation of adenylyl cyclase by G protein beta gamma subunits. *Science* **254,** 1500–1503.

Taylor, J. M., Jacobmosier, G. G., Lawton, R. G., Remmers, A. E., and Neubig, R. R. (1994) Binding of an alpha(2) adrenergic receptor third intracellular loop peptide to G beta and the amino terminus of G alpha. *J. Biol. Chem.* **269,** 27,618–27,624.

Taylor, J. M., Jacobmosier, G. G., Lawton, R. G., Vandort, M., and Neubig, R. R. (1996) Receptor and membrane interaction sites on G beta—a receptor-derived peptide binds to the carboxyl terminus. *J. Biol. Chem.* **271,** 3336–3339.

Tiberi, M. and Caron, M. G. (1994) High agonist-independent activity is a distinguishing feature of the dopamine D1B receptor subtype. *J. Biol. Chem.* **269,** 27,925–27,931.

Turgeon, J. L. and Warring, D. W. (1992) Functional cross-talk between receptors for peptide and steroid hormones. *Trends Endocrinol. Metabol.* **3,** 360–365.

Undie, A. S. and Friedman, E. (1990) Stimulation of a dopamine D1 receptor enhances inositol phosphate formation in rat brain. *J. Pharmacol. Exp. Ther.* **253,** 987–92.

Undie, A. S. and Friedman, E. (1992) Selective dopaminergic mechanism of dopamine and SKF 38393 stimulation of inositol phosphate formation in rat brain. *Eur. J. Pharmacol.* **226,** 297–302.

Undie, A. S., Weinstock, J., Sarau, H. M., and Friedman, E. (1994) Evidence for a distinct D_1-like dopamine receptor that couples to activation of phosphoinositide metabolism in brain. *J. Neurochem.* **62,** 2045–2048.

Vallar, L., Muca, C., Magni, M., Albert, P., Bunzow, J. R., Meldolesi, J., and Civelli, O. (1990) Differential coupling of dopaminergic D2 receptors expressed in different cell types: stimulation of phosphatidylinositol 4,5 biphosphate in Ltk⁻s fibroblasts, hyperpolarization and cytosolic free Ca^{2+} concentration decrease in GH4C1 cells. *J. Biol. Chem.* **265,** 10,320–10,326.

Van Tol, H. H. M. (1994) The dopamine D_4 receptor, in *Dopamine Receptors and Transporters. Pharmacology, Structure, and Function* (Niznik, H. B., ed.), Marcel Dekker, New York, pp. 189–204.

Wade, S. M., Dalman, H. M., Yang, S. Z., and Neubig, R. R. (1994) Multisite interactions of receptors and G proteins: enhanced potency of dimeric receptor peptides in modifying G protein function. *Mol. Pharmacol.* **45,** 1191–1197.

Waddington, J. L. (1989) Functional interactions between D1 and D2 dopamine receptor systems. *J. Psychopharmacol.* **3,** 54–63.

Waddington, J. L. (ed.) (1993) *D1:D2 Dopamine Receptor Interactions.* Academic Press, London.

Waddington, J. L. and Deveney, A. M. (1996) Dopamine receptor multiplicity: 'D1-like'-'D2-like' interactions and 'D1-like' receptors not linked to adenylate cyclase. *Biochem. Soc. Trans.* **24,** 177–182.

Waddington, J. L. and O'Boyle, K. M. (1989) Drugs acting on brain dopamine receptors: a conceptual re-evaluation five years after the first selective D-1 antagonist. *Pharmacol. Ther.* **43,** 1–52.

Waddington, J. L., Daly, S. A., McCauley, P. G., and O'Boyle, K. M. (1994) Levels of functional interaction between D_1-like and D_2-like dopamine receptor systems, in *Dopamine Receptors and Transporters. Pharmacology, Structure, and Function* (Niznik, H. B., ed.), Marcel Dekker, Inc, New York, pp. 511–537.

Waddington, J. L., Daly, S. A., Downes, R. P., Deveney, A. M., McCauley, P. G., and O'Boyle, K. M. (1995) Behavioural pharmacology of 'D1-like' dopamine receptors: further subtyping, new pharmacological probes and interactions with 'D2-like' receptors. *Prog. Neuropsychopharmacol. Biol. Psychiatry* **19**, 811–831.

Wan, F.-J., Taaid, N., and Swerdlow, N. R. (1996) Do D1/D2 interactions regulate prepulse inhibition in rats? *Neuropsychopharmacol.* **14**, 265– 274.

Wang, H. Y., Undie, A. S., and Friedman, E. (1995) Evidence for the coupling of G_q protein to D1-like dopamine sites in rat striatum: possible role in dopamine-mediated inositol phosphate formation. *Mol. Pharmacol.* **48**, 988–994.

Weiner, D., Levey, A., Sunahara, R., Niznik, H., O'Dowd, B., and Brann, M. (1991) Dopamine D1 and D2 receptor mRNA expression in rat brain. *Proc. Natl. Acad. Sci. (USA)* **88**, 1859–1863.

White, F. J. and Hu, X.-T. (1993) Electrophysiological correlates of D_1:D_2 interactions, in *D_1:D_2 Dopamine Receptor Interactions.* (Waddington, J. L., ed.), Academic, London, pp. 79–114.

Wolf, S. S. and Weinberger, D. R. (1996) Schizophrenia: a new frontier in developmental neurobiology. *Israel J. Med. Sci.* **32**, 51–55.

Wong, D. F., Wagner, H. N., Jr., and Tune, L. E. (1986) Positron emission tomography reveals elevated D2 receptors in drug-naive schizophrenics. *Science* **234**, 1558–1563.

Woolverton, W. L. and Johnson, K. M. (1992) Neurobiology of cocaine abuse. *Trends Pharmacol. Sci.* **13**, 193–200.

Wreggett, K. A. and Wells, J. W. (1995) Cooperativity manifest in the binding properties of purified cardiac muscarinic receptors. *J. Biol. Chem.* **270**, 22,488–22,499.

Yamaguchi, I., Walk, S. F., Jose, P. A., and Felder, R. A. (1996) Dopamine D-2L receptors stimulate Na^+/K^+-ATPase activity in murine LTK(–) cells. *Mol. Pharmacol.* **49**, 373–378.

Yang, L., Li, T., Wiese, C., Lannfelt, L., Sokoloff, P., Xu, C. T., Zeng, Z., Schwartz, J. C., Liu, X. H., and Moises, H. W. (1993) No association between schizophrenia and homozygosity at the D3-dopamine receptor gene. *Am. J. Med. Genet.* **48**, 83–86.

Young, S. T., Porrino, L. J., and Iadarola, M. J. (1991) Cocaine induces striatal c-*fos*-immunoreactive proteins via dopaminergic D1 receptors. *Proc. Natl. Acad. Sci. USA* **88**, 1291–1295.

Yu, P. Y., Eisner, G. M., Yamaguchi, I., Mouradian, M. M., Felder, R. A., and Jose, P. A. (1996) Dopamine D-1A receptor regulation of phospholipase C isoform. *J. Biol. Chem.* **271**, 19,503–19,508.

Zhang, X. and Segawa, T. (1989) Selective blockade of dopamine D1 receptor by SCH 23390 affects dopamine agonist binding to [^{3}H]spiperone labeled D2 receptors in rat striatum. *Jap. J. Pharmacol.* **50**, 333–345.

Modulation of G Proteins in Rat Striatum by Neuroleptic Drugs and a Peptidomimetic Analog of Pro-Leu-Gly-NH$_2$

Willard J. Costain, Suresh K. Gupta, Rodney L. Johnson, and Ram K. Mishra

Introduction

Although the dopamine hypothesis of schizophrenia has gained wide acceptance, it has recently become clear that it requires revision (Seeman and Niznik, 1990; Ellison, 1994; Kahn and Davis, 1995). The original hypothesis asserts that schizophrenia is produced by a hyperdopaminergic state, which exists within central dopaminergic neurons, namely, the mesolimbic and mesocortical (A9 and A10) neurons. This hypothesis is supported by several observations:

1. Stimulant-induced psychosis is similar in some ways to schizophrenia;
2. Overmedicated Parkinson's patients exhibit schizophrenia-like symptoms; and
3. The antipsychotic drugs (neuroleptics) used to treat schizophrenia block dopamine receptors (Grace, 1991; Ellison, 1994; Kane and McGlashan, 1995).

Of these, the most compelling line of evidence for a clear dopaminergic component of schizophrenia lies in the obser-

From: *Neuromethods, Vol. 31: G Protein Methods and Protocols*
Ed: R. K. Mishra, G. B. Baker, and A. A. Boulton Humana Press Inc.

vation that neuroleptic efficacy correlates well with their antagonistic affinity toward D2 dopamine receptors (Seeman and Niznik, 1990; Seeman and Van Tol, 1994). One of the primary detractions of the original dopamine hypothesis is its inability to account for the negative symptoms and cognitive deficits considered core symptoms of schizophrenia (Kahn and Davis, 1995). Although neuroleptic therapy is most effective in alleviating the positive (psychotic) symptoms of schizophrenia, it is much less effective with the negative or deficit symptoms (decreased social interaction, avolition, and apathy) (Kahn and Davis, 1995).

Studies of dopamine levels in the brains of schizophrenics have never detected elevations in either dopamine or the dopamine metabolite homovanillic acid (HVA) (Grace, 1991). Although plasma levels of HVA (pHVA) primarily arise from central dopaminergic and peripheral noradrenergic neurons, peripheral dopaminergic and central noradrenergic neurons also contribute. Despite the variety of sources of HVA in the plasma, pHVA is thought to reflect central dopamine function (Kahn and Davis, 1995; Friedhoff and Silva, 1995). Several studies have examined the effect of neuroleptics on pHVA levels, and it has been found that although baseline levels are not different in schizophrenics (compared to controls), chronic neuroleptic therapy significantly lowers pHVA (Kahn and Davis, 1995; Friedhoff and Silva, 1995). The conclusions of Kahn and Davis (1995) suggest that changes in pHVA following chronic neuroleptic treatment are consistent with changes in central dopaminergic function. These results are consistent with the antipsychotic action of neuroleptics being the result of dopamine receptor antagonism.

Postmortem studies that compared striatal dopamine receptor populations in schizophrenics vs controls have reveled that schizophrenia patients have normal D1 dopamine receptor levels, but elevated D2 dopamine receptor levels (Seeman, 1995). Although part of the elevated D2 receptor density may be owing to the patients having received neuroleptic therapy, similar findings in drug-naive schizophrenia patients indicate that this is an inherent feature of the disorder (Seeman, 1995). The observed elevation in D2 dopamine receptor levels supports the hyperdopaminergic theory of schizophrenia and substantiates the belief that the antipsychotic action of neuroleptics is owing to their acting as antagonists at D2 dopamine receptors.

If the pathology of schizophrenia were simply owing to excessive dopamine release in the A9 and A10 neurons or excessive D2 receptor expression, then one would expect the symptoms to be alleviated immediately following D2 receptor blockade. In practice, however, this is not observed. It is known that neuroleptics block D2 receptors shortly after (within hours) the onset of treatment, but no improvement in the patient's symptoms is observed for approx 3–4 wk (Grace, 1991; Kane and McGlashan, 1995). This dilemma presents several questions. Are the D2 antagonistic properties of neuroleptics responsible for their antipsychotic action? Are other uncharacterized actions of neuroleptics responsible for their antipsychotic action? Is D2 receptor blockade the key event in neuroleptic action? Is a homeostatic response to neuroleptics treatment ultimately responsible for the antipsychotic action of neuroleptics? If one assumes that the neuroleptic D2 antagonism is pivotal, then the most likely alternative for the characteristics of neuroleptic antipsychotic action would be a long-term homeostatic compensation in response to the neuroleptics.

Effects of Neuroleptics on Dopamine Receptor Function

Several studies have examined the effects of short- and long-term neuroleptic treatment on dopamine receptor levels in rats. Burt et al. (1977) showed that chronic treatment with several neuroleptics, including haloperidol, increased [3H]-haloperidol binding sites in the striatum. These authors also found that [3H]-haloperidol binding was increased after a short exposure to neuroleptics. Subsequent studies have demonstrated that D2, but not D1, dopamine receptors are upregulated by chronic neuroleptic treatment (Kahn and Davis, 1995; Seeman, 1995). Chronic haloperidol treatment has been shown to change the rate of both synthesis and degradation of D2 dopamine receptors (Pich et al., 1987). However, conflicting results have been obtained in studies that have examined the effects of neuroleptics on dopamine receptor mRNA levels (See and Chapman, 1994). These data indicate that modulation of transcription as well as posttranscriptional alterations may be involved in regulating D2 dopamine receptor expression.

The effect of neuroleptics on D2 receptor expression was critically analyzed by Seeman and Niznik (1990). These authors compared the effects of neuroleptic treatment used in Alzheimer's and Huntington's patients to those of schizophrenia patients. Neuroleptics increased D2 receptor levels in Alzheimer's and Huntington's patients by $\approx$25%, whereas the degree of elevated D2 receptor expression in schizophrenia patients (two- to threefold in some cases) could not be explained as a result of neuroleptic treatment (Seeman and Niznik, 1990).

Another clinically important feature of neuroleptic treatment is the development of dopamine receptor supersensitivity following protracted treatment. Dopamine receptor supersensitivity is manifest on withdrawal of neuroleptic treatment, and presents itself as an enhanced responsiveness to dopaminergic agonists or drugs that promote dopamine release (such as amphetamine) (Kostrzewa, 1993). Although the observed increases in D2 receptor levels is consistent with dopamine receptor supersensitivity, it may not be the only alteration in the signal transduction cascade contributing to the effect. Striatal D2 dopamine receptors are coupled to the enzyme adenylyl cyclase (AC) via the inhibitory G proteins G_i and G_o. Striatal D1 dopamine receptors are coupled to AC via the stimulatory G proteins G_s and G_{olf}. Stimulation of D1 and D2 dopamine receptors results in increased and decreased cAMP levels, respectively. Cyclic AMP (cAMP) activates protein kinase A (PKA), which in turn, modulates the activity of a variety of intracellular proteins. Thus, although dopamine receptors are excellent and logical targets for dopamine supersensitivity, there are several other proteins that are capable of contributing to the process.

Neuroleptics and the Development of Extrapyramidal Side Effects (EPS)

Protracted use of traditional or "typical" neuroleptics often results in the development of EPS, such as dystonia, akinesia, and tardive dyskinesia (Kane and McGlashan, 1995; Meltzer, 1995). To date, three drugs have been classified as "atypical," clozapine, risperidone, and remoxipride (Meltzer, 1995). The use of "atypical" neuroleptics, such as clozapine, is not as strongly associated with the development of EPS. Furthermore, clozapine has been shown to be effective in ≈30–50% of patients who do not respond to typical neuroleptics (Kane and McGlashan, 1995; Meltzer, 1995). These unique qualities of clozapine suggest that it has a mode of action that is distinct from that of typical neuroleptics. This is further supported by the observations of Seeman and Van Tol that clozapine pharmacology stands apart from that of typical neuroleptics (Seeman and Van Tol, 1994; Seeman, 1995).

Not surprisingly, research directed at understanding the mechanism(s) underlying the development of neuroleptic-induced EPS has focused on the effects of neuroleptics on dopamine receptor signaling. The development of neuroleptic-induced dopamine receptor supersensitivity is thought to be the underlying process in EPS (Kostrzewa, 1993). Therefore, researchers have used animal models of dopamine receptor supersensitivity in studies that are ultimately aimed at understanding EPS. Although EPS is clearly the product of a long-term adaptation to protracted neuroleptic treatment, its symptoms are not always alleviated by discontinuation of the neuroleptics (Creese and Sibley, 1981). The lasting effects of EPS are observed despite studies suggesting that D2 dopamine receptor levels return to pretreatment levels following discontinuation of neuroleptic treatment (Creese and Sibley, 1981).

To date there have been few studies that have examined the effect of neuroleptic treatment on dopamine receptor-coupled G protein levels. A recent postmortem study of G protein levels indicated that G_i and G_o were decreased by 42% in schizophrenic brains (Okada et al., 1990). G proteins play an important role in signal transduction, and are a site of signal diversification and amplification. Alterations in G protein levels could affect signal transduction in the absence of altered receptor number or function. Similarly, alterations in G protein function could affect the nature of a response to an agonist as well as the magnitude of the response. Butkerait et al. (1994) demonstrated that $[\alpha\text{-}^{32}P]GTP$ binding in the striatum was increased following reserpine-induced dopamine receptor supersensitivity. Thus, G proteins may be a better candidate for the site of altered function responsible for EPS.

Modulation of Dopaminergic Function by Peptidomimetics

The tripeptide L-prolyl-L-leucyl-glycinamide (PLG) is a hypothalamic factor that was originally found to control the release of melanocyte-stimulating hormone in the anterior pituitary gland (Robertson and Fibiger, 1992). PLG has also been found to have marked modulatory effects on central dopaminergic function (Chiu et al., 1981a,b, 1985; Rajakumar et al., 1987; Srivastava et al., 1988). The effects of PLG on dopaminergic function documented to date include: modulation of D2 receptor affinity states, downregulation of morphine and haloperidol-induced mesolimbic and striatal dopamine receptor supersensitivity, potentiation of behavioral effects of dopaminergic agonists, antagonize oxotremorine-induced tremor, attenuate morphine and neuroleptic-induced catalepsy, and protection from 1-methyl-4-phenyl-1,2,3,6-tetrahydropyridine (MPTP)-induced degeneration of nigrostriatal dopamine neurons (Srivastava et al., 1988).

Effects of Chronic Neuroleptics on G Protein Levels

Haloperidol is one of the most potent and effective neuroleptics available, and is often one of the first drugs used to treat schizophrenia patients (Kane and McGlashan, 1995). Haloperidol's pharmacological and physiological properties are well defined, and it is considered a typical neuroleptic in that it is highly associated with the development of EPS (Kane and McGlashan, 1995; Meltzer, 1995; Seeman, 1995). Although the effects of haloperidol on D2 dopamine receptor expression have been relatively well studied, its effects on G protein levels have not. The atypical neuroleptic clozapine has a distinct pharmacological profile (Seeman and Van Tol, 1994; Seeman, 1995) and has often proven to be effective in patients who do not respond to haloperidol (Kane and McGlashan, 1995; Meltzer, 1995). Furthermore, the development of EPS is much less common in patients who are receiving clozapine (Kane and McGlashan, 1995; Meltzer, 1995). Because of their distinct yet associated effects with respect to schizophrenia, we compared the effects of chronic treatment of these drugs on striatal G protein levels in rats.

Methodology

Animals and Drug Treatment

Male Sprague-Dawley rats (initial weight 200–250 g) were given ip treatments once daily for 14 d. In the studies of neuroleptic effects alone, the animals were given either 3 mg/kg haloperidol (McNeil) or 20 mg/kg clozapine (Sandoz). In studies of the modulatory effects of the PLG analog 3(R)-[N-(L-Prolyl)amino]-2-oxo-1-pyrrolidineacetamide (PAOPA) on haloperidol regulation of G protein expression, the animals were given 2 mg/kg haloperidol, 10 µg/kg PAOPA, or both drugs in combination. Twenty-four hours after the last treatment, the animals were decapitated, the brains were removed, and were placed on ice. The striata were dissected out, frozen immediately in liquid nitrogen, and kept at –80°C until use.

Tissue Preparation

The frozen striata were homogenized with a glass-Teflon homogenizer in an ice-cold homogenization buffer containing: 50 mM Tris (pH 8.0), 6 mM MgCl$_2$, 1 mM EDTA, 3 mM benzamidine, 1 mM dithiothreitol, 5% (w/v) sucrose, 1 µg/mL soybean trypsin inhibitor (Sigma). The protein content was determined according to the method of Lowry et al. (#232) and the protein concentration was adjusted to 1 mg/200 µL homogenization buffer.

G Protein Quantification (Western Blotting)

Quantitation of G protein immunoreactivity was performed as described previously by Gupta and Mishra (1992) and Marcotte et al. (1994). Samples containing 75 µg protein in 40 µL homogenization buffer were incubated at 75°C for 5 min, after which time 20 µL of 100 mM N-ethylmaleimide were added, and the samples were incubated for 15 min at room temperature. Following the addition of 60 µL of sample buffer (40 mM Tris, pH 6.8, 1 mM dithiothreitol, 1% SDS [w/v], 50% glycerol [v/v], and 6% β-mercaptoethanol), the samples were denatured in a boiling water bath for 2 min. The samples were then cooled and loaded on an SDS-PAGE (10% polyacrylamide) gel according to the method of Laemmli (1970). Following separation, proteins were transferred to a nitrocellulose membrane by electrophoresis at 30 V overnight (Towbin et al., 1979). G protein bands were identified with specific antibodies according to the method of McKenzie and Milligan (1990). Briefly, after the transfer, the nitrocellulose papers were washed twice for 10 min each with phosphate-buffered saline (PBS) (pH 7.4) at room temperature. Non-specific sites were blocked by incubating the nitrocellulose papers in 5% (w/v) skim milk dissolved in PBS, pH 7.4, at room temperature for 3 h. The blots were incubated with the primary antibodies (diluted 1:1000) specific for each G protein overnight at 4°C, after which they were washed several times with PBS containing 0.2% NP-40. Certain blots were incubated with a secondary antibody [^{125}I]goat antirabbit IgG (NEN, Dupont, Canada) dissolved in PBS, pH 7.4, + 0.2% NP-

40 at room temperature for 2–3 h. These blots were washed twice for 10 min with PBS followed by two washings with PBS + 0.2% NP-40 and a final wash with PBS. The [^{125}I]-labeled blots were exposed to Kodak XAR film with intensifying screens at –80°C. Other blots were incubated with an HRP-conjugated secondary antibody (donkey antirabbit), followed by peroxidase/chemiluminescence detection (ECL, Amersham) with Kodak X-OMAT film (1–3 min exposure). The intensity of the resultant bands was quantified using an image analyzer (MCID, Image Research, St. Catherines, Ontario, Canada). The optical density of the bands was proportional to the amount of proteins applied to the gel within the range of 20–100 μg protein.

Antibody Specificity

The antibodies toward $G\alpha_s$, $G\alpha_i$, $G\alpha_o$, and β-subunits used in this study were either the gift of Graeme Milligan, Glasgow, UK or bought from Dupont-NEN. Their specificities are as in Table 1.

Results

The polyclonal site-directed antisera CS_1, SG_1, BN_2, and IM_1 produced five bands of varying molecular weight (Fig. 1). The antisera CS_1 recognized two isoforms of $G\alpha_s$ with molecular weights of 45 and 42 kDa. Antisera SG_1, BN_2, and IM_1 each recognized $G\alpha_i$ (41 kDa), $G\alpha_o$ (39 kDa), and β-subunits (36 kDa), respectively. Analysis of the densities of the G protein bands (Fig. 1) revealed that both haloperidol and clozapine modulate the striatal G protein levels in opposite ways. The typical neuroleptic haloperidol significantly decreased $G\alpha_i$, $G\alpha_s$, and β-subunit levels, whereas the atypical neuroleptic clozapine significantly increased the levels of $G\alpha_i$, $G\alpha_s$, and the β-subunits. No significant differences in the levels of $G\alpha_o$ were observed following either haloperidol or clozapine treatment. Figure 2 shows the results of the quantification of the band densities using the image analyzer. The values are expressed as a percentage of the control values (100%).

Table 1
Specificities of Antibodies Toward Various G Protein Sequences

Antiserum identifies	Peptide employed	Corresponding G protein sequences	Antiserum
CS_1	RMHLRQYELL	$G\alpha_s$ 385–394	$G\alpha_s$
SG_1	KENLKDCGLF	Transducin < 341–350	$G_{i\text{-}1}$, $G_{i\text{-}2}$, transducin
BN_2	MSELDQLRQUE	β-subunit	β_1
IM_1	NLKEDGISAAKDVK	G_o 22–35	G_o

Chronic haloperidol treatment significantly decreased the levels of $G\alpha_i$, $G\alpha_s$, and the β-subunits by $\approx$32, 20, and 25%, respectively. Conversely, chronic clozapine significantly increased the levels of $G\alpha_i$, $G\alpha_s$, and the β-subunits by $\approx$78, 34, and 49%, respectively. The levels of $G\alpha_o$ were not significantly altered by chronic treatment with either haloperidol or clozapine.

The modulatory effects of a peptidomimetic analog of PLG on chronic haloperidol-induced downregulation of striatal G proteins were studied (Fig. 3). It was found that the highly active PLG analog PAOPA antagonized the downregulation of the striatal G proteins G_{olf}/G_s and G_i. The effect of PAOPA were highly significant and virtually blocked the effects of haloperidol on G protein levels. The work to determine the effects of PLG and its analogues on clozapine-induced alterations in striatal G proteins is still in progress.

Discussion

The treatment of schizophrenia with neuroleptics is frequently effective in relieving the positive symptoms of the disorder, but much less effective in alleviating negative symptoms (Kahn and Davis, 1995). This selective action of neuroleptics has recently led to their being termed, more accurately, antipsychotic drugs. The primary drawback of neuroleptic use is the development of EPS following protracted use. Atypical

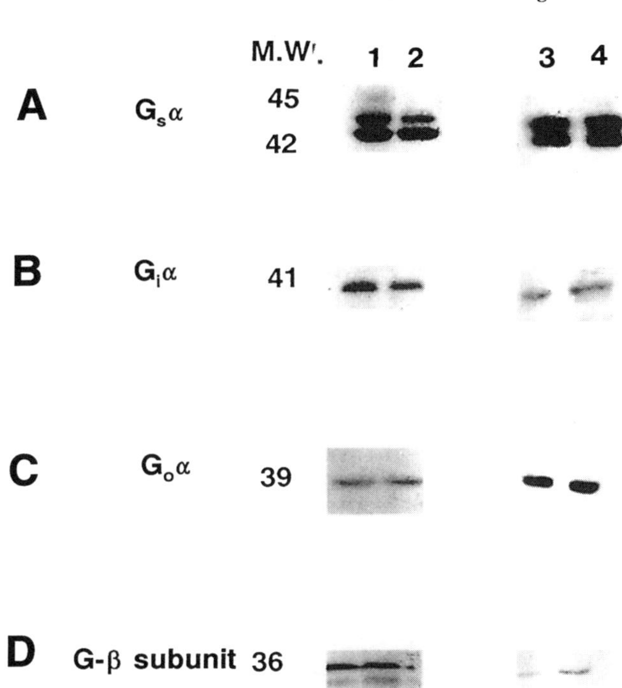

Fig. 1. Autoradiograms showing the effect of haloperidol or clozapine on the levels of G protein immunoreactivity. Rats were treated with vehicle (control) or haloperidol or clozapine for 14 d, after which time striata were isolated and G proteins were separated on 10% SDS-PAGE. Resulting gels were then subjected to immunoblot analysis with specific antisera directed against $G\alpha_s$, $G\alpha_i$, $G\alpha_o$, or β-subunits. Blots were further treated with second antibody goat antirabbit IgG labeled with ^{125}I and autoradiograms were obtained; 1 and 3 represent control, whereas 2 and 4 are haloperidol- and clozapine-treated, respectively. Molecular weight of each G protein is in kilodaltons.

antipsychotic drugs are described as such primarily because of their reduced association with the development of EPS following chronic exposure (Meltzer, 1995). Chronic treatment with typical neuroleptics often induces dopamine receptor supersensitivity, which is believed to be an underlying process in the development of EPS. The upregulation of D2 dopamine receptor expression following chronic neuroleptic treatment is

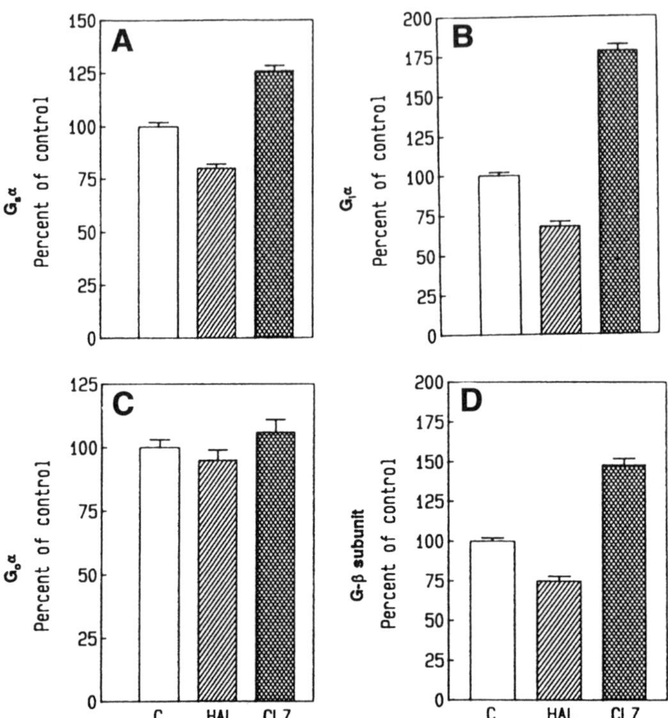

Fig. 2. Quantitative changes in $G\alpha_s$ (**A**), $G\alpha_i$ (**B**), $G\alpha_o$ (**C**), and β-sub-units (**D**) after chronic treatment with haloperidol (HAL) or clozapine (CLZ). Animals were treated as in Fig. 1. Immunoblots were quantified using an image analyzer. Values are presented as percent of control. Values are means of triplicates ± SEM.

consistent with dopamine receptor supersensitivity. Follow-ing along the signal transduction cascade, we observed that chronic neuroleptic treatment is also capable of altering G pro-tein levels in the striatum. Furthermore, we found that the lev-els of G proteins were differentially regulated by the typical neuroleptic haloperidol and the atypical neuroleptic clozapine (Figs. 1 and 2). The observation that chronic neuroleptic treat-ment altered the levels of G protein expression demonstrates that dopamine receptor blockade is capable of altering the expression of downstream mediators and effectors. Similarly, Gupta and Mishra (1993) found that $G\alpha_s$ levels were decreased

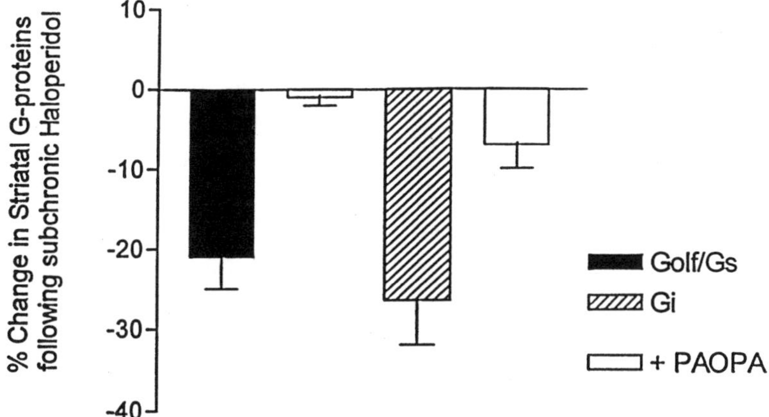

Fig. 3. Effect of protracted treatment with the neuroleptic drug haloperidol and PLG analog PAOPA on striatal G proteins. Various groups of male Sprague-Dawley rats were injected intraperitoneally for 2 wk, with either 2 mg/kg of haloperidol, 10 g/kg PAOPA, or both drugs in combination. Control groups received the vehicle solution. At the end of the last injection, a 3-d wash period was allowed and rats were sacrificed. The striatal tissue was dissected, membranes were prepared, and the G proteins were estimated by Western immunoblotting.

in SK-N-MC cells following prolonged exposure to the D1 agonists dopamine and SKF 38393.

The decreased striatal $G\alpha_i$, $G\alpha_o$ and β-subunit levels following chronic haloperidol treatment are somewhat inconsistent with the development of dopamine receptor supersensitivity. Although increased levels of G proteins could account for the development of supersensitivity, our findings revealed that the opposite occurs. One possible explanation of these data is that the increase in D2 dopamine receptor expression is a compensatory response to haloperidol-induced decreases in $G\alpha_i$ levels (Gupta and Mishra, 1992). Although this hypothesis may explain the alterations in D2 dopamine receptor levels, it does not address the question of receptor supersensitivity. A further question that remains unanswered is: Why are the levels of both stimulatory and inhibitory G proteins decreased following chronic haloperidol treatment? The coregulation of $G\alpha_i$ and $G\alpha_s$ following D2 receptor (which is coupled to $G\alpha_i$, but not $G\alpha_s$) blockade may be owing to interaction between D1 and D2 receptors (Seeman and Niznik, 1990). It is interesting to note that the effects of haloperidol on $G\alpha_s$ and $G\alpha_i$ levels are similar to the effects of antidepressants and lithium (Duman et al., 1989; Li et al., 1991). The findings presented here were substantiated by Shin et al. (1995), who found that $G\alpha_i$ and $G\alpha_{olf}$ were downregulated in the striatum following chronic haloperidol or sulpiride treatment, whereas $G\alpha_q/G\alpha_{11}$ and $G\alpha_o$ were unaffected. Furthermore, these researchers found that chronic haloperidol and sulpiride treatment upregulated the expression of $G\alpha_i$ and $G\alpha_s$ in the hippocampus. Although these authors chose to study sulpiride as a representative atypical neuroleptic, others consider it a typical neuroleptic (Meltzer, 1995).

The effects of the atypical neuroleptic clozapine on striatal G protein levels is completely opposite to the effects of haloperidol (Figs. 1 and 2). Previous research has documented the differential effects of haloperidol and clozapine on brain preprosomatostatin mRNA (Salin et al., 1990) as well as the levels of other neurotransmitters (Hong et al., 1978; Bannon et al., 1987; Compton and Johnson, 1988). Haloperidol and clozapine treatment has also been observed to have differential effects on c-*fos* expression in various brain regions (Robertson and Fibiger, 1992). Consistent with this is the observation that chronic clozapine treatment does not result in dopamine receptor supersensitivity (Seeger et al., 1982; Rupniak et al., 1984). Clearly, haloperidol and clozapine have very different effects on dopamine receptor signaling in the brain. The most likely explanation for the differential effects of haloperidol and clozapine is their distinct pharmacological properties with respect to the D2 dopamine receptor (Seeman, 1995; Meltzer, 1995).

The tripeptide PLG has a variety of effects on the brain, including the ability to modulate central dopaminergic neuronal transmission. The ability of PLG to modulated dopaminergic neuronal function has spurred much research toward the use of PLG as an adjunct in the treatment of disorders, such as depression, schizophrenia, and Parkinson's disease. The data presented here show that the PLG analog PAOPA counteracts the effects of haloperidol on striatal G protein levels. Haloperidol was chosen for this study because its use is associated with the development of EPS. Further studies in our lab will determine the ability of PLG analogs to modulate clozapine-induced changes in G protein expression. PAOPA has been reported to be 1000 times more potent than PLG itself in modulating dopaminergic neurotransmission in the striatum (Mishra et al., 1990; Yu et al., 1998). PAOPA dramatically and significantly antagonized the downregulation of G_{olf}/G_s and G_i and G_i observed following chronic haloperidol treatment (Fig. 3). Similarly, it has previously been shown that PLG is capable of counteracting the development of haloperidol-induced dopamine receptor supersensitivity (Chiu et al., 1981a, 1985; Rajakumar et al., 1987). The present observations demonstrate that PLG or its analogs are capable of antagonizing the downregulation of G protein expression by chronic haloperidol treatment. These observations suggest that PLG may provide some benefit in the neuroleptic treatment of schizophrenia, in that it may decrease the severity of EPS or prevent the development of EPS. Although the precise action of PLG at the molecular level is still being elucidated, it is clear that PLG and its analogs are very influential on central dopaminergic neurons. Figure 4 is a schematic of the G protein cascade and demonstrates the possible sites of action of PLG. Currently, it is believed that PLG and its analogs are acting at the level of G proteins or receptors, by either stabilizing the high affinity state of the receptor (coupled to the GDP-bound G protein) or by increasing the GTPase activity of the $G\alpha_i$ subunit (Fig. 4).

In summary, an understanding of long-term neurological adaptive processes initiated by protracted neuroleptic therapy is important to clinicians and basic scientists alike. One of the primary drawbacks of "typical" neuroleptic therapy is the development of EPS. Although newer "atypical" neuroleptics are not as strongly associated with the development of EPS, the reason for this is not known at this time. The data presented here demonstrate that the typical neuroleptic haloperi-

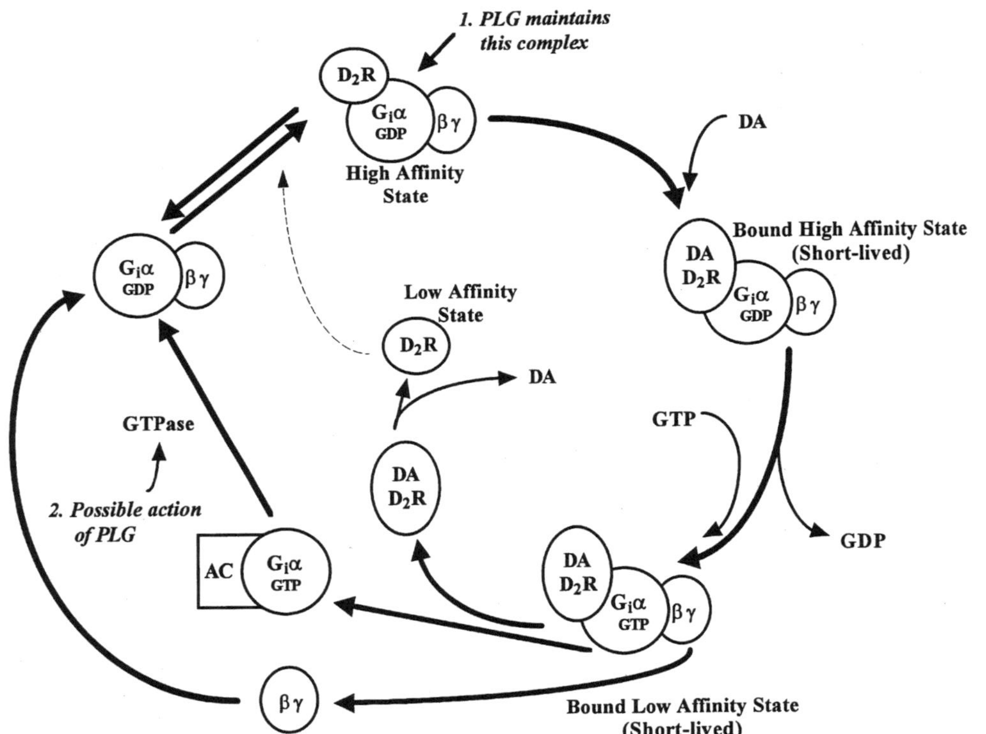

Fig. 4. Schematic representation of PLG's interaction with the D2 receptor–G protein complex. PLG could act through a putative PLG receptor and D2 receptor interactions (1) and/or at the GTPase level (2).

dol and the atypical neuroleptic clozapine have differential effects on G protein levels in the striatum. These observations are similar to the differential effects of haloperidol and clozapine on the expression of brain preprosomatostatin mRNA (Salin et al., 1990), c-*fos* (Robertson and Fibiger, 1992) and other neurotransmitters (Hong et al., 1978; Bannon et al., 1987; Compton and Johnson, 1988). Research directed at understanding the mechanisms involved in the development of neuroleptic-induced EPS suggests that dopamine receptor supersensitivity may play a causative role. Previously, PLG had been found to antagonize the development of haloperidol-induced dopamine receptor supersensitivity (Chiu et al., 1981a, 1985; Rajakumar et al., 1987). Similarly, the present findings indicate that the PLG analog PAOPA antagonizes haloperidol-induced changes in G protein levels. These findings support the role of altered G protein expression in neuroleptic-induced EPS. Furthermore, the observed effects of PLG and its analogs on haloperidol effects present an exciting opportunity for the development of combination therapy aimed at preventing the development of EPS while controlling schizophrenia.

Acknowledgments

The work presented in this manuscript was supported by NIH USA and the Ontario Mental Health Foundation.

References

Bannon, M. J., Elliott, P. J., and Bunney, E. B. (1987) Striatal tachykinin biosynthesis: regulation of mRNA and peptide levels by dopamine agonists and antagonists. *Mol. Brain Res.* **3,** 31–37.

Burt, D. R., Creese, I., and Snyder, S. H. (1977) Antischizophrenic drugs: chronic treatment elevates dopamine receptor binding in brain. *Science* **196,** 326–328.

Butkerait, P., Wang, H.-Y., and Friedman, E. (1994) Increases in guanine nucleotide binding to striatal G proteins is associated with dopamine receptor supersensitivity. *J. Pharmacol. Exp. Ther.* **271,** 422– 428.

Chiu, S., Paulose, C. S., and Mishra, R. K. (1981a) Neuroleptic drug-induced dopamine receptor supersensitivity: antagonism by L-prolyl-L-leucyl-glycinamide. *Science* **214,** 1261, 1262.

Chiu, S., Paulose, C. S., and Mishra, R. K. (1981b) Effect of L-prolyl-L-leucyl-blycinamide (PLG) on neuroleptic-induced catalepsy and dopamine/neuroleptic receptor binding. *Peptides* **2,** 105–111.

Chiu, P., Rajakumar, G., Chiu, S., Johnson, R. L., and Mishra, R. K. (1985) Mesolimbic and striatal dopamine receptor supersensitivity: Prophylactic and reveral effects of L-prolyl-L-leucyl-glycinamide (PLG). *Peptides* **6,** 179–183.

Compton, D. R. and Johnson, K. M. (1988) Effects of acute and chronic clozapine and haloperidol on in vitro release of acetylcholine and dopamine from striatum and nucleus accumbens. *J. Pharmacol. Exp. Ther.* **248,** 521–530.

Creese, I. and Sibley, D. R. (1981) Receptor adaptations to centrally acting drugs. *Ann. Rev. Pharmacol. Toxicol.* **21,** 357–391.

Duman, R. S., Terwlliger, R. Z., and Nestler, E. J. (1989) Chronic antidepressant regulation of $G_s\alpha$ and cyclic AMP-dependent protein kinase. *Pharmacologist* **31,** 182.

Ellison, G. (1994) Stimulant-induced psychosis, the dopamine theory of schizophrenia, and the habenula. *Brain Res. Rev.* **19,** 223–239.

Friedhoff, A. J. and Silva, R. R. (1995) The effects of neuroleptics on plasma homovanillic acid, in *Psychopharmacology: The Fourth Generation of Progress,* (Bloom, F. E. and Kupfer, D. J., eds.), Raven, New York, pp. 1229–1233.

Grace, A. A. (1991) Phasic versus tonic dopamine release and the modulation of dopamine system responsivity: a hypothesis for the etiology of schizophrenia. *Neuroscience* **41,** 1–24.

Gupta, S. K. and Mishra, R. K. (1992) Effects of chronic treatment of haloperidol and clozapine on levels of G protein subunits in rat striatum. *J. Mol. Neurosci.* **3,** 197–201.

Gupta, S. K. and Mishra, R. K. (1993) Desensitization of D1 dopamine receptors down-regulates the $G\alpha_s$ subunit of G protein in SK-N-MC neuroblastoma cells. *J. Mol. Neurosci.* **4,** 117–123.

Hong, J. S., Yang, Y. T., Fratta, W., and Costa, E. (1978) Rat striatal methionine-enkephalin content after chronic treatment with cataleptogenic and noncataleptogenic antischizophrenic drugs. *J. Pharmacol. Exp. Ther.* **205,** 141–147.

Kahn, R. S. and Davis, K. L. (1995) New developments in dopamine and schizophrenia, in *Psychopharmacology: The Fourth Generation of Progress,* (Bloom, F. E. and Kupfer, D. J., eds.), Raven, New York, pp. 1193–1203.

Kane, J. M. and McGlashan, T. H. (1995) Treatment of schizophrenia. *Lancet* **346,** 820–825.

Kostrzewa, R. M. (1993) Dopamine receptor supersensitivity. *Neurosci. Biobehav. Rev.* **19,** 1–17.

Laemmli, U. K. (1970) Cleavage of structural proteins during the assembly of the head of bacteriophage T4. *Nature* **227,** 680–685.

Li, P. P., Tam, Y. K., Young, L. T., and Warsh, J. J. (1991) Lithium decreases G_s, G_{i-1} and $G\alpha_{i2}$ subunit mRNA levels in rat cortex. *Eur. J. Pharmacol.* 206, 165, 166.

Marcotte, E. R., Sullivan, R. M., and Mishra, R. K. (1994) Striatal G proteins: effects of unilateral 6-hydroxydopamine lesions. *Neurosci. Lett.* **169,** 195–198.

McKenzie, F. R. and Milligan, G. (1990) Prostaglandin E_1-mediated, cyclic AMP-independent, down-regulation of $G_s\alpha$ in neuroblastoma X glioma hybrid cells. *J. Biol. Chem.* **265,** 17,084–17,093.

Meltzer, H. Y. (1995) Atypical antipsychotic drugs, in *Psychopharmacology: The Fourth Generation of Progress* (Bloom, F. E. and Kupfer, D. J., eds.), Raven, New York, pp. 1277–1286.

Mishra, R. K., Srivastava, L. K., and Johnson, R. L. (1990) Modulation of high-affinity CNS dopamine D2 receptor by L-pro-L-leu-gly-cinamide (PLG) analogue 3(*R*)-(*N*-L-prolylamino)-2-oxo-1-pyrrolidi-neacetamide. *Prog. Neuro-Psychopharmacol. Biol. Psychiat.* **14,** 821–827.

Pich, E. M., Benfenati, F., Farabegoli, C., Fuxe, K., Meller, E., Aronsson, M., Goldstein, M., and Agnati, L. F. (1987) Chronic haloperidol affects striatal D2-dopamine receptor reappearance after irreversible receptor blockade. *Brain Res.* **435,** 147–152.

Rajakumar, G., Naas, F., Johnson, R. L., Chiu, S., Yu, K. L., and Mishra, R. K. (1987) Down-regulation of haloperidol-induced striatal dopamine receptor supersensitivity by active analogues of L-prolyl-L-leucyl-glycinamide (PLG). *Peptides* **8,** 855–861.

Robertson, G. S. and Fibiger, H. C. (1992) Neuroleptics increase c-*fos* expression in the forebrain: contrasting effects of haloperidol and clozapine. *Neuroscience* **46,** 315–328.

Rupniak, N. M. J., Kipatrick, G., Hall, M. D., Jennen, P., and Marsden, C. D. (1984) Differential alterations in striatal dopamine receptor sensitivity induced by repeated administration of clinically equivalent doses of haloperidol, sulpiride, or clozapine in rats. *Psychopharmacology* **84,** 512–519.

Salin, P., Mercugliano, M., and Chesselet, M.-F. (1990) Differential effects of chronic treatment with haloperidol and clozapine on the level of pre-prosomatostatin mRNA in the striatum nucleus accumbens and frontal cortex of rats. *Cell Mol. Neurobiol.* **10,** 27–144.

See, R. E. and Chapman, M. A. (1994) The consequences of long-term antipsychotic drug administration on basal ganglia neuronal function in laboratory animals. *Critical Rev. Neurobiol.* **8,** 85–124.

Seeger, T. F., Thol, L., and Gardner, E. (1982) Behavioural and biochemical aspects of neuroleptic-induced dopaminergic supersensitivity: studies with chronic clozapine and haloperidol. *Psychopharmacology* **76,** 182–187.

Seeman, P. (1995) Dopamine receptors: clinical correlates, in *Psychopharmacology: The Fourth Generation of Progress* (Bloom, F. E. and Kupfer, D. J., eds.), Raven, New York, pp. 295–302.

Seeman, P. and Niznik, H. B. (1990) Dopamine receptors and transporters in Parkinson's disease and schizophrenia. *FASEB J.* **4,** 2737–2744.

Seeman, P. and Van Tol, H. H. M. (1994) Dopamine receptor pharmacology. *Trends Pharmacol. Sci.* **15,** 264–270.

Shin, C. J., Kim, Y. S., Park, J.-B., and Juhnn, Y.-S. (1995) Changes in G protein levels in the hippocampus and the striatum of rat brain after chronic treatment with haloperidol and sulperide. *Neuropharmacology* **34,** 1335–1338.

Srivastava, L. K., Bajwa, S. B., Johnson, R. L., and Mishra, R. K. (1988) Interaction of L-prolyl-L-leucyl glycinamide with dopamine D2 receptor: evidence for modulation of agonist affinity states in bovine striatal membranes. *J. Neurochem.* **50,** 960–968.

Towbin, H., Staehelin, T., and Gordon, J. (1979) Electrophoretic transfer of proteins from polyacrylamide gels to nitrocellulose sheets: procedure and some applications. *Proc. Natl. Acad. Sci. USA* **76,** 4350–4354.

Yu, K.-L., Rajakumar, G., Srivastava, L. K., Mishra, R. K., and Johnson, R. L. (1988) Dopamine receptor modulation by conformationally constrained analogues of pro-leu-gly-NH$_2$. *J. Med. Chem.* **31,** 1430–1436.

G Proteins and Animal Models
of Parkinson's Disease

Eric R. Marcotte and Ram K. Mishra

Introduction

G protein-linked receptor function and regulation have received considerable attention owing to the pivotal role of these receptors in mediating cellular responses to chemical transmitters. Typically, alterations in neurotransmitter status can either result in reduced receptor activity, also known as receptor desensitization, or increased receptor activity, referred to as receptor supersensitivity. Receptor desensitization has perhaps been best characterized in the β-adrenergic receptor system, where prolonged treatment with β-receptor agonists results in the decreased number and sensitivity of β-adrenergic receptors (Hadcock and Malbon, 1993). Although the role of G proteins and other signal transduction components in regulating receptor desensitization has long been appreciated, the study of receptor supersensitivity has generally focused more narrowly on changes in receptor levels and affinity. Receptor supersensitivity can be brought about in a number of ways, most commonly by receptor antagonist treatment or through denervation (Srivastava and Mishra, 1994). In both cases, the dopamine receptor system has received a great deal of attention and is generally accepted as the prototype for the study of receptor supersensitivity. The purpose of this chapter is to discuss the potential role of G proteins in mediating dopamine receptor supersensitivity.

From: *Neuromethods, Vol. 31: G Protein Methods and Protocols*
Ed: R. K. Mishra, G. B. Baker, and A. A. Boulton Humana Press Inc.

The general focus on dopamine neurotransmission is owing in part to its pivotal role in a large number of neuropsychiatric and motor-related processes. For example, the standard treatment of schizophrenia, which is estimated to affect 1% of the population worldwide, is the use of dopamine receptor antagonists known as antipsychotics or neuroleptics (Leiberman and Koreen, 1993). Many of the side effects and long-term adverse effects of neuroleptics are thought to result from dopamine receptor supersensitivity (Hyman and Nestler, 1993; Sunahara et al., 1993). Dopamine also plays a key role in the development of Parkinson's disease, where degeneration of substantia nigra neurons results in the depletion of striatal dopamine and corresponding supersensitivity of dopamine receptors (Korczyn, 1995). As in the case of schizophrenia, many of the long-term adverse consequences of dopamine replacement therapy in Parkinson's disease are believed to result from dopamine receptor supersensitivity.

The most obvious and direct explanation for these related supersensitivity syndromes is receptor upregulation due to increased receptor synthesis or reduced degradation. Changes in dopamine receptor levels and gene expression, as a result of receptor blockade or denervation, have been extensively examined (Srivastava and Mishra, 1994). Despite the heuristic value of such an approach, it has increasingly been recognized that signal transduction elements may play a role in mediating changes in receptor sensitivity. Increased G protein activity could result in a comparable level of supersensitivity to that produced by increased receptor number. In pharmacological terms, this is generally depicted as a leftward shift of the dose–response curve, as shown in Fig. 1A. Moreover, G proteins also offer the potential advantage of increasing the maximum physiological response to receptor stimulation, thus "raising the ceiling" of the dose–response curve, as shown in Fig. 1B (Ross, 1992). G proteins may also modulate the "steepness" of the curve, allowing for greater adaptive control over receptor signal transduction.

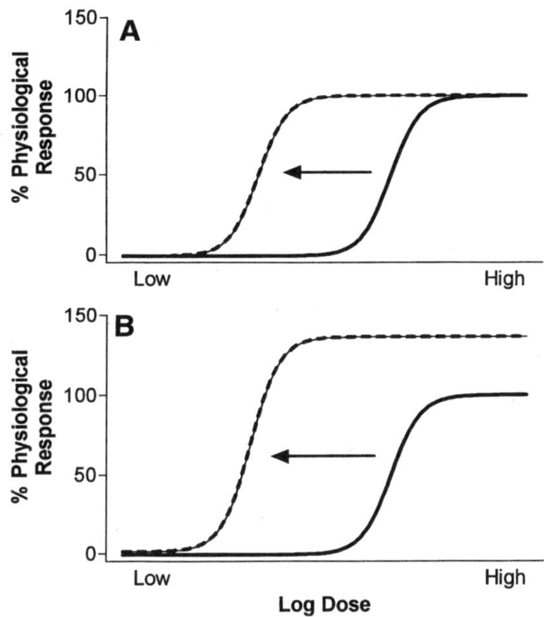

Fig. 1. Dose–response curves for a hypothetical G protein-linked receptor system. **(A)** With increasing drug dose, a greater physiological response is elicited up to a certain maximum (100% physiological response in the figure). Receptor upregulation (dotted line) results in increased sensitivity of the receptor system, and thus produces a comparable physiological response at a lower dose (i.e., higher potency), as demonstrated by the leftward shift of the dose–response curve. Note that the maximum physiological response remains unaltered, however. **(B)** G protein upregulation (dotted line) produces a similar increase in drug potency (leftward shift of the dose–response curve), but also increases the maximum physiological response. Other parameters of the dose–response relationship, such as the range of the dose response (i.e., "steepness" of the curve), may also be altered by G protein manipulation.

In contrast, a preferential advantage of modulation at the receptor level is the maintenance of signal specificity. This is achieved by the selective upregulation of receptor subtypes with their correspondingly distinct signal transduction systems. Alterations at the G protein level would be expected to result in a corresponding loss of selectivity for a given receptor system. However, the integration of diverse receptor-mediated transduction systems through common G proteins may prove advantageous under certain circumstances. Consider the example of a simplified model involving receptors for two different neurotransmitters colocalized on the same cell, coupled in a similar manner to the same G protein. In response to the loss of one transmitter, upregulation of the common G protein subunit would allow for increased sensitivity to both neurotransmitter signals. In this fashion, the loss of one neurotransmitter may be offset by increased signaling through the common G protein. G proteins may thus serve as potential sources of redundancy and integration among diverse neurotransmitter systems. In this chapter, the potential role of G proteins in mediating dopamine receptor supersensitivity in Parkinson's disease in particular will be described.

Molecular Biology of Dopamine Receptors and Their G Proteins

Dopamine receptors are members of the seven-transmembrane domain receptor superfamily of G protein-coupled receptors (O'Dowd, 1993). They fall into two main categories, initially classified as D1 and D2 based on their differing pharmacological profiles and opposite effects on adenylyl cyclase (Kebabian and Calne, 1979). With the advent of modern molecular cloning techniques, this characterization has broaden to include several subtypes, but the general D1/D2 subcategories still hold (Sibley and Monsma, 1992). The D2 receptor subfamily can be divided into a long and short form of the D2 receptor, D2L and D2S, that differ in the length of the third cytoplasmic loop, as well as D3 and D4 receptors. The D1 subfamily consists of just two members to date, D1 and D5 receptors (Sokoloff and Schwartz, 1995). Although the regional distributions of the mRNAs of these various subtypes have been mapped out, their potentially different roles in signal transduction have not been clearly established because of a relative lack of specific pharmacological agents. For the purpose of this discussion, only the prototypical D1 and D2 receptor subtypes will be considered, since these appear to be the predominant forms in the striatum (Civelli et al., 1993).

Based on an α-subunit classification system for G proteins, there are three major categories that are relevant for consideration here. These include the G_s family, which stimulates adenylyl cyclase, the G_i family, which inhibits adenylyl cyclase, and the G_q family, which stimulates phospholipases (Birnbaumer, 1993). All three signaling systems may play a role in mediating dopamine receptor signal transduction. In the striatum, two subtypes of stimulatory G proteins have been observed, G_s and G_{olf}, which are believed to couple to D1 receptors. G_{olf}, so named because it was initially discovered in the olfactory tubercle (Jones and Reed, 1989), is now believed to be the predominant stimulatory G protein in the striatum (Hervé et al., 1993). The G_i family includes multiple subtypes including G_i proper, which is involved in the inhibition of adenylyl cyclase, and G_o, which stimulates K^+ channels and inhibits Ca^{2+} channels (Birnbaumer, 1993). Both subtypes are believed to be coupled to D2 receptors in the striatum. The G_q family consists of several subtypes involved in the stimulation of phospholipases C and D. Although G_q is not generally believed to play a significant role in dopamine signal transduction, this view has recently been challenged for the D1 receptor system (Wang et al., 1995). Despite the general reliance on the α-subunit nomenclature, βγ-subunits have been shown to have direct stimulatory effects on a variety of second messenger systems, including both stimulatory and inhibitory effects on adenylyl cyclase isoforms (Clapham and Neer, 1993). Whether these factors play a significant role in mediating dopamine receptor signal transduction in the striatum remains to be determined.

G Proteins in the Basal Ganglia

Structure and Function of the Basal Ganglia

The basal ganglia are a collection of subcortical nuclei that are involved in the control of movement (Cote and Crutcher, 1991). They are not directly connected to the spinal cord, but instead receive the majority of their input from the cerebral cortex and direct their output back to the motor cortex through the thalamus. As illustrated in Fig. 2, the basal ganglia consist of the substantia nigra (pars compacta and pars reticulata), striatum (caudate and putamen), globus pallidus (external and internal segments), subthalamic nucleus, and the nuclei of the thalamus. Despite their apparent segregation from the corticospinal and extrapyramidal motor pathways, attempts to separate out the specific contributions of the basal ganglia in motor control have proven difficult owing to the extensive interconnections of this network (Cote and Crutcher, 1991).

The precise cellular localization of receptors and G proteins needs to be addressed in any meaningful discussion of basal ganglia function. If D1 and D2 receptors were colocalized on the same neurons, then increased signaling through G_i would be expected to inhibit G_s-stimulated adenylyl cyclase, because of direct competition of G_i for adenylyl cyclase and thermodynamic inhibition of G_s as a result of the greater levels of inhibitory $\beta\gamma$-subunits released by activated G_i (Birnbaumer, 1993). This latter point stems from the fact that G_i levels are higher than G_s in virtually all cell types. There are also several isoforms of adenylyl cyclase that are differentially regulated by G protein α- and $\beta\gamma$-subunits (Mons and Cooper, 1995). This diversity of adenylyl cyclase isoforms has given rise to the suggestion that they may play a role in integrating coincident signals from different receptor systems (Lustig et al., 1993; Mons and Cooper, 1995). A similar possibility exists for the various G protein subtypes already described, as postulated in the Introduction.

The majority of available evidence suggests, however, that D1 and D2 receptors are localized on discrete populations of striatal output neurons (Gerfen, 1992a). This segregation of D1 and D2 output pathways is also in agreement with the observation that activation of both D1 and D2 receptors is often required to elicit behavioral responses—a feature referred to as D1/D2 synergism (Robertson, 1992b). Nevertheless, there have been reports of a significant amount of

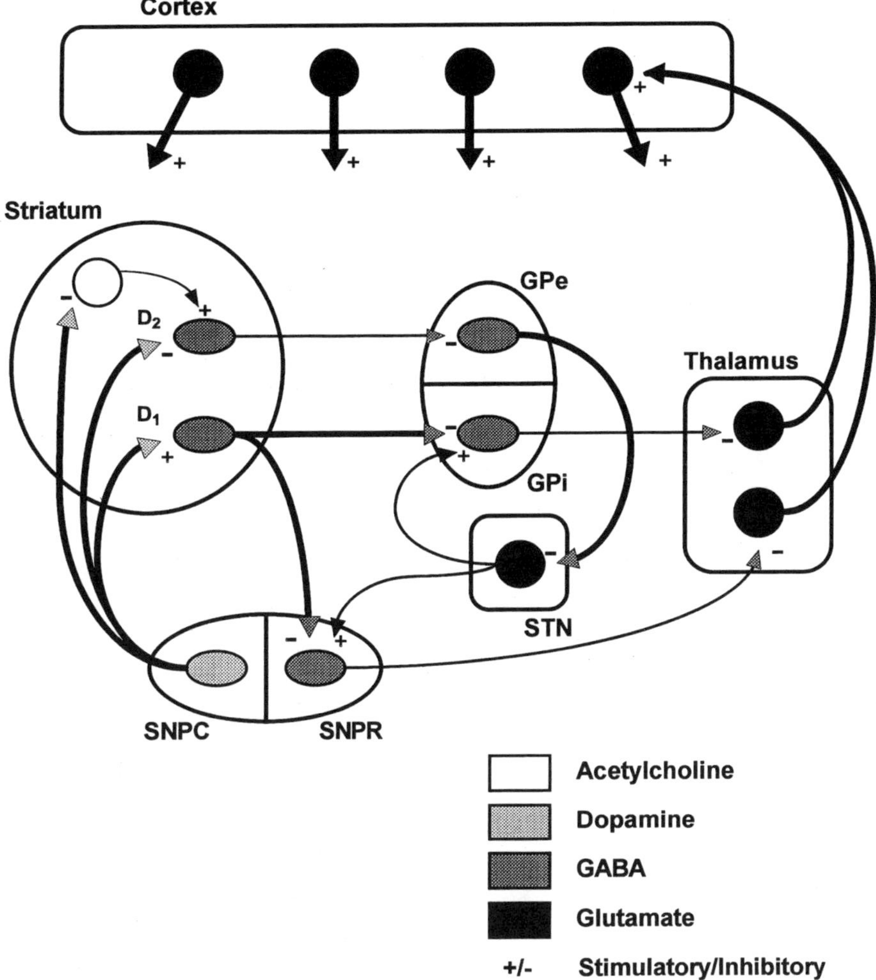

Fig. 2. This figure depicts the main neurochemical connections between basal ganglia nuclei. Thickness of the arrows indicates relative output strength of each group of neurons, with +/− indicating whether the effect is stimulatory (i.e., excitatory) or inhibitory. Glutamatergic input from the cortex is received by a virtually all nuclei depicted in this figure, with the exception of the globus pallidus, and is thus omitted for the sake of clarity. Similarly, thalamic glutamatergic input to the striatum is also omitted. STN refers to the subthalamic nucleus, GPe and GPi refer to the external and internal segments of the globus pallidus, and SNPC and SNPR refer to the substantia nigra pars compacta and pars reticulata, respectively. Note that the ultimate effect of dopamine release in the striatum is a relatively low level of inhibition of excitatory thalamic neurons.

colocalization of D1 and D2 receptors on striatal neurons, based mainly on electrophysiological and immunohistochemical techniques (Surmeier et al., 1993). Despite this contradictory data, the standard model of basal ganglia function consisting of distinct D1 and D2 output pathways, as presented in Fig. 2, is generally accepted (Albin et al., 1989; Gerfen, 1992a).

Under resting conditions, the activity of striatal GABAergic neurons is generated by excitatory glutamatergic inputs from the cortex (omitted from Fig. 2 for the sake of clarity) (Gerfen, 1992b). Release of dopamine from presynaptic terminals of substantia nigra neurons in this area modulates the activity of these output neurons through D1 and D2 receptors. D1 receptors are believed to be located primarily on GABAergic neurons that project to substantia nigra pars reticulata and the internal segment of the globus pallidus—the so-called direct pathway, since these regions then project topographically directly to the thalamus. The thalamus in turn provides excitatory feedback to the frontal cortex and striatum. The second striatal output pathway consists of GABAergic neurons containing D2 receptors that project to the external globus pallidus. This pathway is termed the indirect pathway, since it is connected to the substantia nigra pars reticulata and internal globus pallidus through the subthalamic nucleus. These striatal output pathways are also often characterized by the neuropeptides they synthesize and release; the direct output neurons contain substance P and dynorphin, and the indirect output neurons contain met-enkephalin and possibly neurotensin (Gerfen et al., 1991). There are also a large number of cholinergic interneurons within the striatum that contain somatostatin and neuropeptide Y, and may modulate dopamine/glutamate interactions in this area (Graybiel, 1990; Kawaguchi et al., 1995).

Integration of Signal Transduction Systems

In addition to the dopaminergic and glutamatergic input into the striatum already described, a large number of neurotransmitters and their receptors have been detected in this area. This confluence of receptor systems in the striatum may thus lead to the interaction and integration of diverse signaling pathways at the level of G proteins, effectors, second messengers, or immediate-early genes. Dopamine is known to have direct interactions with many neurotransmitter systems in the striatum, including glutamate (Di Chiara et al., 1994), adenosine (Ferre et al., 1993, 1996), acetylcholine (De Klippel et al., 1993; Imperato et al., 1994; Odagaki and Fuxe, 1995), opiate (Tirone et al., 1985; Groppetti et al., 1990; Dourmap and Costentin, 1994), GABA (Hossain and Weiner, 1995), and benzodiazepine (Griffiths et al., 1994) systems. Many of these interactions are described in detail in the accompanying chapters of this book.

Given the complex nature of the various signal transduction systems present in the striatum, alterations in G proteins could potentially have dramatic effects on neurotransmission. As previously mentioned, G proteins provide a great degree of flexibility in the regulatory control of signal transduction. Upregulation of G protein levels or function could thus result in a compensatory effect for the loss of one transmitter system by increasing the sensitivity to other concurrent inputs. As will be demonstrated shortly, this concept is especially appealing in the discussion of Parkinson's disease.

Parkinson's Disease

Parkinson's disease is one of the most prevalent neurodegenerative conditions in the elderly population today. Clinically, it is characterized by the related motor syndromes of akinesia, bradykinesia, muscle rigidity, and tremor at rest. It results from degeneration of dopaminergic neurons in the substantia nigra pars compacta, with subsequent depletion of striatal dopamine (Marsden, 1994; Korczyn, 1995). The discovery that L-DOPA, the metabolic precursor to dopamine, can reverse the clinical signs of Parkinson's disease in early stages has stimulated decades of intensive research into dopamine neurotransmission. Despite this effort, the cause(s) of dopaminergic cell death in idiopathic Parkinson's disease remains unknown (Rajput, 1992). Although many researchers favor an oxidative stress hypothesis of cell loss, the precise mechanisms through which this may occur remains controversial.

In terms of basal ganglia circuitry, loss of dopamine in the striatum would alter the output of the both the direct and indirect motor pathways, as presented in Fig. 3. Dopamine depletion would be expected to increase the activity of the direct pathway and decrease the activity of the indirect pathway, resulting in corresponding changes throughout the circuit. The end result of these alterations would be a decrease in the thalamic output to cortical and striatal areas. Decreased "thalamic drive" has been proposed to account for the clinical manifestations of akinesia and bradykinesia in Parkinson's disease. As appealing as the model presented in Figure 3 may be, however, it is important to remember that it is ultimately an oversimplification of the rich interconnections between basal ganglia nuclei.

Despite the tremendous research effort that has been focused on the cause and treatment of Parkinson's disease, the design of more effective therapeutic agents has been hampered by the lack of precise knowledge of the cellular mechanisms underlying this disorder. Although dopamine receptor changes in Parkinson's disease have been well documented (Seeman and Niznik, 1990), little is known about alterations in other postsynaptic signal transduction components. Integration of multiple receptor systems at the G protein level may have beneficial compensatory effects on the loss of dopamine in the striatum. Interestingly, Parkinson's disease patients are asymptomatic for the disorder until a critical threshold of approx 80% degeneration occurs. Although this most likely owing to presynaptic mechanisms that maintain relatively constant dopamine levels in the striatum, increased responsiveness of G proteins could play a role. In order to help resolve these issues, we have examined G protein levels in two popular animal models of Parkinson's disease, the 6-hydroxy-dopamine (6-OHDA) lesioned rat and 1-methyl-4-phenyl-1,2,3,6-tetrahydropyridine (MPTP) treated mouse.

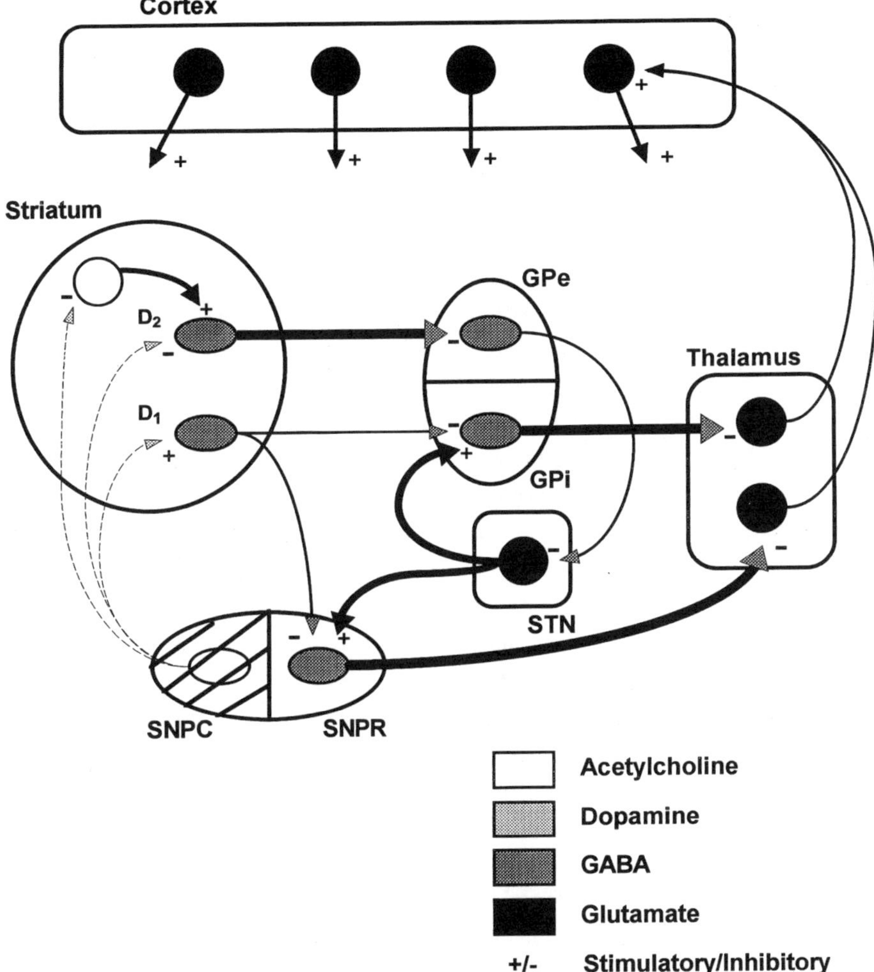

Fig. 3. This figure depicts the neurochemical alterations in basal ganglia nuclei in Parkinson's disease. Thickness of the arrows indicates relative output strength of each group of neurons, with +/− indicating whether the effect is stimulatory (i.e., excitatory) or inhibitory. STN refers to the subthalamic nucleus, GPe and GPi refer to the external and internal segments of the globus pallidus, and SNPC and SNPR refer to the substantia nigra pars compacta and pars reticulata, respectively. The loss of SNPC neurons and the corresponding depletion of striatal dopamine result in a reversal of the relative strength of the direct and indirect pathways, culminating in a greatly increased inhibition of thalamic neurons. This in turns reduces the excitatory output of both thalamic and cortical glutamatergic neurons, and may underlie the clinical manifestations of akinesia and bradykinesia in Parkinson's disease.

6-OHDA Lesions

Unilateral injections 6-OHDA into the substantia nigra pars compacta have long been used as a model of hemi-Parkinsonism in the rat (Herrera-Marschitz and Ungerstedt, 1984; Robertson, 1992a). 6-OHDA is toxic for catecholaminergic neurons, and although its precise mechanism of action is not clear, it is believed to involve auto-oxidation and free radical production (Oertel and Kupsch, 1993). Lesioning of the substantia nigra with 6-OHDA results in the corresponding depletion of dopamine levels in the ipsilateral striatum. The resulting supersensitivity of dopamine receptors in the depleted striatum can easily be detected by the administration of a dopamine receptor agonist. This results in contralateral rotational behaviour (i.e., rotation away from the side of the lesion) owing to the hemispheric imbalance in dopamine receptor sensitivity. The pharmacological characteristics of this rotational behavior suggest two distinct forms, corresponding to the D1 and D2 output pathways (Herrera-Marschitz and Ungerstedt, 1984).

Although receptor upregulation is the most obvious mechanism by which this may occur, it does not appear to be adequate in this case. The reported increases in D2 receptor levels are generally small, on the order of 20–50% (Srivastava and Mishra, 1994). There are conflicting reports for D1 receptors, with some studies showing a transient or persistent increase, but most finding either no change or a decrease in D1 receptors (Blunt et al., 1992; LaHoste and Marshall, 1992; Srivastava and Mishra, 1994). The reason for this discrepancy in D1 receptor studies is unclear, but may reflect differing methodologies, choice of ligands, or temporal changes in receptor expression. Interestingly, this pattern is not observed in chronic receptor blockade studies, where both selective D1 and D2 antagonists have consistently been reported to increase dopamine receptor levels in the striatum (McGonigle et al., 1989). In any case, dopamine receptor supersensitivity in the 6-OHDA lesioned rat does not appear to be maintained by an increase at the receptor level alone, at least not for the D1 system. This latter result is of particular interest, since cAMP levels in the lesioned striatum are elevated, and the ability of dopamine or D1 agonists to stimulate cAMP is enhanced (Mishra et al., 1980).

Alterations in G protein levels or function could account for this increased receptor sensitivity, given the lack of change in receptor levels. In support of this hypothesis, studies from our laboratory (Marcotte et al., 1994) and others (Hervé et al., 1993) have demonstrated a significant increase in stimulatory G proteins in rats who received unilateral 6-OHDA lesions. As shown in Fig. 4, both G_s and G_{olf} levels are significantly elevated in the lesioned striatum within 8 d of a 6-OHDA lesion, with little change in inhibitory G proteins. This upregulation of stimulatory G proteins is on the order of 30–60%, and persists for at least 6 wk (Hervé et al., 1993; Marcotte et al., 1994). Given the nature of G protein signal amplification, this could prove highly significant in increasing the sensitivity to dopamine agonists. In a similar fashion, subchronic neuroleptic treatment has been shown in our lab and others to increase certain G protein levels (Gupta and Mishra, 1992; Schettini et al., 1992) (*see* Chapter 5). Similarly, the monoamine-depleting agent reserpine, which has been suggested to cause a reversible denervation of dopaminergic systems, has also been shown to increase G protein levels and activity in the striatum (Butkerait and Friedman, 1993; Butkerait et al., 1994). Taken together, these studies clearly indicate that G protein levels in the rat striatum are capable of being modulated in response to denervation.

MPTP Lesions

The neurotoxin MPTP has been shown to produce a severe Parkinsonian-like state in humans and several animal species (Gerlach et al., 1991; Oertel and Kupsch, 1993). MPTP is highly selective for the dopaminergic neurons of the substantia nigra, and closely mimics the pattern of neurotransmitter depletion seen in Parkinson's disease. The mechanism of action of MPTP is not entirely understood, but is believed to involved oxidative stress through the biotransformation of

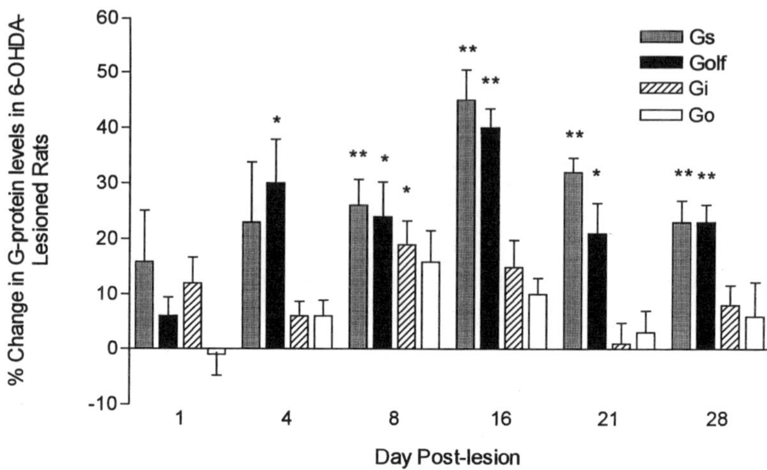

Fig. 4. This graph represents the percent change in G protein levels in 6-OHDA lesioned rats over a 28-d time-course. Individual bars refer to the change in G proteins in the lesioned hemisphere relative to the nonlesioned hemisphere for each animal (mean ± SEM) as measured by Western blotting. Specific G-protein antibodies were purchased from NEN (G_s, G_i, G_o) or synthesized by Research Genetics (G_{olf}) according to established protocols (Hervé et al., 1993). Statistical significance is based on comparison of lesioned groups ($n = 6$) to sham controls ($n = 4$) using Student's *t*-test, two-tailed. *$p < 0.05$, **$p < 0.01$.

MPTP into the cytotoxic metabolite 1-methyl-4-phenylpyridinium ion (MPP+) by monoamine oxidase B. Although much interest has focused on MPTP and MPTP-like compounds as potential causes of idiopathic Parkinson's disease, it needs to be stressed that not all of the structural and motor abnormalities of Parkinson's disease are reproduced by MPTP (Gerlach et al., 1991). Moreover, animal models tend to show behavioral recovery with time, in contrast to the progressive degeneration seen in humans.

One popular animal model of MPTP lesioning involves the systemic administration of MPTP to C57 black mice (BL/6) (Sundstrom et al., 1990). Dopamine receptor supersensitivity can be difficult to establish in this animal model, however, beacause of the bilateral lesioning of the substantia nigra. Nevertheless, there are several reports of increased dopamine sensitivity in these animals (Lau and Fung, 1986; Lange, 1990; Fredriksson et al., 1994). Corresponding increases in D1 or D2 receptors are not generally found (Ogawa et al., 1987; Camps et al., 1989; Lange, 1990), and even those that report positive findings generally show only a very small and transient increase in D2 receptors (Peroutka et al., 1985; Lau and Fung, 1986). Similar findings have also been reported in MPTP-lesioned monkeys (Graham et al., 1993). The similarity of the dopamine receptor response in MPTP and 6-OHDA lesion models suggests a comparable change in G protein levels may occur. However, preliminary data from our lab (Marcotte et al., 1995) has shown that stimulatory G protein levels are differentially regulated in MPTP-lesion mice, with reduced levels of G_s and G_{olf} acutely following the lesion, and increased levels during recovery. The significance of these findings are unclear, but it suggests that G proteins may be regulated in a more complicated fashion than indicated by the simple model presented above.

Drug-Induced Dyskinesia

As previously mentioned, dopamine receptor supersensitivity is believed to play a role in mediating some of the long-term adverse effects of treatment in schizophrenia and Parkinson's disease. The chronic use of neuroleptics in schizophrenia, or L-DOPA in Parkinson's disease, is associated with the development of abnormal, involuntary movements—the so-called dyskinesias (Tanner, 1986). By definition, both of these are "tardive" in as much as they only manifest on chronic treatment, but the term tardive dyskinesia is generally reserved for neuroleptic treatment (Jeste and Caligiuri, 1993). Both forms of dyskinesia are characterized by chorea of the face and extremities, akathisia, dystonia, and occasionally even myoclonic jerks (Tanner, 1986; Nutt, 1990). Prevalence estimates for these drug-induced dyskinesias vary greatly; they have been reported to develop in 20–80% of all L-DOPA-treated patients and 1–55% of neuroleptic-treated patients (Tanner, 1986; Bedard et al., 1992). Clearly, this represents a significant concern in the long-term treatment with dopaminergic agents. Interestingly, the development of tardive dyskinesia, although clearly associated with neuroleptic treatment, is not generally well correlated with dose, duration, or type of neuroleptic used (Tanner, 1986). In contrast, L-DOPA dyskinesia is well correlated with the severity of degeneration and the corresponding higher doses of L-DOPA used. Both forms of dyskinesia can frequently be reduced by attenuating the dose of neuroleptic or L-DOPA, but generally only at the cost of increased disability (Tanner, 1986; Bedard et al., 1992).

A possible rationale for L-DOPA-induced dyskinesia can be presented in terms of the alterations in basal ganglia function reported in Parkinson's disease (Fig. 3). The short-lived effects of L-DOPA on striatal neurons have been proposed to result in brief bursts of activity of the supersensitive D1-mediated direct output pathway (Yoshida, 1991). This in turn would lead to corresponding bursts of activity in the tonically inhibited thalamus, potentially producing the excessive motor movements associated with L-DOPA-induced dyskinesia. Despite the oversimplification implied by this model, recent evidence of increased D1 receptor levels in the caudal striatum of dyskinetic monkeys supports this theory (Graham et al., 1993). Alternatively, there is also evidence for increased inhibition of the D2-mediated indirect pathway (Crossman, 1990). In either case, the imbalance between D1 and D2 striatal output pathways is often presented as the common mechanism in the development of dyskinesias. Delineating the specific pathways involved has proven difficult, however, and has led to the suggestion that a functional redundancy may exist within the system (Blanchet et al., 1994).

Future Directions

In addition to the changes in G protein levels discussed here, other components of the signal transduction cascade may also show long-term adaptations to denervation. In particular, adenylyl cyclases have been suggested to represent critical sites of signal integration (Anholt, 1994). At least eight isoforms of adenylyl cyclase have been described that are differentially regulated by G proteins. In addition, adenylyl cyclase isoforms have been shown to respond to a wide range of factors, including second messengers (Mons and Cooper, 1995). One particular adenylyl cyclase isoform, type V, is enriched in the striatum and the olfactory tubercle. This is reminiscent of the expression of G_{olf}, which is localized in the same brain areas (Hervé et al., 1993). Also of interest, a G protein γ-subunit isoform, γ7, shares the same specific pattern of expression (Watson et al., 1994). The significance of these findings is unclear, but the restricted colocalization of three specific signal transduction components warrants further investigation.

An alternative approach to delineating the role of G proteins in the basal ganglia is the selective blockade, or "knockdown," of G proteins using locally administered antisense oligonucleotides. Recently, antisense oligonucleotides have successfully been used to block protein expression in vivo, most notably for the dopamine D2 receptor in 6-OHDA lesioned mice (Weiss et al., 1993; Zhang and Creese, 1993; Silvia et al., 1994; Qin et al., 1995). Based on the increased levels of stimulatory G proteins in 6-OHDA lesioned rats (Hervé et al., 1993; Marcotte et al., 1994), and the demonstration of the effectiveness of G protein antisense oligonucleotides in reducing G protein levels in vitro (Chen et al., 1994; Buckley et al., 1995; Tang et al., 1995) and in vivo (Raffa et al., 1994; Plata-Salaman et al., 1995), we are currently pursuing studies involving G protein antisense knockdown in this animal model.

In conclusion, further elucidation of the role of G proteins in mediating dopamine receptor supersensitivity may have significant implications for the study of receptor signaling in general. Ultimately, this knowledge may lead in turn to a greater understanding of the mechanisms involved in mediating, and compensating for, alterations in neurotransmission brought about by related disease processes.

Acknowledgements

The work described in this chapter was supported by the Parkinson's Foundation of Canada and the NIH (grant NS20036).

References

Albin, R. L., Young, A. B., and Penney, J. B. (1989) The functional anatomy of basal ganglia disorders. *Trends Neurosci.* **12,** 366–375.

Anholt, R. R. H. (1994) Signal integration in the nervous system: adenylate cyclases as molecular coincidence detectors. *Trends Neurosci.* **17,** 37–41.

Bédard, P. J., Mancilla, B. G., Blanchette, P., Gagnon, C., and Di Paolo, T. (1992) Levodopa-induced dyskinesia: facts and fancy. What does the MPTP monkey model tell us? *Can. J. Neurol. Sci.* **19,** 134–137.

Birnbaumer, L. (1993) Heterotrimeric G proteins: molecular diversity and functional correlates. *J. Recept. Res.* **13,** 19–26.

Blanchet, P. J., Boucher, R., and Bedard, P. J. (1994) Excitotoxic lateral pallidotomy does not relieve L-DOPA-induced dyskinesia in MPTP parkinsonian monkeys. *Brain Res.* **650,** 32–39.

Blunt, S. B., Jenner, P., and Marsden, C. D. (1992) Autoradiographic study of striatal D1 and D2 dopamine receptors in 6-OHDA-lesioned rats receiving foetal ventral mesencephalic grafts and chronic treatment with L-DOPA and carbidopa. *Brain Res.* **582,** 299–311.

Buckley, N. J., ffrench-Mullen, J., and Caulfield, M. (1995) Use of antisense oligodeoxynucleotides and monospecific antisera to inhibit G protein gene expression in cultured neurons. *Biochem. Soc. Trans.* **23,** 137–141.

Butkerait, P. and Friedman, E. (1993) Repeated reserpine increases striatal dopamine receptor and guanine nucleotide binding protein mRNA. *J. Neurochem.* **60,** 566–571.

Butkerait, P., Wang, H.-Y., and Friedman, E. (1994) Increases in guanine nucleotide binding to striatal G proteins is associated with dopamine receptor supersensitivity. *J. Pharmacol. Exp. Ther.* **271,** 422–428.

Camps, M., Ambrosio, S., Ballarin, M., Reiriz, J., Blesa, R., and Mahy, N. (1989) Dopamine D1 and D2 receptors visualized in MPTP treated C57 mice by in vitro autoradiography: lack of evidence of receptor modifications in parkinsonian mice. *Pharmacol. Toxicol.* **65,** 169–174.

Chen, Y., Baez, M., and Yu, L. (1994) Functional coupling of the 5-HT2C serotonin receptor to G proteins in Xenopus oocytes. *Neurosci. Lett.* **179,** 100–102.

Civelli, O., Bunzow, J. R., and Grandy, D. K. (1993) Molecular diversity of the dopamine receptors. *Annu. Rev. Pharmacol. Toxicol.* **32,** 281–307.

Clapham, D. E. and Neer, E. J. (1993) New role for G protein βgamma-dimers in transmembrane signalling. *Nature* **365**, 403–406.

Cote, L. and Crutcher, M. D. (1991) The basal ganglia, in *Principles of Neural Science*, 3rd ed. (Kandel, E. K., Schwartz, J. H., and Jessell, T. M., eds.), Elsevier, New York, pp. 648–659.

Crossman, A. R. (1990) A hypothesis on the pathophysiological mechanisms that underlie levodopa- or dopamine agonist-induced dyskinesia in Parkinson's disease: implications for future strategies in treatment. *Mov. Disord.* **5**, 100–108.

De Klippel, N., Sarre, S., Ebinger, G., and Michotte, Y. (1993) Effect of M1- and M2-muscarinic drugs on striatal dopamine release and metabolism: an in vivo microdialysis study comparing normal and 6-hydroxydopamine-lesioned rats. *Brain Res.* **630**, 57–64.

Di Chiara, G., Morelli, M., and Consolo, S. (1994) Modulatory functions of neurotransmitters in the striatum: ACh/dopamine/NMDA interactions. *Trends Neurosci.* **17**, 228–232.

Dourmap, N. and Costentin, J. (1994) Involvement of glutamate receptors in the striatal enkephalin-induced dopamine. *Eur. J. Pharmacol.* **253**, R9–R11.

Ferre, S., O'Connor, W. T., Fuxe, K., and Ungerstedt, U. (1993) The striopallidal neuron: a main locus for adenosine-dopamine interactions in the brain. *J. Neurosci.* **13**, 5402–5406.

Ferre, S., Popoli, P., Tinner-Staines, B., and Fuxe, K. (1996) Adenosine A_1 receptor-dopamine D1 receptor interaction in the rat limbic system: modulation of dopamine D1 receptor antagonist binding sites. *Neurosci. Lett.* **208**, 109–112.

Fredriksson, A., Plaznik, A., Sundstrom, E., and Archer, T. (1994) Effects of D1 and D2 agonists on spontaneous motor activity in MPTP treated mice. *Pharmacol. Toxicol.* **75**, 36–41.

Gerfen, C. R. (1992a) The neostriatal mosaic: multiple levels of compartmental organization in the basal ganglia. *Annu. Rev. Neurosci.* **15**, 285–320.

Gerfen, C. R. (1992b) The neostriatal mosaic: multiple levels of compartmental organization. *Trends Neurosci.* **15**, 133–139.

Gerfen, C. R., McGinty, J. F., and Young III, W. S. (1991) Dopamine differentially regulates dynorphin, substance P, and enkephalin expression in striatal neurons: in situ hybridization histochemical analysis. *J. Neurosci.* **11**, 1016–1031.

Gerlach, M., Riederer, P., Przuntek, H., and Youdim, M. B. H. (1991) MPTP mechanisms of neurotoxicity and their implications for Parkinson's disease. *Eur. J. Pharmacol. Mol. Pharmacol.* **208**, 273–286.

Graham, W. C., Sambrook, M. A., and Crossman, A. R. (1993) Differential effect of chronic dopaminergic treatment on dopamine D1 and D2 receptors in the monkey brain in MPTP-induced parkinsonism. *Brain Res.* **602**, 290–303.

Graybiel, A. M. (1990) Neurotransmitters and neuromodulaters in the basal ganglia. *Trends Neurosci.* **13,** 244–254.

Griffiths, P. D., Perry, R. H., and Crossman, A. R. (1994) A detailed anatomical analysis of neurotransmitter receptors in the putamen and caudate in Parkinson's disease and Alzheimer's disease. *Neurosci. Lett.* **169,** 68–72.

Groppetti, A., Ceresoli, G., Mandelli, V., and Parenti, M. (1990) Role of opiates in striatal D-1 dopamine receptor supersensitivity induced by chronic L-DOPA treatment. *J. Pharmacol. Exp. Ther.* **253,** 950–956.

Gupta, S. K. and Mishra, R. K. (1992) Effects of chronic treatment of haloperidol and clozapine on levels of G protein subunits in rat striatum. *J. Mol. Neurosci.* **3,** 197–201.

Hadcock, J. R. and Malbon, C. C. (1993) Agonist regulation of gene expression of adrenergic receptors and G proteins. *J. Neurochem.* **60,** 1–9.

Herrera-Marschitz, M. and Ungerstedt, U. (1984) Evidence that striatal efferents relate to different dopamine receptors. *Brain Res.* **323,** 269–278.

Hervé, D., Levi-Strauss, M., Marey-Semper, I., Verney, C., Tassin, J.-P., Glowinski, J., and Girault, J.-A. (1993) G_{olf} and G_s in rat basal ganglia: possible involvement of G_{olf} in the coupling of Dopamine D1 receptor with adenylyl cyclase. *J. Neurosci.* **13,** 2237–2248.

Hossain, M. A. and Weiner, N. (1995) Interactions of dopaminergic and GABAergic neurotransmission: impact of 6-hydroxydopamine lesions into the substantia nigra of rats. *J. Pharmacol. Exp. Ther.* **275,** 237–24.

Hyman, S. E. and Nestler, E. J. (1993) *The Molecular Foundations of Psychiatry.* American Psychiatric Press, Inc., Washington, DC.

Imperato, A., Obinu, M. C., and Gessa, G. L. (1994) Does dopamine exert a tonic inhibitory control on the release of striatal acetylcholine in vivo? *Eur. J. Pharmacol.* **251,** 271–279.

Jeste, D. V. and Caligiuri, M. P. (1993) Tardive dyskinesia. *Schizophrenia Bull.* **19,** 303–315.

Jones, D. T. and Reed, R. R. (1989) Golf: an olfactory neuron specific-G protein involved in odorant signal transduction. *Science* **244,** 790–795.

Kawaguchi, Y., Wilson, C. J., Augood, S. J., and Emson, P. C. (1995) Striatal interneurones: chemical, physiological and morphological characterization. *Trends Neurosci.* **18,** 527–535.

Kebabian, J. W. and Calne, D. B. (1979) Multiple receptors for dopamine. *Nature* **277,** 93–96.

Korczyn, A. D. (1995) Parkinson's disease, in: *Pyschopharmacology: The Fourth Generation of Progress,* (Bloom, F. E. and Kupfer, D. J. eds.), Raven, New York, pp. 1479–1484.

LaHoste, G. J. and Marshall, J. F. (1992) Dopamine supersensitivity and D1/D2 synergism are unrelated to changes in striatal receptor density. *Synapse* **12,** 14–26.

Lange, K. W. (1990) Behavioural effects and supersensitivity in the rat following intranigral MPTP and MPP+ administration. *Eur. J. Pharmacol.* **175,** 57–61.

Lau, Y.-S. and Fung, Y. K. (1986) Pharmacological effects of 1-methyl-4-phenyl-1,2,3,6-tetrahydropyridine (MPTP) on striatal dpamine receptor system. *Brain Res.* **369,** 311–315.

Leiberman, J. A. and Koreen, A. R. (1993) Neurochemistry and neuroendocrinology of schizophrenia: A selective review. *Schizophrenia Bull.* **19,** 371–429.

Lustig, K. D., Conklin, B. R., Hermark, P., Taussig, R., and Bourne, H. R. (1993) Type II adenylylcyclase integrates coincident signals from G_s, G_i, and G_q. *J. Biol. Chem.* **268,** 13,900–13,905.

Marcotte, E. R., Sullivan, R. M., and Mishra, R. K. (1994) Striatal G proteins: effects of unilateral 6-hydroxydopamine lesions. *Neurosci. Lett.* **169,** 195–198.

Marcotte, E. R., Chugh, A., Barlas, C., and Mishra, R. K. (1995) G protein expression in 6-hydroxydopamine lesioned rats and 1-methyl-4-phenyl-1,2,3,6-tetraphydropyridine treated mice. *Soc. Neurosci.* **21,** (Abstract).

Marsden, C. D. (1994) Parkinson's disease. *J. Neurol. Neurosurg. Psychiatry* **57,** 672–681.

McGonigle, P., Boyson, S. J., Reuter, S., and Molinoff, P. B. (1989) Effects of chronic treatment with selective and nonselective antagonists on the subtypes of dopamine receptors. *Synapse* **3,** 74–82.

Mishra, R. K., Marshall, A. M., and Varmuza, S. L. (1980) Supersensitivity in rat caudate nucleus: effects of 6-hydroxydopamine on the time course of dopamine receptor and cyclic AMP changes. *Brain Res.* **200,** 47–57.

Mons, N. and Cooper, D. M. F. (1995) Adenylate cyclases: critical foci in neuronal signaling. *Trends Neurosci.* **18,** 536–542.

Nutt, J. G. (1990) Levodopa-induced dyskinesia: review, observations, and speculations. *Neurology* **40,** 340–345.

Odagaki, Y. and Fuxe, K. (1995) Functional coupling of dopamine D2 and muscarinic cholinergic receptors to their respective G proteins assessed by agonist-induced activation of high affinity GTPase activity in rat striatal membranes. *Biochem. Pharmacol.* **50,** 325–335.

O'Dowd, B. F. (1993) Structures of dopamine receptors. *J. Neurochem.* **60,** 804–816.

Oertel, W. H. and Kupsch, A. (1993) Pathogenesis and animal studies of Parkinson's disease. *Curr. Opinion Neurol. Nuerosurg.* **6,** 323–332.

Ogawa, N., Mizukawa, K., Hirose, Y., Kajita, S., Ohara, S., and Watanabe, Y. (1987) MPTP-induced parkinsonian model in mice: biochemistry, pharmacology and behaviour. *Eur. Neurol.* **26(Suppl. 1),** 16–23.

Peroutka, S. J., DeLanny, L., Irwin, I., Ison, P.J., Ricaurte, G., Schlegel, J. R., and Langston, J. W. (1985) 1-Methyl-4-phenyl-1,2,3,6-tetraphydropyridine (MPTP) induced dopamine D2 receptor hypersensitivity in the mouse is transient. *Res. Commun. Chem. Pathol. Pharmacol.* **48,** 163–171.

Plata-Salaman, C. R., Wilson, C. D., Sonti, G., Borkoski, J. P., and ffrench-Mullen, J. M. H. (1995) Antisense oligodeoxynucleotides to G protein α-subunit subclasses identify a transductional requirement for the modulation of normal feeding dependent on Gs_{OA} subunit. *Brain Res. Mol. Brain Res.* **33,** 72–78.

Qin, Z.-H., Zhou, L.-W., Zhang, S. P., Wang, Y., and Weiss, B. (1995) D2 dopamine receptor antisense oligonucleotide inhibits the synthesis of a functional pool of D2 dopamine receptors. *Mol. Pharmacol.* **48,** 730–737.

Raffa, R. B., Martinez, R. P., and Connelly, C. D. (1994) G protein antisense oligodeoxyribonucleotides and μ-opioid supraspinal antinociception. *Eur. J. Pharmacol.* **258,** R5–R7.

Rajput, A. H. (1992) Frequency and cause of Parkinson's disease. *Can. J. Neurol. Sci.* **19,** 103–107.

Robertson, H. A. (1992a) Synergistic interactions of D1- and D2-selective dopamine agonists in animal models for Parkinson's disease: sites of action and implications for the pathogenesis of dyskinesias. *Can. J. Neurol. Sci.* **19,** 147–152.

Robertson, H. A. (1992b) Dopamine receptor interactions: some implications for the treatment of Parkinson's disease. *Trends Neurosci.* **15,** 201–205.

Ross, E. M. (1992) G proteins and receptors in neuronal signalling, in *An Introduction to Molecular Neurobiology* (Hall, Z. W., ed.), Sinauer Associates, Sunderland, pp. 181–207.

Schettini, G., Ventra, C., Florio, T., Grimaldi, M., Meucci, O., and Marino, A. (1992) Modulation by GTP of basal and agonist-stimulated striatal adenylate cyclase activity following chronic blockade of D1 and D2 dopamine receptors: involvement of G proteins in the development of receptor supersensitivity. *J. Neurochem.* **59,** 1667–1674.

Seeman, P. and Niznik, H. B. (1990) Dopamine receptors and transporters in Parkinson's disease and schizophrenia. *FASEB J.* **4,** 2737–2744.

Sibley, D. R. and Monsma, F. J., Jr. (1992) Molecular biology of dopamine receptors. *Trends Pharmacol. Sci.* **13,** 61–69.

Silvia, C. P., King, G. R., Lee, Xue, Z.-Y., Caron, M. G., and Ellinwood, E. H. (1994) Intrastriatal administration of D2 dopamine receptor antisense oligodeoxynucleotides establishes a role for nigrostriatal D2 autoreceptors in the motor actions of cocaine. *Mol. Pharmacol.* **46,** 51–57.

Sokoloff, P. and Schwartz, J.-C. (1995) Novel dopamine receptors half a decade later. *Trends Pharmacol. Sci.* **16,** 269–285.

Srivastava, L. K. and Mishra, R. K. (1994) Dopamine receptor gene expression: effects of neuroleptics, denervation, and development, in

Dopamine Receptor Function and Pharmacology (Niznik, H. B., ed.), Marcel Dekker, New York, N.Y. pp. 437–457.

Sunahara, R. K., Seeman, P., Van Tol, H. H. M., and Niznik, H. B. (1993) Dopamine receptors and antipsychotic drug response. *Br. J. Psychiatry* **163 (Suppl. 22)**, 31–38.

Sundstrom, E., Fredriksson, A., and Archer, T. (1990) Chronic neurochemical and behavioral changes in MPTP-lesioned C57BL/6 mice: a model for Parkinson's disease. *Brain Res.* **528**, 181–188.

Surmeier, D. J., Reiner, A., Levine, M. S., and Ariano, M. A. (1993) Are neostriatal dopamine receptors co-localized? *Trends Neurosci.* **16**, 299–305.

Tang, T., Kiang, J. G., Cote, T. E., and Cox, B. M. (1995) Antisense oligodeoxynucleotide to the Gi_2 protein α subunit sequence inhibits an opioid-induced increase in the intracellular free calcium concentration in ND8-47 neuroblastoma × dorsal root ganglion hybrid cells. *Mol. Pharmacol.* **48**, 189–193.

Tanner, M. T. (1986) Drug-induced movement disorders (tardive dyskinesia and dopa-induced dyskinesia), in *Etrapyramidal Disorders,* 49th ed. (Vinken, P. J., Bruyn, G. W., and Klawans, H. L., eds.), Elsevier Science Publishers, Amsterdam pp. 185–203.

Tirone, F., Parenti, M., and Groppetti, A. (1985) Opiate and dopamine stimulate different GTPase in striatum: evidence for distinct modulatory mechanisms of adenylate cyclase. *J. Cyclic Nucl. Protein Phos. Res.* **10**, 327–339.

Wang, H.-Y., Undie, A. S., and Friedman, E. (1995) Evidence for the coupling of G_q protein to D1-like dopamine sites in rat striatum: possible role in dopamine-mediated inositol phosphate formation. *Mol. Pharmacol.* **48**, 988–994.

Watson, J. B., Coulter II, P. M., Margulies, J. E., de Lecea, L., Danielson, P. E., Erlander, M. G., and Sutcliffe, J. G. (1994) G protein gamma7 subunit is selectively expressed in medium-sized neurons and dendrites of the rat neostriatum. *J. Neurosci. Res.* **39**, 108–116.

Weiss, B., Zhou, L. W., Zhang, S. P., and Qin, Z. H. (1993) Antisense oligodeoxynucleotide inhibits D2 dopamine receptor-mediated behavior and D2 messenger RNA. *Neuroscience* **55**, 607–612.

Yoshida, M. (1991) The neuronal mechanism underlying parkinsonism and dyskinesia: differential roles of the putamen and caudate nucleus. *Neurosci. Res.* **12**, 31–40.

Zhang, M. and Creese, I. (1993) Antisense oligodeoxynucleotide reduces brain dopamine D2 receptors: behavioral correlates. *Neurosci. Lett.* **161**, 223–226.

Protein Kinase C as a Modulator of 5-HT1A and Dopamine-D2 Receptor Signaling

Paul R. Albert, Paola M. C. Lembo,
and Stephen J. Morris

Introduction

Since its identification as a family of protein kinases that is activated in the presence of calcium, phosphatidyl serine, and diacylglycerol (DAG), protein kinase C (PKC) has been regarded as one of the early steps in receptor signaling (Nishizuka, 1988, 1995; Berridge, 1993; Rasmussen et al., 1995). The generation of the second messengers DAG and calcium by receptor-mediated phospholipase C (PLC) activation leads to rapid activation of several isoforms of PKC (Newton, 1995), its translocation to the membrane (Kiley et al., 1995), and the phosphorylation of numerous proteins. With the finding by Castagna et al. (1982) that phorbol esters (e.g., 12-*O*-tetradecanoyl 4β-phorbol 13-acetate [TPA]) are potent and selective activators of multiple PKC isoforms, this pharmacological tool was used extensively to document the importance of PKC in a variety of stimulatory events: lymphocyte activation, contractile and secretory responses, gene transcription, and cell proliferation (Nishizuka, 1988, 1995; Berridge, 1993; Rasmussen et al., 1995). In addition to these stimulatory actions, PKC activation was also found to have a negative feedback action to inhibit the activity of receptors that coupled to PLC to augment DAG and calcium levels. Among the receptors that are inhib-

From: *Neuromethods, Vol. 31: G Protein Methods and Protocols*
Ed: R. K. Mishra, G. B. Baker, and A. A. Boulton Humana Press Inc.

ited by PKC are the 5-HT1A and dopamine-D2 receptors (Albert, 1994) where feedback inhibition by PKC is remarkably selective to inhibit receptor-mediated PLC activation, and does not block other pathways (e.g., receptor coupling to adenylyl cyclase). The mechanism of PKC-induced negative feedback regulation is discussed, focusing on recent experiments using site-directed mutagenesis to identify the crucial sites required for PKC action. These findings are related to the mechanisms by which 5-HT1A and dopamine-D2 receptors couple to multiple signaling pathways, and the role of phosphorylation in specifying receptor coupling.

Receptor Signaling: G_i/G_o Protein Selectivity for Multiple Pathways

The binding of agonists to G protein-coupled receptors initiates a series of intracellular second messenger changes that depends on the selection of heterotrimeric G proteins to which the receptor couples (Birnbaumer, 1992; Clapham and Neer, 1993; Neer, 1995). Specific G proteins mediate stimulation or inhibition of effector enzymes (e.g., adenylyl cyclase, PLC) or ion channels (e.g., potassium, sodium, or calcium channels) that regulate the levels of second messengers, such as cyclic AMP (cAMP), DAG inositol trisphosphate, or calcium ion ($[Ca^{2+}]_i$) (Exton, 1994; Hille, 1994; Wickman and Clapham, 1995; Nishizuka, 1995). In particular, activation of the 5-HT1A (or dopamine-D2) receptor in neuronal and pituitary cells (Civelli et al., 1993; Albert, 1994; Raymond, 1995; Albert et al., 1996) initiates multiple inhibitory actions, including opening of potassium channels (Innis and Aghajanian, 1987; Andrade and Nicoll, 1987; Vallar et al., 1990; Clapham and Neer, 1993; Penington et al., 1993a,b), closing of calcium channels (Penington and Kelly, 1990; Liu and Albert, 1991; Seabrook et al., 1994) and inhibition of basal and stimulated adenylyl cyclase activity (Vallar et al., 1990; Albert et al., 1990a,b; Liu and Albert, 1991; Seabrook et al., 1994). Opening of potassium channels hyperpolarizes the membrane potential, leading to reductions in action potential frequency and decreased neurotransmission (Andrade and Nicoll, 1987; Albert et al., 1996). Inhibition of calcium channel activation would decrease $[Ca^{2+}]_i$ attenuating calcium-dependent processes, such as presynaptic release of neurotransmitters (Albert et al., 1996). Inhibition of cAMP levels reduces the activity of protein kinase PKA to inhibit cAMP-dependent processes. The cumulative effect of 5-HT1A or dopamine-D2 receptor activation in endocrine cells is an inhibition of prolactin (PRL) secretion (Albert et al., 1990a; Albert and Raquidan, 1995), PRL gene transcription (Elsholtz et al., 1991), and cell proliferation (Florio et al., 1992; Senogles, 1994; Albert, 1995). All of these actions are blocked by pretreatment with PTX, a toxin that selectively ADP-ribosylates G_i/G_o proteins to inactivate coupling (Birnbaumer, 1992; Clapham and Neer, 1993; Neer, 1995). Thus, coupling to G_i/G_o proteins mediates these inhibitory actions of 5-HT1A and dopamine-D2 receptors (Albert, 1994).

However, the activation of 5-HT1 or dopamine-D2 receptors expressed in mesenchymal cell types (such as Ltk$^-$, Balb/c-3T3, CHO, or HeLa cells) induced a paradoxical increase in PLC activity to mobilize intracellular calcium stores resulting in increased $[Ca^{2+}]_i$ (Fargin et al., 1989; Vallar et al., 1990; Albert et al., 1990b; Liu and Albert, 1991; Abdel-Baset et al., 1992; Lajiness et al., 1993; Chio et al., 1994; Lew et al., 1994; Albert, 1995; Lew and Elsholtz, 1995). These receptor-induced actions occurred in the absence of inhibition of basal cAMP levels, although inhibition of forskolin- or G_s-stimulated cAMP was observed. The stimulatory response is surprising, since the action of the 5-HT1A receptor is generally "inhibitory" as described above, whereas stimulation of PLC enhances $[Ca^{2+}]_i$ and PKC activity. These stimulatory changes were associated with MAP kinase activation (Lajiness et al., 1993; Chio et al., 1994; Dhanasekaran et al., 1995), increased PRL gene transcription (Lew et al., 1994; Lew and Elsholtz, 1995), enhanced DNA synthesis (Abdel-Baset et al., 1992; Varrault et al., 1992; Lajiness et al., 1993; Chio et al., 1994), and cellular transformation (Abdel-Baset et al., 1992; Varrault et al., 1992). All of these actions were blocked by pretreatment with PTX, indicating that these receptors initiate multiple signaling pathways by interacting with a single family of closely related G proteins, the G_i/G_o family.

Other receptor subtypes are even more promiscuous in activating across G protein families: for example, the α2-C10 adrenergic receptor couples to both G_s and G_i/G_o proteins (Eason et al., 1992), whereas the thyrotropin receptor couples to G_s, G_q, and G_i (Laugwitz et al., 1996). With the observation of this multiplicity of receptor-initiated signals, several groups have addressed the mechanisms underlying coupling specificity (Albert, 1994; Albert and Morris, 1994). In the case of "inhibitory" receptors, such as dopamine-D2 or 5-HT1A receptors, several groups have examined the identity of the G proteins that mediate the diverse second messenger changes initiated by these receptors (Lledo et al., 1990; Kleuss et al., 1991, 1992, 1993; Baertschi et al., 1992; Liu et al., 1994). Using antisense approaches in lactotroph pituitary cells, the mRNAs encoding individual G_i or G_o α-subunits have been specifi-cally depleted to examine the consequences for receptor sig-naling. Dopamine-D2, 5-HT1A, somatostatin, muscarinic, and other G_i/G_o-coupled receptors have been shown to require G_o proteins (G_oA or G_oB) to couple to inhibition of cal-cium channel opening (Kleuss et al., 1991). Similarly, the GABAB and other inhibitory receptors appear to couple via G_o to inhibition of calcium channels in other neuronal and pituitary cell types (Albert and Morris, 1994). The coupling of muscarinic, somatostatin, and dopamine-D2 receptors to opening of potassium channels in pituitary cells appears to involve G_i3 (Lledo et al., 1990, 1992). It appears to be βγ-sub-units that associate with $α_i3$ that induce potassium channel opening, based on data from muscarinic receptor coupling in cardiac tissue (Clapham and Neer, 1993; Hille, 1994; Wickman and Clapham, 1995). Receptor coupling to inhibition of ade-nylyl cyclase is mediated primarily via $G_iα$ proteins (Tang and Gilman, 1991), the selection of which is receptor-dependent

(Kleuss et al., 1993; Albert and Morris, 1994). For example (Kleuss et al., 1993), in α_i2-antisense cells, the dopamine-D2L, muscarinic and somatostatin receptors were entirely uncoupled from inhibition of adenylyl cyclase, whereas the dopamine-D2S receptor remained coupled to inhibition of G_s-stimulated adenylyl cyclase. By contrast, receptor coupling to inhibition of basal or stimulated cAMP levels was unimpaired in $G\alpha_o$ antisense clones. These results indicate at least two levels of specification in receptor signaling: (1) Receptor–G protein interactions that determine the receptor-specific priority of G protein heterotrimers. For example, the dopamine-D2L receptor has a greater dependence on G_i2 than does the D2S receptor, and (2) G protein–effector interactions that establish a hierarchy of G protein-mediated functions. For example, $G\alpha_i$ couples to inhibition of adenylyl cyclase, but not to closing of calcium channels, whereas G_o couples to inhibition of calcium channels without mediating inhibition of adenylyl cyclase.

The structural basis for the G protein specificity of receptors remains to be elucidated and awaits crystallographic determination of receptor topology. Various indirect approaches (e.g., chimeric receptors, point and deletion mutagenesis of receptors) have led to the identification of coupling domains in the intracellular loop domains (i2, i3), C-terminal domains of the receptor, and in the C-terminal portion of $G\alpha$-subunits (Ostrowski et al., 1992; Strader et al., 1994). Although no conserved consensus sequences have been identified, these domains provide clear targets to examine the specific amino acids that may be involved in determining receptor-G protein specificity, as discussed below.

The basis for conditional stimulation of calcium mobilization by 5-HT1A and dopamine-D2 receptors in mesenchymal cell types (such as Ltk⁻ cells) appears to involve coupling by multiple G_i/G_o subtypes to PLC. The finding that certain subtypes of PLC (e.g. PLC-β2, PLC-β3) are sensitive to $\beta\gamma$ dimers derived from G_i/G_o heterotrimers (Clapham and Neer, 1993; Exton, 1994; Neer, 1995) suggests a mechanism by which "inhibitory" 5-HT and dopamine receptors coupled to stimulation of this enzyme. Preliminary evidence suggests that this pathway involves Gβγ-mediated activation of PLC-β2 to generate IP3, which mobilizes calcium from intracellular stores (Lew and Elsholtz, 1995). For 5-HT1A and dopamine-D2 receptors, the PLC-mediated signaling pathway is selectively and completely blocked by acute (1–2 min) preactivation of PKC, whereas receptor-mediated inhibition of cAMP accumulation is not affected by PKC (Liu and Albert, 1991; Liu et al., 1992; Lembo and Albert, 1994, 1995). It is the mechanism of this pathway-selective uncoupling of 5-HT1A and dopamine-D2 receptors by PKC that is discussed further.

Receptor Desensitization by Protein Kinase Activation

Receptor desensitization has been defined as the loss of responsiveness in the sustained presence of agonist, and has been most intensively studied in the β2-adrenergic-G_s-adenylyl cyclase paradigm (Lefkowitz et al., 1990; Kobilka, 1992; Sterne-Marr and Benovic, 1995). Homologous desensitization involves several processes that occur sequentially on binding of agonist to the receptor: uncoupling (seconds to minutes), a phosphorylation-mediated event where the receptor no longer activates G proteins; internalization (minutes to hours) of the receptor in acidic, clathrin-coated vesicles and dissociation of the receptor-ligand complex; and downregulation (hours to days) where internalized vesicles fuse with lysosomes that proteolytically degrade the receptor. In addition, receptor desensitization can be homologous (agonist-mediated) or heterologous (kinase-mediated). These two mechanisms overlap in that for both homologous and heterologous regulation, activation of protein kinases has been implicated. The elegant studies of Lefkowitz and coworkers have identified two primary pathways that mediate homologous desensitization: (1) via activation of PKA and (2) via activation of receptor kinases (GRKs). Activation of β2-adrenergic receptors leads to the production of cAMP, which activates PKA. In addition to mediating downstream actions of cAMP, PKA acts in a negative feedback loop to reduce the signaling of the β2-adrenergic receptor by phosphorylating key sites on the receptor. Heterologous desensitization via activation of other receptors that increase cAMP utilizes the same mechanism. In contrast, GRKs act independent of second messengers, like cAMP, to phosphorylate agonist-induced receptors, and thus, cannot mediate heterologous desensitization. The consensus substrate sites of the two kinases are different and are located in different parts of the β2-adrenergic receptor. The PKA sites are located near the transmembrane domains in the i3 loop and C-terminal, regions that are required for receptor-G protein interactions and coupling. The GRK sites are on the C-terminal tail and are removed from the coupling domains of the receptor. Site-directed mutagenesis of phosphate-acceptor serine residues on the i3 loop and C-terminal tail of the β2-adrenergic receptor prevented agonist-induced uncoupling of the receptor from the adenylyl cyclase pathway. A partial effect was observed if only the PKA sites were removed, whereas a complete block of desensitization occurred if both PKA and GRK sites were eliminated. Thus, for the β2-adrenergic receptor, multiple sites of phosphorylation are required for receptor uncoupling, which results in an inactivation of receptor signaling.

Another example of kinase-mediated receptor uncoupling is observed for receptors that activate PLC (Nishizuka, 1988, 1995; Berridge, 1993; Rasmussen et al., 1995). Receptor-mediated generation IP3 mobilizes intracellular calcium stores to increase $[Ca^{2+}]_i$, whereas DAG activates PKC. Preactivation of PKC (e.g., using TPA) results in a loss of receptor-mediated PI turnover and calcium mobilization. Thus, PKC acts in a negative feedback loop, such that PLC activation generates a product (DAG) that inhibits phospholipase activation via PKC. Several receptors are regulated by this mechanism, including G_q- and G_i/G_o-coupled receptors, such as neurokinin receptor types I and II, gastrin-releasing peptide, and mu opioid receptors (Barr and Watson, 1994; Alblas et al.,

1995; Benya et al., 1995; Mestek et al., 1995). However, the site at which PKC acts to inhibit receptor coupling to PLC has not been identified. PKC has been shown to phosphorylate receptors (e.g., the 5-HT1A receptor [Raymond, 1991]), G proteins (e.g., G_i2, G_z [Bushfield et al., 1990; Yatomi et al., 1992; Strassheim and Malbon, 1994; Fields and Casey, 1995]), and PLC-β (Ryu et al., 1990). Activation of PKC is associated with inhibition of G_i2 and PLC activity in vitro. It was unclear which component of the receptor G protein PLC-β signaling system was the crucial target of PKC to block this receptor signaling pathway. Although each signaling component has been demonstrated biochemically to be phosphorylated by PKC, these studies did not address which component(s) play the predominant role in an intact cell. For example, receptor phosphorylation by PKC may be sufficient to inhibit signaling at physiological levels of PKC activation, whereas other phosphorylation events may operate at supra-physiologic levels of PKC activation. A different approach was needed to address the site of PKC action.

Mechanisms of PKC-Induced Receptor Uncoupling

Site of PKC Action

The most direct way to identify the site of PKC action is by site-directed mutagenesis of consensus sites for PKC-induced phosphorylation located on putative substrate proteins. The biochemical identification of PKC phosphorylation sites in a variety of substrate proteins has yielded peptide consensus sequences (Kemp and Pearson, 1990; Kennelly and Krebs, 1991). Elimination of the phosphate acceptor hydroxyl side chain by conversion of the serine/threonine residues to structurally similar aliphatic amino acids, such as alanine or glycine, prevents the phosphorylation event without greatly altering protein structure. Site-directed mutagenesis or chimeric substitution has been used to identify the roles of PKC sites in modulation of β2-adrenergic, GABA-A, and hepatocyte growth factor receptors (Bouvier et al., 1991; Gandino et al., 1994; Krishek et al., 1994; Yuan et al., 1994) and of voltage-gated calcium channels (Stea et al., 1995). This approach has the advantage of selectively modifying an individual target site of PKC action, without altering the global activities of the kinase. Strategies involving nonselective pharmacological inhibitors that block multiple actions of PKC that may influence signaling at a variety of sites, and often the inhibitors impair the activity of other kinases. Biochemical strategies to demonstrate phosphorylation are correlative and do not address the importance of phosphorylation events in intact cell signaling systems. We therefore used the approach of site-directed mutagenesis as a selective approach to address the role of receptor phosphorylation in PKC-induced receptor uncoupling (Lembo and Albert, 1995).

Putative PKC phosphorylation sites were identified on the 5-HT1A receptor, which is the most sensitive to TPA-induced uncoupling (Liu and Albert, 1991; Liu et al., 1992). The 5-HT1A receptor was chosen for initial studies in part because of biochemical evidence that the receptor was phosphorylated on multiple sites following acute activation of PKC (Raymond, 1991). We identified four consensus phosphorylation sites on the receptor, three in the i3 loop domain and one in the i2 loop domain (Lembo and Albert, 1995). These sites are conserved in human, rat, and murine 5-HT1A receptors. Each site was mutagenized to a nonphosphorylatable residue (glycine or alanine), and receptors with single mutations, one double mutant (two i3 sites), and the triple mutant (three i3 sites) were transfected into Ltk⁻ fibroblast cells. All of the i3 loop receptor mutants were functional, except for the one threonine mutant (T229A) in which 5-HT-mediated calcium mobilization was reduced. For each of the single-mutant receptors, pretreatment with 100 n*M* TPA remained effective, completely blocking 5-HT-mediated calcium mobilization (Fig. 1). However, the double mutant was approx 50% resistant to TPA, whereas the triple mutant was nearly 75% resistant to preactivation of PKC. These results provide the first evidence that uncoupling of the receptor G protein PLC-β pathway that is induced by PKC requires multiple phosphorylation sites on the receptor. At least in Ltk⁻

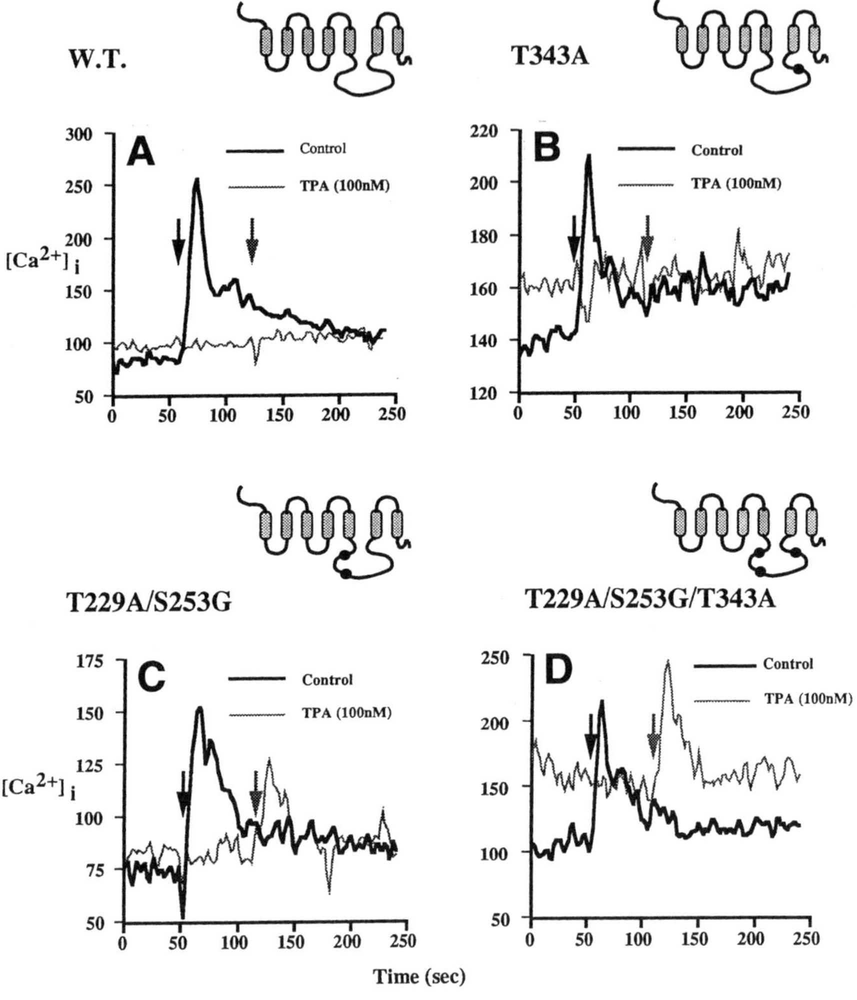

Fig. 1. Calcium mobilization by PKC site mutants of the 5-HT1A receptor transfected in Ltk⁻ cells. Dark tracing represents control samples: 5-HT (100 nM) was added at 60 s. Light tracing represents TPA-treated samples: TPA (100 nM) was added at time 0, and 5-HT (100 nM) at 120 s. **(A)** Wildtype receptor, in TPA-treated samples (light tracing) the serotonin-mediated calcium response was completely inhibited. **(B)** T343A (TPA completely blocked the 5-HT response). **(C)** T229A/S253G, post-treatment with TPA, 50% of the 5-HT-induced calcium response was recovered. **(D)** T229A/S253G/T343A, 74% of 5-HT-mediated calcium response was recovered post-TPA. All curves were generated by a computer and were from a single experiment, which was repeated at least three times and with at least two independent clones, which gave similar results. Reprinted with permission from Lembo and Albert, 1995, Fig. 2.

cells, 75% of the effect of PKC is mediated at the receptor, rather than by phosphorylation at the G protein or PLC-β. It should be noted that at very high concentrations of TPA (>1 μM), receptor-induced calcium mobilization was attenuated even in the triple mutant, indicating that phosphorylation at other sites plays a secondary role in acute uncoupling and occurs only at pharmacological levels of PKC activation. The importance of receptor phosphorylation sites in PKC-induced uncoupling supports the hypothesis that receptor phosphorylation plays the predominant role in TPA-induced uncoupling of the 5-HT1A receptor from stimulation of PI turnover and calcium mobilization. The isoforms of PKC that mediate receptor uncoupling remain to be identified.

Determinants of Receptor Sensitivity to PKC

Pseudosubstrate Domains

We and others had shown previously that in addition to cloned 5-HT1A receptors, both dopamine D2 receptor subtypes also mediate PTX-sensitive PI turnover and calcium mobilization when expressed in a variety of mesenchymal cell lines (Vallar et al., 1990; Liu et al., 1992; Lew and Elsholtz, 1995). Acute pretreatment (1–2 min) with TPA to activate PKC resulted in a complete inhibition of receptor-mediated calcium mobilization for the transfected 5-HT1A (Liu and Albert, 1991) and dopamine-D2S receptors, whereas the dopamine-D2L receptor was resistant (Fig. 2). The EC$_{50}$ values for sensitivity to acute TPA pretreatment were <10 nM, 100 nM, and >1 μM for 5-HT1A, dopamine-D2S, and D2L receptors, respectively. Calcium mobilization induced by ATP (via endogenously expressed G$_q$-coupled P2-purinergic receptors) or thrombin (endogenous G$_i$/G$_o$/G$_q$-coupled) were also blocked by TPA. The resistance of the D2L receptor subtype to TPA pretreatment was not owing to a general insensitivity of the cell line to TPA, since TPA inhibited thrombin-mediate calcium mobilization in both D2S- and D2L-transfected cell lines with equivalent affinity (Liu et al., 1992). These results indi-

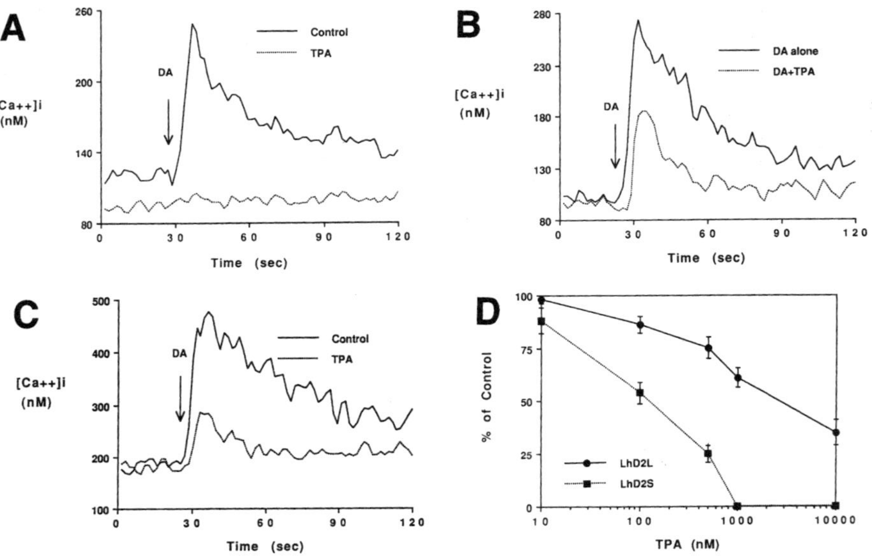

Fig. 2. TPA inhibits the dopamine-induced increase in $[Ca^{2+}]_i$ in LhD2L and LhD2S cells. All curves were from a typical experiment that has been repeated six times with the similar results. The cells were incubated with 1 μ*M* TPA for 5 min before recording. 100 n*M* dopamine were added as indicated. **(A)** TPA (1 μ*M*) completely abolishes the dopamine induced-increase in $[Ca^{2+}]_i$ in LhD2S cells. **(B)** TPA (1 μ*M*) partially inhibits the dopamine-induced increase in $[Ca^{2+}]_i$ in LhD2L cells. **(C)** LhD2L cells were incubated as above with 100 μ*M* TPA and treated with 100 n*M* dopamine as indicated. **(D)** Concentration dependency of TPA-induced inhibition of the maximal increase in $[Ca^{2+}]_i$ in LhD2L and LhD2S cells. The error bars are the SD calculated from six independent experiments performed in duplicate and in parallel with D2S and D2L cells. Reprinted with permission from Liu et al., 1992, Fig. 3

cate that the additional 29 amino acid insert present in the D2L receptor subtype confers resistance to the action of PKC. There is a weak pseudosubstrate domain present in the i3 loop of the D2S receptor that is lacking in the 5-HT1A receptor (Fig. 3). The presence of a vicinal i3 loop pseudo-substrate domain could act as an intramolecular inhibitor of PKC-induced receptor phosphorylation at substrate sites located in the i2 and i3 loops of the receptor. Partial inhibition of PKC by this domain could account for the 10-fold higher EC_{50} for TPA at the D2S receptor. In the D2L receptor, the inclusion of intronic sequence creates a strong consensus pseudosubstrate domain that might explain the resistance of the D2L receptor to TPA-induced uncoupling.

There is a precedent for intramolecular inhibition of PKC activity by pseudosubstrate domains. A pseudosubstrate domain is present in all forms of PKC identified to date that is lodged in the catalytic domain and renders the kinase inactive (Newton, 1995). The binding of a variety of activators (including DAG, TPA, or polyamines) to noncatalytic domains of the kinase alters the protein conformation to displace the pseudo-substrate domain and allow interaction of the catalytic domain with substrate proteins. Our hypothesis that pseudo-substrate domains present in substrate proteins may inhibit the activity of PKC at adjacent phosphorylation sites is an extension of these observations in PKC. It is of interest that the pseudosubstrate domains of typical (α, β, δ, ϵ) and atypical (λ, ζ) PKC subtypes are not functionally interchangeable, suggesting that differences in D2 receptor pseudosubstrate domains may confer differential sensitivity to PKC. However, further mutagenesis experiments are required to distinguish the role of these putative pseudosubstrate domains in receptor sensitivity to modulation by PKC.

Sculpting Receptor Signaling by Selective Uncoupling

The uncoupling produced by PKC was of interest, because the selective impairment in receptor signaling com-

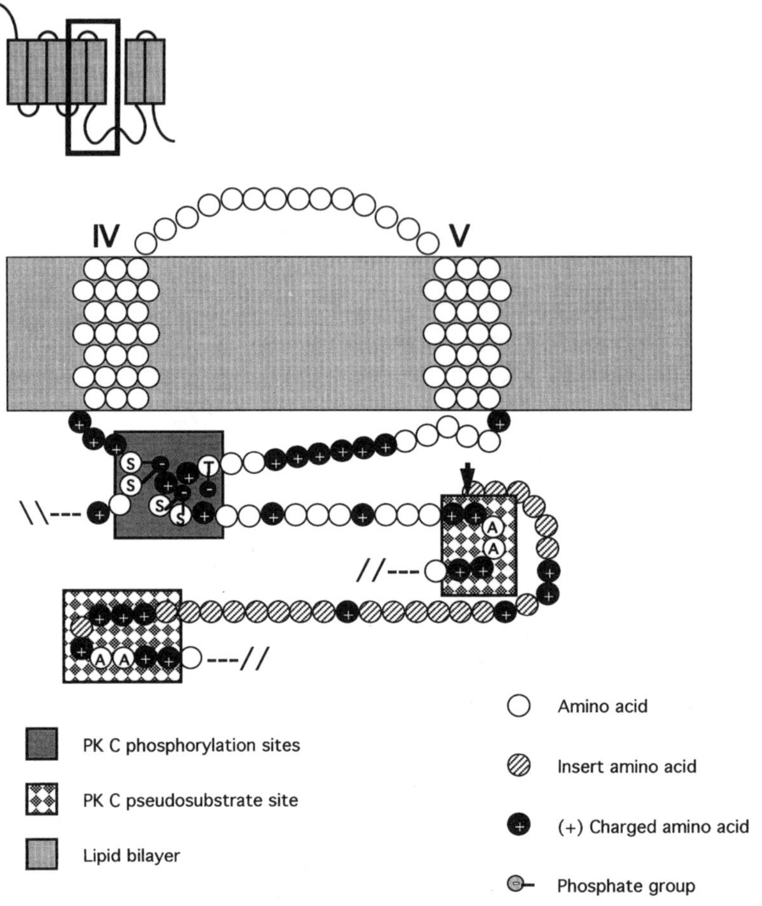

Fig. 3. Model of the dopamine D2S and D2L receptors: presence of PKC substrate and pseudosubstrate sequences in the intracellular loops. The inset (upper right) shows a model of the dopamine-D2 receptor, with the expanded region boxed. The expanded region includes the fourth and fifth transmembrane domains, and portions of the intracellular second (i2) and third loops (i3). The 29 amino acid intron of the D2L subtype inserts the residues that are shaded and creates a stronger putative pseudosubstrate domain. This domain consists of alanine residues **(A)** flanked by multiple positively charged amino acids as indicated. Multiple putative PKC substrate sites analogous to those identified in the 5-HT1A receptor are indicated as negatively charged phosphorylated serine and threonine (**S** and **T**) residues.

pletely blocked agonist-induced calcium mobilization, whereas there was no effect on receptor-mediated inhibition of cAMP accumulation (Liu and Albert, 1991; Liu et al., 1992; Lembo and Albert, 1995). We have previously proposed the concept of pathway-selective uncoupling, whereby desensitization can lead to a sculpting of receptor signaling, preserving or enhancing some pathways while blocking others (Albert, 1994). In the case of the 5-HT1A and dopamine-D2S receptors in Ltk⁻ fibroblast cells, preactivation of PKC blocks the stimulatory pathway (PI turnover and calcium mobilization), but permits coupling to inhibition of forskolin- or G_s-stimulated cAMP accumulation. However, the mechanism by which TPA inactivated certain pathways selectively was unresolved: For example, TPA might act by differential phosphorylation of G proteins or effectors. In the experiments described above, both the non-mutated 5-HT1A receptor and the triple mutant inhibited cAMP accumulation induced by prostaglandin E1, which activates a G_s-coupled receptor. The EC_{50} values for 5-HT-induced inhibition of cAMP in the presence or absence of TPA (1 μM) were indistinguishable (Lembo and Albert, 1995). These results indicate that receptor phosphorylation uncouples the receptor selectively from activation of PLC, but allows coupling to inhibition of adenylyl cyclase. It appears that modification of the receptor by phosphorylation can selectively disrupt interactions with G proteins that couple to calcium mobilization, while retaining interactions with G_i-proteins that couple to inhibition of cAMP.

It is unclear why the PLC pathway, but not the adenylyl cyclase pathway, is selectively blocked by receptor phosphorylation. As presented above, evidence suggesting that multiple G_i/G_o proteins couple the receptor to PLC whereas a single subtype can couple to cAMP suggests that a partial uncoupling of receptor–G protein interaction may be sufficient to block calcium mobilization. However, for some receptors (e.g., the muscarinic-m2 receptor) a single G protein (although overexpressed) may be sufficient to couple to all pathways (Hunt et al., 1994). Another explanation is that since there is a higher dependence of the PLC signal on receptor number (Albert, 1994), partial phosphorylation of the receptor population would preferentially inactivate this pathway. However, the cAMP pathway was not blocked at high (1–10 μM) TPA concentrations that saturate 5-HT1A receptor phosphorylation (Raymond, 1991). This argues against the latter "receptor reserve" hypothesis, and for a selective uncoupling of receptor–G protein interactions by PKC. It may be possible to test this directly using the Sf-9 insect cell system to examine the action of PKC on interaction with individual $G\alpha$-subunits or G protein heterotrimers (Butkerait et al., 1995). This could also provide a system to identify which PKC isoforms mediate receptor uncoupling by introducing each isoform individually.

G Protein Contact Sites: Role of the i2 Loop

The role of the i2 loop phosphorylation site (threonine-149) in PKC regulation of 5-HT1A receptor uncoupling was not be addressed because this mutation uncoupled the receptor from calcium mobilization in Ltk⁻ fibroblasts (Lembo et al., 1997). Interestingly, the receptor mutant still coupled to inhibition of cAMP generation. When transfected into GH4 pituitary cells, the T149A mutant also coupled to both inhibition of basal and G_s-stimulated cAMP formation. However, coupling to inhibition of calcium channel activation in this cell line was blocked. These results indicate that the conversion of the i2 loop threonine to an alanine residue was sufficient to uncouple the receptor selectively from calcium mobilization. This suggests that the modification of T149 by phosphorylation may also mediate selective uncoupling of the receptor. An alternate interpretation of resistance of the triple mutant to TPA could be that mutations have generated multiple weak pseudosubstrate sites and that these sites reduce phosphorylation at T149. This would be consistent with the partial resistance to TPA of the double mutant in which only two pseudosubstrate sites are generated. Since the i2 loop threonine is widely invariant among receptors that couple to stimulate

PLC (either via G_i/G_o or via G_q), the hypothesis that this residue plays a key role in PKC-mediated receptor uncoupling is appealing, but needs further testing. It should be noted that this domain has been implicated in coupling to both PLC and adenylyl cyclase (Varrault et al., 1994; Blin et al., 1995; Liu et al., 1995).

PKC In Agonist-Induced Homologous Desensitization

Agonist-induced receptor desensitization is a key element in receptor regulation. Although the above studies indicate that heterologous desensitization by PKC activation occurs, they do not address the role of PKC in homologous desensitization of 5-HT1A or dopamine-D2 receptors. In cell types that lack the PLC pathway (e.g., neuronal or pituitary cells), the receptor would not activate PKC-mediated feedback inhibition (but *see* Senogles, 1994). In cells (such as Ltk⁻ fibroblasts) in which the 5-HT1A receptor signals via PLC activation to increase DAG and activate PKC, PKC-induced uncoupling of this pathway may play an important role in homologous desensitization. The above studies do not directly address this possibility. Transfected 5-HT1A and dopamine-D2 receptors have been observed to undergo agonist-induced desensitization with respect to inhibition of adenylyl cyclase in a variety of mesenchymal or insect cells (Bates et al., 1990; Raymond, 1991; van Huizen et al., 1993; Harrington et al., 1994; Nebigil et al., 1995), but the mechanisms involved remain unclear. It should noted that the priority of kinases involved in homologous desensitization has been found to be cell-specific (Shih and Malbon, 1994), and thus, homologous desensitization is best studied in cells that endogenously express the receptor of interest.

Physiological Relevance

The above studies have addressed the mechanism by which PKC uncouples receptor signaling. However, the role of such uncoupling in the brain has not been addressed. Recent studies in dissociated raphe neurons indicate that PKC activation can induce pathway-selective modulation in the brain. Following acute (minutes) pretreatment with TPA, the 5-HT1A receptor was uncoupled from inhibition of calcium channel opening, but remained coupled to opening of potassium channels (Chen and Penington, 1996). It is tempting to speculate that similar mechanisms of receptor phosphorylation may be involved in this modulation of receptor function. The loss of calcium channel inhibition would preclude a presynaptic inhibition of transmitter release. However the retention of potassium channel opening would sustain membrane hyperpolarization and inhibition of raphe firing rate. Activation of PKC by receptors (e.g., metabotropic glutamatergic) that couple to PLC would produce selective alterations in the function of inhibitory receptors that are sensitive to PKC. Interestingly, the dopamine-D2S is sensitive to PKC, but the D2L isoform is not. Thus, alterations in the relative levels of these subtypes, could influence the signaling and desensitization of the D2 receptors.

Summary

Modulation of receptor responsiveness by PKC is a potentially important mechanism of receptor signaling plasticity. Rather than completely uncoupling the receptor, the action of PKC is to uncouple 5-HT1A, dopamine-D2, and other "inhibitory" receptors selectively from stimulatory coupling to PLC, but not from inhibitory coupling to adenylyl cyclase. This selective uncoupling may extend to other pathways (e.g., inhibition of calcium influx vs opening of potassium channels). The pathway selectivity of PKC action suggests that these phosphorylation events result in a selective uncoupling of the receptor from one or more G protein subtypes.

The mechanism of acute PKC action to uncouple the 5-HT1A receptor selectively involves phosphorylation at multiple sites on the receptor. These sites are located in or proximal to intracellular receptor domains (e.g., the i2 and i3 loops) proposed to be crucial G protein contact sites. The sensitivity of a given receptor to PKC activation correlates inversely with the presence of putative pseudosubstrate sites in these receptor domains. It remains to be determined whether receptor phosphorylation, as opposed to phosphorylation at downstream points (e.g., G proteins or effectors), predominates as a mechanism for uncoupling receptors in general.

References

Abdel-Baset, H., Bozovic, V., Szyf, M., and Albert, P. R. (1992) Conditional transformation mediated via a pertussis toxin-sensitive receptor signalling pathway. *Mol. Endocrinol.* **6,** 730–740.

Albert, P. R. (1994) Heterologous expression of G protein-linked receptors in pituitary and fibroblast cell lines. *Vitam. Horm.* **48,** 59–109.

Albert, P. R. (1995) Dopamine-D2 receptors mediate inhibition of DNA synthesis in GH pituitary cells via G_i/G_o proteins. Endocrine Soc. Mtg., Washington, DC, #P3–93.

Albert, P. R. and Morris, S. J. (1994) Antisense knockouts: molecular scalpels for the dissection of signal transduction. *Trends Pharmacol. Sci.* **15,** 250–254.

Albert, P. R. and Raquidan, D. (1995) Dopamine-D2 receptor routing through multiple G proteins for inhibition of TRH-stimulated prolactin secretion. Endocrine Society Mtg., Washington, DC, #P1–3.

Albert, P. R., Neve, K., Bunzow, J., and Civelli, O. (1990a) Coupling of a rat dopamine D2 receptor to inhibition of adenylyl cyclase and prolactin secretion. *J. Biol. Chem.* **265,** 2098–2104.

Albert, P. R., Zhou, Q. Y., VanTol, H. H. M., Bunzow, J., and Civelli, O. (1990b) Cloning, functional expression and mRNA tissue distribution of the rat 5-HT1A receptor gene. *J. Biol. Chem.* **265,** 5825–5832.

Albert, P. R., Lembo, P., Storring, J. M., Charest, A., and Saucier, C. (1996) The 5-HT1A receptor: signalling, desensitization, and gene regulation. *Neuropsychopharmacology* **14,** 19–25.

Alblas, J., van Etten, I., Khanum, A., and Moolenaar, W. H. (1995) C-terminal truncation of the neurokinin-2 receptor causes enhanced and sustained agonist-induced signaling. Role of receptor phosphorylation in signal attenuation. *J. Biol. Chem.* **270,** 8944–8951.

Andrade, R. and Nicoll, R. A. (1987) Pharmacologically distinct actions of serotonin on single pyramidal neurones of the rat hippocampus recorded *in vitro. J. Physiol. (Lond.)* **394,** 99–124.

Baertschi, A. J., Audigier, Y., Lledo, P.-M., Israel, J.-M., Bockaert, J., and Vincent, J.-D. (1992) Dialysis of lactotropes with antisense oligonucleotides assigns guanine nucleotide binding protein subtypes to their channel effectors. *Mol. Endocrinol.* **6,** 2257–2265.

Barr, A. J. and Watson, S. P. (1994) Protein kinase C mediates delayed inhibitory feedback regulation of human neurokinin type 1 receptor activation of phospholipase C in UC11 astrocytoma cells. *Mol. Pharmacol.* **46,** 266–273.

Bates, M. D., Senogles, S. E., Bunzow, J. R., Liggett, S. B., Civelli, O., and Caron, M. G. (1990) Regulation of responsiveness at D2 dopamine receptors by receptor desensitization and adenylyl cyclase sensitization. *Mol. Pharmacol.* **39,** 55–63.

Benya, R. V., Kusui, T., Battey, J. F., and Jensen, R. T. (1995) Chronic desensitization and down-regulation of the gastrin-releasing peptide receptor are mediated by a protein kinase C-dependent mechanism. *J. Biol. Chem.* **270,** 3346–3352.

Berridge, M. J. (1993) Inositol trisphosphate and calcium signalling. *Nature (Lond.)* **361,** 315–325.

Birnbaumer, L. (1992) Receptor-to-effector signaling through G proteins: roles for βγ dimers as well as α-subunits. *Cell* **71,** 1069–1072.

Blin, N., Yun, J., and Wess, J. (1995) Mapping of single amino acid residues required for selective activation of $G_q/11$ by the m3 muscarinic acetylcholine receptor. *J. Biol. Chem.* **270,** 17,741–17,748.

Bouvier, M., Guibault, N., and Bonin, H. (1991) Phorbol-ester induced phosphorylation of the β2-adrenergic receptor decreases its coupling to G_s. *FEBS Lett.* **279,** 243–248.

Bushfield, M., Murphy, G. J., Lavan, B. E., Parker, P. J., Hruby, V. J., Milligan, G., and Houslay, M. D. (1990) Hormonal regulation of G_i2 α-subunit phosphorylation in intact hepatocytes. *Biochem. J.* **268,** 449–457.

Butkerait, P., Aheng, Y., Hallak, H., Graham, T. E., Miller, H. A., Burris, K. D., Molinoff, P. B., and Manning, D. R. (1995) Expression of the human 5-hydroxytryptamine-1A receptor in Sf9 cells. Reconstitution of a coupled phenotype by co-expression of mammalian G protein subunits. *J. Biol. Chem.* **270,** 18,691–18,699.

Castagna, M., Takai, Y., Kaibuchi, K., Sano, K., Kikkawa, U., and Nishizuka, Y. (1982) Direct activation of calcium-activated, phospholipid-dependent protein kinase by tumor-promoting phorbol esters. *J. Biol. Chem.* **257,** 7847–7851.

Chen, Y. and Penington, N. J. (1996) Differential effects of PKC activation on 5-HT1A receptor coupling to calcium and potassium currents in serotonergic neurons. *J. Physiol.* **496.1,** 129–137.

Chio, C. L., Drong, R. F., Riley, D. T., Gill, G. S., Slightom, J. L., and Huff, R. M. (1994) D4 dopamine receptor-mediated signaling events determined in transfected Chinese hamster ovary cells. *J. Biol. Chem.* **269,** 11,813–11,819.

Civelli, O., Bunzow, J. R., and Grandy, D. K. (1993) Molecular diversity of the dopamine receptors. *Ann. Rev. Pharmacol. Toxicol.* **32,** 281–307.

Clapham, D. E. and Neer, E. J. (1993) New roles for G protein βγ-dimers in transmembrane signalling. *Nature (Lond.)* **365,** 403–406.

Dhanasekaran, N., Heasley, L. E., and Johnson, G. L. (1995) G protein-coupled receptor systems involved in cell growth and oncogenesis. *Endocrin. Rev.* **16,** 259–270.

Eason, M. G., Kurose, H., Holt, B. D., Raymond, J. R., and Liggett, S. B. (1992) Simultaneous coupling of alpha 2-adrenergic receptors to two G proteins with opposing effects. Subtype-selective coupling of alpha 2C10, alpha 2C4, and alpha 2C2 adrenergic receptors to G_i and G_s. *J. Biol. Chem.* **267,** 15,795–15,801.

Elsholtz, H. P., Lew, A. M., Albert, P. R., and Sundmark, V. C. (1991) Inhibitory control of prolactin and Pit-1 gene promoters by dopamine. Dual signaling pathways required for D2 receptor-regulated expression of the prolactin gene. *J. Biol. Chem.* **266,** 22,919–22,925.

Exton, J. H. (1994) Phosphoinositide phospholipases and G proteins in hormone action. *Annual Rev. Physiol.* **56,** 349–369.

Fargin, A., Raymond, J. R., Regan, J. W., Cotecchia, S., Lefkowitz, R. J., and Caron, M. G. (1989) Effector coupling mechanisms of the cloned 5-HT1A receptor. *J. Biol. Chem.* **264,** 14,848–14,852.

Fields, T. A. and Casey, P. J. (1995) Phosphorylation of Gz alpha by protein kinase C blocks interaction with the beta gamma complex. *J. Biol. Chem.* **270,** 23,119–23,125.

Florio, T., Pan, M., Newman, B., Hershberger, R. E., Civelli, O., and Stork, P. J. S. (1992) Dopaminergic inhibition of DNA synthesis in pituitary tumor cells is associated with phosphotyrosine phosphatase activity. *J. Biol. Chem.* **267,** 24,169–24,172.

Gandino, L., Longati, P., Medico, E., Prat, M., and Comoglio, P. M. (1994) Phosphorylation of serine 985 negatively regulates the hepatocyte growth factor receptor kinase. *J. Biol. Chem.* **269,** 1815–1820.

Harrington, M.A., Shaw, K., Zhong, P., and Ciaranello, R. D. (1994) Agonist-induced desensitization and loss of high-affinity binding sites of stably expressed human 5-HT1A receptors. *J. Pharmacol. Exp. Ther.* **268,** 1098–1106.

Hille, B. (1994) Modulation of ion-channel function by G protein-coupled receptors. *Trends Neurosci.* **17,** 531–536.

Hunt, T. W., Carroll, R. C., and Peralta, E. G. (1994) Heterotrimeric G proteins containing G alpha i3 regulate multiple effector enzymes in the same cell. Activation of phospholipases C and A2 and inhibition of adenylyl cyclase. *J. Biol. Chem.* **269,** 29,565–29,570.

Innis, R. B. and Aghajanian, G. H. (1987) Pertussis-toxin blocks 5-HT1A receptor and GABAB receptor-mediated inhibition of serotonergic neurons. *Eur. J. Pharmacol.* **143,** 195–204.

Kemp, B. E. and Pearson, R. B. (1990) Protein kinase recognition sequence motifs. *Trends Biochem. Sci.* **15,** 342–346.

Kennelly, P. J. and Krebs, E. G. (1991) Consensus sequences as substrate specificity determinants for protein kinases and protein phosphatases. *J. Biol. Chem.* **266,** 15,555–15,558.

Kiley, S. C., Jaken, S., Whelan, R., and Parker, P. J. (1995) Intracellular targeting of protein kinase C isoenzymes: functional implications. *Biochem. Soc. Transact.* **23,** 601–605.

Kleuss, C., Hescheler, J., Ewel, C., Rosenthal, W., Schultz, G., and Wittig, B. (1991) Assignment of G protein subtypes to specific receptors inducing inhibition of calcium currents. *Nature (Lond.)* **353,** 43–48.

Kleuss, C., Scherübl, H., Hescheler, J., Schultz, G., and Wittig, B. (1992) Different β-subunits determine G protein interaction with transmembrane receptors. *Nature (Lond.)* **358,** 424–426.

Kleuss, C., Scherübl, H., Hescheler, J., Schultz, G., and Wittig, B. (1993) Selectivity in signal transduction determined by γ subunits of heterotrimeric G proteins. *Science* **259,** 832–834.

Kobilka, B. (1992) Adrenergic receptors as models for G protein-coupled receptors. *Annu. Rev. Neurosci.* **15,** 87–114.

Krishek, B. J., Xie, X., Blackstone, C., Huganir, R. L., Moss, S. J., and Smart, T. G. (1994) Regulation of GABAA receptor function by protein kinase C phosphorylation. *Neuron* **12,** 1081–1095.

Lajiness, M. E., Chio, C. L., and Huff, R. M. (1993) D2 dopamine receptor stimulation of mitogenesis in transfected chinese hamster ovary cells: relationship to dopamine stimulation of tyrosine phosphorylations. *J. Pharm. Exp. Ther.* **267,** 1573–1581.

Laugwitz, K. L., Allgeier, A., Offermanns, S., Spicher, K., Van Sande, J., Dumont, J. E., and Schultz, G. (1996) The human thyrotropin receptor: a heptahelical receptor capable of stimulating members of all four G protein families. *Proc. Natl. Acad. Sci. USA* **93,** 116–120.

Lefkowitz, R. J., Hausdorff, W. P., and Caron, M. G. (1990) Role of phosphorylation in desensitization of the β-adrenoceptor. *Trends Pharmacol. Sci.* **11,** 190–194.

Lembo, P. and Albert, P. R. (1994) 5-HT1B receptors mediate a stimulatory calcium signaling opossum kidney (OK) cells: negative regulation by protein kinase C. *Can. J. Physiol. Pharmacol.* **72 (Suppl. 1)** 536.

Lembo, P. M. C. and Albert, P. R. (1995) Multiple phosphorylation sites are required for pathway-selective uncoupling of the 5-hydroxytryptamine 1A receptor by protein kinase C. *Mol. Pharmacol.* **48,** 1024–1029.

Lembo, P. M. C., Ghahremani, M. H., Morris, S. J., and Albert, P. R. (1997) A conserved threonine residue in the second cytoplasmic loop of the 5-HT1A receptor directs pathway selective signaling. *Mol. Pharmacol.,* in press.

Lew, A. M. and Elsholtz, H. P. (1995) A dopamine-responsive domain in the N-terminal sequence of Pit-1. Transcriptional inhibition in endocrine cell types. *J. Biol. Chem.* **270,** 7156–7160.

Lew, A. M., Yao, H., and Elsholtz, H. P. (1994) G(i) alpha 2- and G(o) alpha-mediated signaling in the Pit-1-dependent inhibition of the prolactin gene promoter. Control of transcription by dopamine D2 receptors. *J. Biol. Chem.* **269,** 12,007–12,013.

Liu, J., Conklin, B. R., Blin, N., Yun, J., and Wess, J. (1995) Identification of a receptor/G protein contact site critical for signaling specificity and G protein activation. *Proc. Natl. Acad. Sci. USA* **92,** 11,642–11,646.

Liu, Y. F. and Albert, P. R. (1991) Cell-specific signalling of the 5-HT1A receptor. Modulation by PKC and PKA. *J. Biol. Chem.* **266,** 23,689–23,697.

Liu, Y. F., Civelli, O., Zhou, Q. Y., and Albert, P. R. (1992) Differential sensitivity of the short and long human dopamine-D2 receptor subtypes to protein kinase C. *J. Neurochem.* **59,** 2311–2317.

Liu, Y. F., Jakobs, K. H., Rasenick, M. M., and Albert, P. R. (1994) G protein specificity in receptor-effector coupling. Analysis of the roles of Go and Gi2 in GH4C1 pituitary cells. *J. Biol. Chem.* **269,** 13,880–13,886.

Lledo, P.-M., Legendre, P., Israel, J.-M., and Vincent, J.-D. Dopamine inhibits two characterized voltage-dependent calcium currents in identified rat lactotroph cells. *Endocrinology* **127,** 990–1001.

Lledo, P.-M., Homburger, V., Bockeart, J., and Vincent, J.-D. (1992) Differential G protein-mediated coupling of D2 dopamine receptors to K^+ and Ca^{2+} currents in rat anterior pituitary cells. *Neuron* **8,** 455–463.

Mestek, A., Hurley, J. H., Bye, L. S., Campbell, A. D., Chen, Y., Tian, M., Liu, J., Schulman, H., and Yu, L. (1995) The human mu opioid receptor: modulation of functional desensitization by calcium/calmodulin-dependent protein kinase and protein kinase C. *J. Neurosci.* **15,** 2396–2406.

Nebigil, C. G., Garnovskaya, M. N., Casanas, S. J., Mulheron, J. G., Parker, E. M., Gettys, T. W., and Raymond, J. R. (1995) Agonist-induced desensitization and phosphorylation of human 5-HT1A receptor expressed in Sf9 insect cells. *Biochemistry* **34,** 11,954–11,962.

Neer, E. J. (1995) Heterotrimeric G proteins: organizers of transmembrane signals. *Cell* **80,** 249–257.

Newton, A. C. (1995) Protein kinase C: structure, function, and regulation. *J. Biol. Chem.* **270,** 28,495–28,498.

Nishizuka, Y. (1988) The molecular heterogeneity of protein kinase C and its implications for cellular regulation. *Nature (Lond.)* **334,** 661–665.

Nishizuka, Y. (1995) Protein kinase C and lipid signaling for sustained cellular responses. *FASEB J.* **9,** 484–496.

Ostrowski. J., Kjelsberg M. A., Caron, M. G., and Lefkowitz, R. J. (1992) Mutagenesis of the β2-adrenergic receptor: How structure elucidates function. *Annu. Rev. Pharmacol. Toxicol.* **32,** 167–183.

Penington, N. J. and Kelly, J. S. (1990) Serotonin receptor activation reduces calcium current in an acutely dissociated adult central neuron. *Neuron* **4,** 751–758.

Penington N. J., Kelly, J. S., and Fox, A. P. (1993a) Whole-cell recordings of inwardly rectifying K+ currents activated by5-HT1A receptors on dorsal raphe neurones of the adult rat. *J. Physiol. (Lond.)* **469,** 387–405.

Penington, N. J., Kelly, J. S., and Fox, A. P. (1993b) Unitary properties of potassium channels activated by 5-HT in acutely isolated rat dorsal raphe neurones. *J. Physiol. (Lond.)* **469,** 407–426.

Rasmussen H., Isales C. M., Calle R., Throckmorton D., and Anderson M., Gasalla-Herraiz J., McCarthy R. (1995) Diacylglycerol production, Ca2+ influx, and protein kinase C activation in sustained cellular responses. *Endocr. Rev.* **16,** 649–681.

Raymond, J. R. (1991) Protein kinase C induces phosphorylation and desensitization of the human 5-HT1A receptor. *J. Biol. Chem.* **266,** 14,747–14,753.

Raymond, J. R. (1995) Multiple mechanisms of receptor-G protein signaling specificity. *Am. J. Physiol.* **269,** F141–F158.

Ryu, S. H., Kim, U., Wahl, M. I., Brown, A. B., Carpenter, G., Huang, K., and Rhee, S. G. (1990) Feedback regulation of PLC-β by protein kinase C. *J. Biol. Chem.* **265,** 17,941–17,945.

Seabrook, G. R., Knowles, M., Brown, N., Myers, J., Sinclair, H., Patel, S., Freedman, S. B., and McAllister (1994) Pharmacology of high-threshold calcium currents in GH4C1 pituitary cells and their regulation by activation of human D2 and D4 receptors. *Br. J. Pharmacol.* **112,** 728–734.

Senogles, S. E. (1994) The D2 dopamine receptor mediates inhibition of growth in GH4ZR7 cells: involvement of protein kinase-Cε. *Endocrinology* **134,** 783–789.

Shih, M. and Malbon, C. C. (1994) Oligodeoxynucleotides antisense to mRNA encoding protein kinase A, protein kinase C, and beta-adrenergic receptor kinase reveal distinctive cell-type-specific roles in agonist-induced desensitization. *Proc. Natl. Acad. Sci. USA* **91,** 12,193–12,197.

Stea, A., Soong, T. W., and Snutch, T. P. (1995) Determinants of PKC-dependent modulation of a family of neuronal calcium channels. *Neuron* **15,** 929–940.

Sterne-Marr, R. and Benovic, J. L. (1995) Regulation of G protein-coupled receptors by receptor kinases and arrestins. *Vitam. Horm.* **51,** 193–234.

Strader, C. D., Fong, T. M., Tota, M. R., Underwood, D., and Dixon, R. A. (1994) Structure and function of G protein-coupled receptors. *Ann. Rev. Biochem.* **63,** 101–132.

Strassheim, D. and Malbon, C. C. (1994) Phosphorylation of G_i2 attenuates inhibitory adenylyl cyclase in neuroblastoma/glioma (NG 108-15) cells. *J. Biol. Chem.* **269,** 14,307–14,313.

Tang, W.-J. and Gilman, A. G. (1991) Type-specific regulation of adenylyl cyclase by G protein βγ subunits. *Science* **254,** 1500–1503.

Vallar, L., Claudia, M., Magni, M., Albert, P., Bunzow, J., Meldolesi, J., and Civelli, O. (1990) Differential coupling of dopaminergic D2 receptor expressed in different cell types. *J. Biol. Chem.* **265,** 10,320–10,326.

van Huizen, F., Bansse, M., and Stam, N. J. (1993) Agonist-induced down-regulation of human 5-HT1A and 5-HT2 receptors in Swiss 3T3 cells. *Neuroreport* **4**, 1327–1330.

Varrault A., Bockaert J., and Waeber C. (1992) Activation of 5-HT1A receptors expressed in NIH-3T3 cells induces focus formation and potentiates EGF effect on DNA synthesis. *Mol. Biol. Cell* **3**, 961–969.

Varrault, A., Nguyen, D. L., McClue, S., Harris, B., Jouin, P., and Bockaert, J. (1994) 5-hydroxytryptamine 1A receptor synthetic peptides: Mechanisms of adenylyl cyclase inhibition. *J. Biol. Chem.* **269**, 16,720–16,725.

Wickman, K. and Clapham, D. E. (1995) Ion channel regulation by G proteins. *Physiol. Rev.* **75**, 865–885.

Yatomi, Y., Arata, Y., Tada, S., Kume, S., and Ui, M. (1992) Phosphorylation of the inhibitory guanine-nucleotide-binding protein as a possible mechanism of inhibition by protein kinase C of agonist-induced Ca^{+2} mobilization in human platelet. *Eur. J. Biochem.* **205**, 1003–1009.

Yuan, N., Friedman, J., Whaley, B. S., and Clark, R. B. (1994) cAMP-dependent protein kinase and protein kinase C consensus site mutations of the β2-adrenergic receptor. *J. Biol. Chem.* **269**, 23,032–23,038.

Molecular Mechanisms of Opiate Addiction in the Nucleus Accumbens

Morphine Regulation of G Proteins

Katherine L. Widnell and Eric J. Nestler

Introduction

Addiction exacts enormous costs on society in terms of both its clinical and social ramifications. Addiction, defined as the compulsive use of a drug despite adverse consequences, is thought to develop as a result of adaptive changes in specific brain areas. The persistence of these progressively developing adaptations in brain function suggests that long-lasting changes in the brain are important in mediating addictive phenomena. The adaptations associated with compulsive drug use are defined by the terms tolerance, sensitization, dependence, and withdrawal. Tolerance develops when repeated administration of a drug produces a decreased effect or when increasing doses must be administered to obtain the effect observed with the original dose. Sensitization describes the opposite situation, wherein repeated drug administration elicits increasing effects. Dependence refers to an altered physiological state that necessitates the continued administration of the abused drug to prevent the appearance of a withdrawal syndrome. Withdrawal is characterized by psychological and perhaps physical disturbances when the abused drug is withdrawn. The goal of this chapter is to review the evidence for a role of G proteins in these phenomena.

From: *Neuromethods, Vol. 31: G Protein Methods and Protocols*
Ed: R. K. Mishra, G. B. Baker, and A. A. Boulton Humana Press Inc.

Brain Regions Implicated
as Mediators of Drug Addiction

The study of the cellular and molecular mechanisms underlying drug addiction has been facilitated by the availability of animal models of drug reinforcement and addiction. There is now compelling evidence that the mesolimbic dopamine system, which consists of dopaminergic neurons projecting from the ventral tegmental area (VTA) to the nucleus accumbens (NAc), plays an important role in the acute reinforcement produced by opiates and many other drugs of abuse (Bozarth, 1986; Wise, 1990; Koob, 1992; Dworkin and Smith, 1993). For example, animals will self-administer opiates directly into the VTA and NAc, whereas lesions of these regions attenuate systemic drug self-administration. These and other behavioral tests have provided a consistent body of evidence with regard to the involvement of these brain regions in drug reinforcement. Moreover, long-term adaptations within these same brain regions in response to repeated drug exposure are thought to contribute to changes in reinforcement mechanisms that underlie drug addiction (Koob, 1992; Self and Nestler, 1995).

Cellular Adaptations to Morphine
in the Mesolimbic Dopamine System

Regulation of G Proteins by Morphine in the NAc

The mechanisms by which extracellular agents, such as neurotransmitters, neuropeptides, and pharmacologic agents, act on plasma membrane receptors to alter intracellular processes have been a subject of intense research. Increasing evidence has demonstrated that a class of heterotrimeric guanine nucleotide binding proteins, termed G proteins, function to couple most types of plasma membrane receptors to the generation of intracellular second messengers. Receptor conformation is altered by ligand binding, causing the receptor to associate with the α-subunit of the G protein. This in turn alters the conformation of the α-subunit, generating a free α-

subunit bound to guanosine triphosphate (GTP), which is biologically active and can regulate the functional activity of effector proteins within the cell. G protein $\beta\gamma$-subunits are also generated by this process, and they, too, can regulate specific effector proteins. The cycle is thought to be reset when GTP is hydrolyzed to guanosine diphosphate (GDP), which leads to the reassociation of the α- and $\beta\gamma$-subunits.

Distinct α-subunits are defined by their ability to regulate effector proteins differentially. One of these effectors is the enzyme adenylyl cyclase, which, when activated, forms adenosine-3',5'-cyclic monophosphate (cAMP) by creating a cyclic phosphodiester bond with the α-phosphate group of ATP. $G\alpha_s$ activates adenylyl cyclase, whereas $G\alpha_i$ inhibits the enzyme. Alterations in levels of cAMP specifically affect the activity of cAMP-dependent protein kinase, thus altering phosphorylation of numerous downstream target proteins.

Opiates exert their acute effects by binding to opioid receptors on the cell surface. This leads to inhibition of cAMP formation via G_i (and perhaps G_o) proteins. Activation of G_i and G_o also leads to direct regulation of specific ion channels, including activation of inwardly rectifying K^+ channels and inhibition of voltage-dependent Ca^{2+} channels.

In contrast to the acute inhibitory actions of morphine on the cAMP pathway, chronic morphine regulates multiple second messenger proteins in the NAc in a manner predicted to increase the functional activity of the cAMP pathway (*see* Fig. 1). In the chronic morphine paradigm, morphine pellets are implanted subcutaneously into Sprague-Dawley rats for 5 d; this results in profound states of tolerance and dependence (Guitart et al., 1992b). Chronic morphine treatment decreases levels of $G\alpha_i$ specifically in the NAc, but does not alter levels of other G protein subunits, including $G\alpha_o$ and $G\beta$ (Terwilliger et al., 1991; Nestler, 1992). Chronic opiate administration also increases levels of adenylyl cyclase and cAMP-dependent protein kinase in the NAc, but not in several other brain regions, such as the caudate-putamen and frontal cortex, that are generally not implicated in drug reinforcement mechanisms (Terwilliger et al., 1991; Nestler, 1992). The upregulation of the

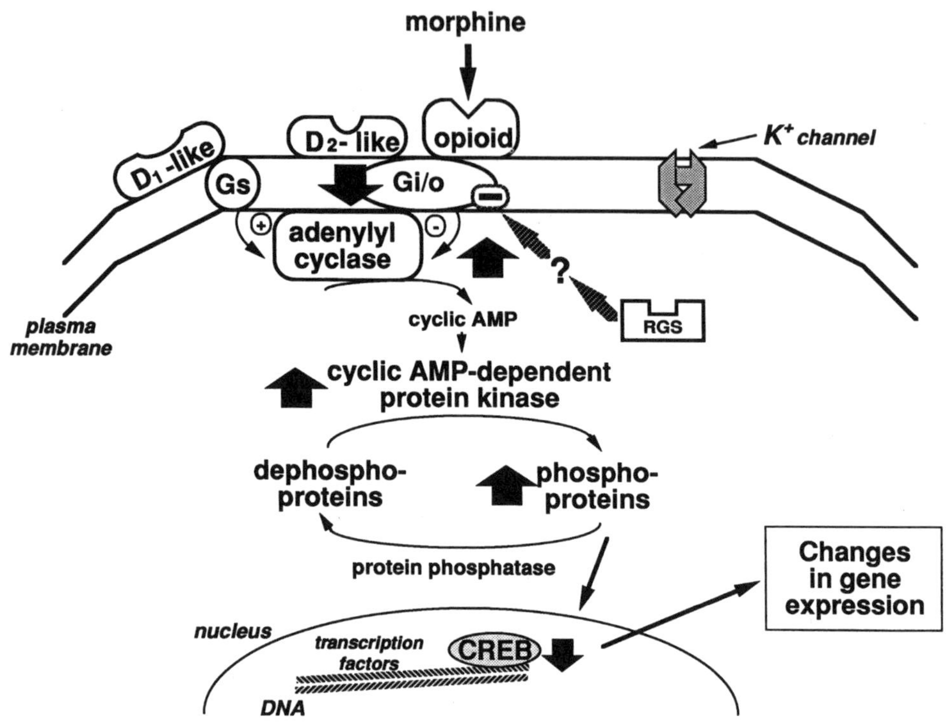

Fig. 1. Scheme illustrating biochemical parameters in the NAc associated with aspects of the "drug-addicted" and "drug-preferring" state. The figure depicts an NAc neuron. Shown in the NAc cell are dopamine receptors (D1, D2), opioid receptors, K$^+$ channel, G proteins ($G_{i/o}$ and G_s), a novel mammalian protein that could play a role in the regulation of G protein signaling (RGS), components of the intracellular cAMP system (adenylyl cyclase, cAMP-dependent protein kinase), and CREB, a known substrate for the kinase that functions as a nuclear transcription factor. Bold arrows denote the direction of alteration in protein levels in the chronic morphine-treated state: levels of G_i (which after activation by D2-like or opioid receptors inhibits adenylyl cyclase activity) and of CREB are decreased, and levels of adenylyl cyclase and of cAMP-dependent protein kinase levels are increased. Bold arrows also denote baseline strain differences in the NAc in Lewis compared to Fischer 344 rats. These biochemical parameters are hypothesized to mediate certain aspects of the addictive state, as well as individual vulnerability for addiction.

cAMP pathway could be viewed as a mechanism of tolerance and dependence in this brain region. The degree of downregulation of $G\alpha_{i1/2}$, a 20% decrease from control, is likely to be functionally important, since small changes in levels of G proteins have significant functional consequences. This has been demonstrated by administration of pertussis toxin (PTX), which inhibits $G\alpha_i$ (and $G\alpha_o$) by catalyzing the ADP-ribosylation of a specific cysteine residue in the protein. A 10–15% inhibition of $G\alpha_i$ and $G\alpha_o$ in dorsal raphe neurons decreases by about 40–50% the ability of neurotransmitters to produce their electrophysiological effect, and a 50% inhibition completely abolishes electrophysiological responses (Innis et al., 1988).

The finding that G proteins can be downregulated in response to agonist exposure is not unique to NAc neurons. Chronic opiate exposure has been shown to decrease $G\alpha_i$ levels in other neuronal systems (Attali and Vogel, 1989; Nestler, 1992; Eriksson et al., 1992). Moreover, such a response is not limited to opioid receptors. In rat adipocytes, adenosine A1 receptors are coupled to adenylyl cyclase via inhibitory $G\alpha_{i2}$ proteins. Prolonged treatment of cultured rat adipocytes with adenosine results in a 50% reduction in levels of $G\alpha_{i2}$, with a downregulation of $G\alpha_{i1}$ and $G\alpha_{i3}$ also observed (Green et al., 1990). However, downregulation of $G\alpha_i$ is not a general response to opiate treatment. Some cell types do not show such changes, and others exhibit adaptations in other G protein subunits (Van Vliet et al., 1993; Parolaro et al., 1993; Rubino et al., 1994). The factors that determine regulation of a G protein subunit in a particular cell type and the mechanisms by which such regulation occurs remain unknown.

It is important to emphasize that regulation of levels of G proteins probably represents only one of many mechanisms that can influence the specificity of interactions between receptors and G proteins (Raymond, 1995). Exciting evidence points to a newly identified class of molecules that may prove to play a role in the desensitization of G protein-mediated signaling. This family of mammalian proteins, termed Regulators of G protein Signaling (RGS), has been demonstrated to regulate G protein-linked signaling pathways (Druey et al., 1996). It is possible that a novel form of desensitization may occur as a result of the induction of RGS family members in response to receptor binding and G protein activation (Druey et al., 1996; Roush, 1996). RGS proteins (and their homologs) have been shown to desensitize $G\alpha_i$-linked signals in cultured cells (De Vries et al., 1995; Druey et al., 1996). One of the RGS family members, RGS4, is highly expressed in mammalian brain (Druey et al., 1996). These findings raise the interesting possibility that RGSs could play a role in the dampening of the inhibitory $G\alpha_i$ signal in a complementary fashion to the observed morphine-induced downregulation of $G\alpha_i$ levels. This hypothesis is demonstrated schematically in Fig. 1.

Behavioral Consequences
of Cellular Adaptations in G Proteins in the NAc

Direct evidence for the involvement of G proteins in the NAc in acute drug reinforcement has been provided by assessment of an animal's self-administration of heroin following injections of PTX. The self-administration paradigm provides a direct test of a drug's reinforcing property, or the effect of pharmacologic interventions on reinforcement, by measuring the rate at which an animal self-administers the drug. It is believed that an increase in the rate of self-administration reflects a decrease in the reinforcing effects produced by the drug, causing the animal to shorten the time between injections of the drug. Intra-NAc administration of PTX increases the rate of heroin self-administration in a manner identical to reducing the unit dose per injection of the drug or to pretreating animals with an opioid receptor antagonist (Self et al., 1994; Self and Nestler, 1995). This effect is not observed following injection of heat-inactivated PTX. The increase in drug self-administration did not appear to result from a neurotoxic effect of PTX or from a nonspecific increase in lever pressing. These findings, combined with the observed chronic morphine regulation of G proteins in the NAc, are consistent with the idea that long-term drug-induced reductions in G_i and G_o proteins in the NAc may be implicated in some of the motivational changes underlying drug addiction (Self and Nestler, 1995). Specifically, these findings support the hypothesis stated above that the morphine-induced decrease in $G\alpha_i$ in the NAc (which would mimic the experimental effects of PTX) represents a mechanism of tolerance and dependence.

Strain-Selective Differences in Biochemical and Behavioral Parameters

Further evidence for the involvement of the G protein–cAMP pathway in addictive processes has been provided by studies of two inbred rat strains that differ in their preference for several drugs of abuse. The Lewis inbred rat strain more readily self-administers opiates, alcohol, and cocaine compared with the Fischer 344 inbred rat strain (Suzuki et al., 1988; George and Goldberg, 1989). Lewis rats also show a higher degree of morphine- and cocaine-induced conditioned place preference (Guitart et al., 1992a). Significant differences in levels of G protein subunits were found in drug-naive Fischer and Lewis rats. Levels of $G\alpha_{i1/2}$, $G\alpha_{i3}$, and $G\beta$ were significantly lower in the NAc of drug-naive Lewis rats compared to drug-naive Fischer 344 rats (Guitart et al., 1993). Strain differences in levels of G proteins were also demonstrated in the locus ceruleus, but were not generalized throughout the brain. Other strain differences in activity of the cAMP pathway have been found. Lewis rats display significantly higher levels of adenylyl cyclase activity and of cAMP-dependent protein kinase activity in the NAc compared to the Fischer rat (Guitart et al., 1993). No strain differences in activities of these enzymes were detected in the caudate-putamen, cerebral cortex, or hippocampus. The pattern of intracellular proteins in drug-naive Lewis rats (compared to drug-naive Fischer 344 rats) resembles the pattern produced by chronic morphine treatment of outbred Sprague-Dawley rats (Fig. 1). These prominent differences in levels of G protein–cAMP intracellular messenger proteins in the two inbred strains suggest that these biochemical parameters could underlie some of the inherent strain differences in drug preference (Nestler et al., 1993).

Molecular Mechanisms Underlying G Protein Regulation in the NAc

Regulation of the Transcription Factor cAMP Response Element Binding Protein (CREB) by Morphine in the NAc

Chronic drug treatment has been found to involve alterations in total amounts of second messenger proteins as outlined above. Although both pre- and posttranslational mechanisms could account for these changes, most attention to date has focused on drug regulation of gene expression as a result of several demonstrated cases of equivalent changes in protein and mRNA (Nestler, 1992). Unique combinations of transcription factors bind to specific DNA sequences in the promoter of a given gene to regulate cooperatively that gene's transcription. One well-characterized transcription factor is cAMP response element binding protein (CREB). CREB belongs to a family of leucine zipper transcription factors that share certain structural motifs, bind DNA as dimers, and demonstrate alterations in transcriptional activity following phosphorylation (Meyer and Habener, 1993). Regulation of CREB expression has been observed in vitro and in vivo (Widnell et al., 1994, 1996; Brecht et al., 1994; Walker et al., 1995).

Based on its central role in mediating the effects of the cAMP pathway on gene expression, we have considered a role for CREB in long-term adaptations to morphine in the NAc. Indeed, we have demonstrated that chronic, but not acute, morphine treatment downregulates CREB immunoreactivity in the NAc, but does not alter CREB levels in the caudate-putamen or frontal cortex (Widnell et al., 1996). This morphine effect on CREB immunoreactivity in the NAc is blocked by concomitant administration of naltrexone, an opioid receptor antagonist, indicating that it is mediated via opioid receptors (Widnell et al., 1996).

Role of CREB in G Protein Downregulation in the NAc

It has proven difficult in vivo to determine the role of a given transcription factor in the regulation of any given target gene, particularly in chronic treatment paradigms. Much work to date has focused on the use of "knockout" mice in which the transcription factor is either mutated or disrupted completely, although these studies have been complicated by developmental compensations (Struthers et al., 1991; Hummler et al., 1994). In addition, there are no traditional pharmacologic agents that inhibit transcription factor function. Therefore, in order to investigate the functional role of the morphine-induced downregulation of CREB in the NAc and its possible involvement as a mediator of alterations in putative downstream target proteins, such as $G\alpha_{i1/2}$ or cAMP-dependent protein kinase, an antisense oligonucleotide strategy was developed. This approach has been used successfully to dissect some of the complex interactions in signal transduction pathways (Albert and Morris, 1994). For example, it has proven possible to link specific receptors and assign cellular function to individual G protein subtypes (Kleuss et al., 1991; Wang et al., 1992). Antisense oligonucleotides represent small (single-stranded sequences of DNA typically 15–25 bp in length) that are complementary to a region of mRNA derived from a gene of interest. As shown in Fig. 2, antisense oligonucleotides are believed to work either by (1) forming a physical barrier to translation initiation and/or mRNA processing or (2) forming a substrate for RNase H, an endogenous enzyme that cleaves RNA at the site of an RNA/DNA heteroduplex (Bischofberger and Shea, 1992; Milligan et al., 1993). However, antisense oligonucleotides can also be nonspecific or toxic either by (1) interfering directly with ribosomes or (2) hybridizing nonspecifically to other mRNAs. In addition, concerns have been raised about the degree of oligonucleotide uptake by a target cell, generalized spread of

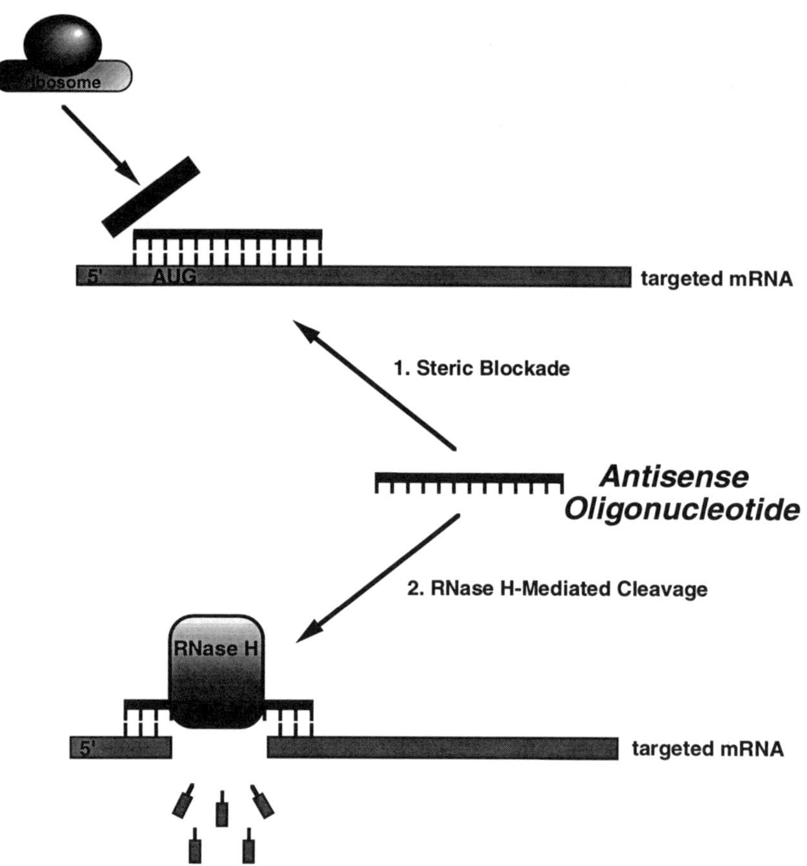

Fig. 2. Schematic illustration of the sequence-specific mechanisms of action of antisense oligonucleotides. The figure depicts an antisense oligonucleotide binding to its complementary sequence at the translation start site (AUG) on the targeted mRNA. In case 1, the oligonucleotide prevents ribosomal binding, thus disrupting translation and/or mRNA processing. In case 2, the DNA oligonucleotide/RNA heteroduplex serves as a substrate for RNase H, an endogenous enzyme that has been shown to cleave RNA when bound to DNA.

the antisense oligonucleotide to surrounding tissues, and oligonucleotide susceptibility to nucleases (Stein and Chen, 1993). The use of phosphorothioate linkages in the backbone of the oligonucleotide has been shown to increase the stability of the oligonucleotide (and possibly also increase its transport across the membrane) (Woolf et al., 1990). However, totally phosphorothioate-modified oligonucleotides have been associated with an increase in toxicity.

We have devised a series of controls considered essential for the appropriate interpretation of antisense experiments (Russell et al., 1996). A selective reduction in levels of the targeted protein by the antisense oligonucleotide must be demonstrated. Antisense oligonucleotide effects (either biochemical or behavioral) should not be observed with a sense oligonucleotide (complementary to the antisense oligonucleotide) or with a scrambled antisense oligonucleotide (a sequence containing the same GC and phosphorothioate content as the original antisense oligonucleotide). Observed antisense oligonucleotide effects should be fully reversible following discontinuation of oligonucleotide treatment. Finally, there should be no degenerative changes assessed by histological techniques.

By taking advantage of sequence differences among CREB and other CREB-like transcription factors, it is possible to use antisense oligonucleotides to decrease levels of CREB selectively. This results in an inhibition of CREB-dependent transcriptional activation without an effect on other CREB-like transcription factor-dependent activity. Since we were interested in the long-term effects of CREB downregulation, we administered partially phosphorothioate-modified CREB antisense oligonucleotide by constant infusions at a rate of 20 μg/d (0.5 μL/h) via Alzet minipumps for a period of 5 d into the NAc. As shown in Table 1, this produced a significant degree of CREB downregulation, similar in magnitude to that observed following chronic morphine treatment. This effect is specific based on our control criteria. As shown in Table 2, there was no effect of CREB sense of CREB scrambled oligonucleotide compared to sham treatment (PBS infusion). Furthermore, there appeared to be little diffusion of the CREB oligonucleotide outside of the NAc, since 5-d infusion of CREB antisense oligonucleotide into the NAc did not affect levels of CREB immunoreactivity in a 0.5-mm thick ring of tissue sur-

Table 1

Comparison of Chronic Morphine Treatment
and Chronic CREB Antisense Treatment in Rat NAc[a]

	Chronic morphine treatment	Chronic CREB antisense oligonucleotide treatment
CREB	$-48 \pm 4\%$ (16)[b,d]	$-40 \pm 4\%$ (16)[b,d]
$G\alpha_{i1/2}$	$-19 \pm 6\%$ (6)[c,d]	$-21 \pm 3\%$ (21)[b,d]
PKA	$27 \pm 8\%$ (8)[c,e]	$-27 \pm \%$ (5)[b,d]
	$42 \pm 7\%$ (8)[c,f]	

[a]Values are given as percent change from control SEM (*N*).

[b]From reference Widnell et al. (1996). $p < 0.01$ by unpaired Student's *t*-test.

[c]From reference Terwilliger et al. (1991). $p < 0.05$ by χ^2-test.

[d]Values represent levels of immunoreactivity, as assessed by immunoblotting.

[e]Values present cAMP-dependent protein kinase activity in particulate fraction.

[f]Values represent cAMP-dependent protein kinase activity in soluble fraction.

Table 2

Lack of Effect of Chronic CREB Sense or Scrambled Oligonucleotide[a]

	Chronic CREB sense oligonucleotide treatment	Chronic scrambled CREB oligonucleotide treatment
CREB	$3 \pm 5\%$ (6)	$4 \pm 8\%$ (5)
$G\alpha_{i1/2}$	$4 \pm 4\%$ (6)	$2 \pm 9\%$ (8)
PKA	$9 \pm 13\%$ (5)	$4 \pm 15\%$ (7)

[a]From reference Widnell et al. (1996). $p < 0.01$ by unpaired Student's *t*-test. Values are given as percent change from control $\pm$ SEM (*N*).

rounding the NAc (Widnell et al., 1996). There also was no detectable toxicity of the antisense oligonucleotide on Cresyl violet stains of rat brain sections containing the NAc (Widnell et al., 1996). Finally, the antisense oligonucleotide effect was reversible (Widnell et al., 1996).

Having demonstrated that the CREB antisense oligonucleotide effect on levels of CREB immunoreactivity was specific, we turned our attention to the effect of downregulation

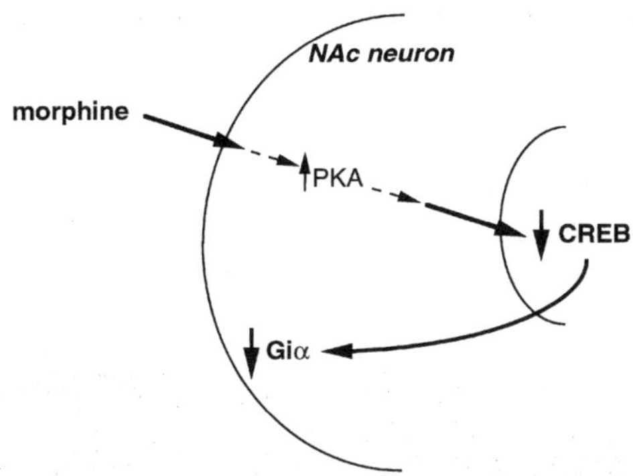

Fig. 3. Hypothetical scheme illustrating a role for CREB in mediating morphine regulation of $G\alpha_i$ in the NAc. CREB antisense studies in the rat NAc demonstrate that sustained (5-d) downregulation of CREB protein levels is sufficient to downregulate $G\alpha_i$. The antisense-induced downregulation of CREB and of $G\alpha_i$ are equivalent in magnitude to the downregulation of these proteins seen following chronic morphine treatment, consistent with the possibility that decreased levels of CREB mediate the morphine effect on $G\alpha_i$. This downregulation of $G\alpha_i$, in turn, has been shown to alter drug reinforcement mechanisms in the NAc. Morphine downregulation of CREB may be mediated, directly or indirectly, via the morphine-induced increase in cAMP-dependent protein kinase (PKA), which would appear to occur via a CREB-independent mechanism. Reprinted with permission from Widnell et al. (1996).

of CREB on target proteins. Selective morphine regulation of CREB levels in the NAc, but not in brain areas where morphine-induced alterations in components of the cAMP signaling pathway have not been observed, suggested a possible role for CREB in mediating morphine regulation of these intracellular messenger proteins. Indeed, CREB antisense oligonucleotide infusion decreased levels of $G\alpha_{i1/2}$ in the NAc to about the same extent as that seen after chronic morphine treatment (Table 1). This supports a scheme (shown in Fig. 3) whereby the morphine-induced downregulation of $G\alpha_i$ is mediated by the downregulation of CREB. This effect was specific, using the same criteria as described for the antisense oligonucleotide effect on CREB immunoreactivity (*see* Table 2) (Widnell et al., 1996). Furthermore, $G\alpha_i$ regulation occurred in the absence of any changes in levels of $G\alpha_o$ or $G\beta$. Of note, cAMP-dependent protein kinase was regulated in an opposite fashion by CREB antisense oligonucleotide and by chronic morphine treatment, suggesting that CREB downregulation by morphine does not mediate increases in the protein kinase. Finally, there was no observed effect of CREB antisense oligonucleotide on levels of a number of other protein kinases that have been shown to phosphorylate and transcriptionally activate CREB (Widnell et al., 1996).

The mechanism(s) by which decreases in levels of CREB downregulate $G\alpha_{i1/2}$ and cAMP-dependent protein kinase remain unknown. The antisense oligonucleotide studies also demonstrate that long-term adaptations to drugs of abuse in the NAc are likely mediated by more than one transcription factor. Nevertheless, the finding that decreases in levels of CREB can regulate levels of a G protein subunit shown to play an important role in drug reinforcement suggest that morphine regulation of CREB is likely to be functionally significant. Further studies along these lines promise to establish the precise molecular steps by which chronic exposure to morphine or other drugs of abuse produce long-lasting changes in the brain that underlie addiction.

References

Albert, P. R. and Morris, S. J. (1994) Antisense knockouts: molecular scalpels for the dissection of signal transduction. *Trends Pharmacol. Sci.* **15,** 250–254.

Attali, B. and Vogel, Z. (1989) Long-term opiate exposure leads to reduction of the alpha i-1 subunit of GTP-binding proteins. *J. Neurochem.* **53,** 1636–1639.

Bischofberger, N. and Shea, R. G. (1992) Oligonucleotide-based therapeutics, in *Nucleic Acid Targeted Drug Design* (Probpst, C. L. and Perun T. J., eds.), Marcel Dekker, New York, pp. 579–612.

Bozarth, M. A. (1986) Neural basis of psychomotor stimulant and opiate reward: evidence suggesting the involvement of a common dopaminergic system. *Behav. Brain Res.* **22,** 107–116.

Brecht, S., Gass, P., Anton, F., Bravo, R., Zimmerman, M., and Herdegen, T. (1994) Induction of c-Jun and suppression of CREB transcription factor proteins in axotomized neurons of substantia nigra and covariation with tyrosine hydroxylase. *Mol. Cell. Neurosci.* **5,** 431–441.

De Vries, L., Mousli, M., Wurmser, A., and Farquhar, M. G. (1995) GAIP, a protein that specifically interacts with the trimeric G protein $G\alpha_{i3}$, is a member of a protein family with a highly conserved core domain. *Proc. Natl. Acad. Sci. USA* **92,** 11,916–11,920.

Druey, K. M., Blumer, K. J., Kang, V. H., and Kehrl, J. H. (1996) Inhibition of G protein mediated MAP kinase activation by a new mammalian gene family. *Nature* **379,** 742–746.

Dworkin, S. I. and Smith, J. E. (1993) Opiates/opioids and reinforcement, in *Biological Basis of Substance Abuse* (Korenman, S. G. and Barchas, J. D., eds.), Oxford University Press, New York, pp. 327–338.

Eriksson, P. S., Carlsson, B., Isaksson, O. G. P., Hansson, E., and Ronnback, L. (1992) Altered amounts of G protein mRNA and cAMP accumulation after long-term opioid receptor stimulation of neurons in primary culture from the rat cerebral cortex. *Mol. Brain Res.* **14,** 317–325.

George, F. R. and Goldberg, S. R. (1989) Genetic approaches to the analysis of addiction processes. *Trends Pharmacol. Sci.* **10,** 78–83.

Green, A., Johnson, J. L., and Milligan, G. (1990) Down-regulation of Gi subtypes by prolonged incubation of adipocytes with an A1 adenosine receptor agonist. *J. Biol. Chem.* **265,** 5206–5210.

Guitart, X., Beitner-Johnson, D., and Nestler, E. J. (1992a) Fischer and Lewis rat strains differ in basal levels of neurofilament proteins and in their regulation by chronic morphine in the mesolimbic dopamine system. *Synapse* **12,** 242–253.

Guitart, X., Thompson, M. A., Mirante, C. K., Greenberg, M. E., and Nestler, E. J. (1992b) Regulation of cyclic AMP response element-binding protein (CREB) phosphorylation by acute and chronic morphine in the rat locus coeruleus. *J. Neurochem.* **58,** 1168–1171.

Guitart, X., Kogan, J. H., Berhow, M., Terwilliger, R. Z., Aghajanian, G. K., and Nestler, E. J. (1993) Lewis and Fischer rat strains display differences in biochemical, electrophysiological and behavioral parameters: studies in the nucleus accumbens and locus coeruleus of drug naive and morphine-treated animals. *Brain Res.* **611,** 7–17.

Hummler, E., Cole, T. J., Blendy, J. A., Ganss, R., Aguzzi, A., Schmid, W., Beermann, F., and Schütz, G. (1994) Targeted mutation of the CREB gene: compensation within the CREB/ATF family of transcription factors. *Proc. Natl. Acad. Sci. USA* **91,** 5647–5651.

Innis, R. B., Nestler, E. J., and Aghajanian, G. K. (1988) Evidence for G protein mediation of serotonin- and $GABA_B$-induced hyerpolarization of rat dorsal raphe neurons. *Brain Res.* **459,** 27–36.

Kleuss, C., Hescheler, J., Ewel, C., Rosenthal, W., Schultz, G., and Wittig, B. (1991) Assignment of G protein subtypes to specific receptors inducing inhibition of calcium currents. *Nature* **353,** 43–47.

Koob, G. F. (1992) Drugs of abuse: anatomy, pharmacology and function of reward pathways. *Trends Pharmacol. Sci.* **13,** 177–184.

Meyer, T. E. and Habener, J. F. (1993) Cyclic adenosine 3′,5′-monophosphate response element binding protein (CREB) and related transcription-activating deoxyribonucleic acid-binding proteins. *Endocr. Rev.* **14,** 269–290.

Milligan, J. F., Mateucci, M. D., and Martin, J. C. (1993) Current concepts in antisense drug design. *J. Med. Chem.* **36,** 1923–1937.

Nestler, E. J. (1992) Molecular mechanisms of drug addiction. *J. Neurosci.* **12,** 2439–2450.

Nestler, E. J., Hope, B. T., and Widnell, K. L. (1993) Drug addiction: a model for the molecular basis of neural plasticity. *Neuron* **11,** 995–1006.

Parolaro, D., Rubino, T., Gori, E., Massi, P., Bendotti, C., Patrini, G., Marcozzi, C., and Parenti, M. (1993) In situ hybridization reveals specific increases in $G\alpha_s$ and $G\alpha_o$ mRNA in discrete brain regions of morphine-tolerant rats. *Eur. J. Pharmacol. Mol. Pharmacol.* **244,** 211–222.

Raymond, J. R. (1995) Multiple mechanisms of receptor-G protein signaling specificity. *Am. J. Physiol.* **269,** F141–158.

Roush, W. (1996) Regulating G protein signaling. *Science* **271,** 1056–1058.

Rubino, T., Massi, P., Patrini, G., Venier, I., Giagnoni, G., and Parolaro, D. (1994) Effect of chronic exposure to naltrexone and opioid selective agonists on G protein mRNA levels in the rat nervous system. *Mol. Brain. Res.* **23,** 333–337.

Russell, D. S., Widnell, K. L., and Nestler, E. J. (1996). Use of antisense oligonucleotides to investigate brain function. *The Neuroscientist,* **2,** 79–82.

Self, D. W. and Nestler, E. J. (1995) Molecular mechanisms of drug reinforcement and addiction. *Annu. Rev. Neurosci.* **18,** 463–495.

Self, D. W., Terwilliger, R. Z., Nestler, E. J., and Stein, L. (1994) Inactivation of G_i and G_o proteins in nucleus accumbens reduces both cocaine and heroin reinforcement. *J. Neurosci.* **14,** 6239–6247.

Stein, C. A. and Chen, Y.-C. (1993) Antisense oligonucleotides as therapeutic agents—is the bullet really magical? *Science* **261,** 1004–1012.

Struthers, R. S., Vale, W. W., Arias, C., Sawchenko, P. E., and Montminy, M. R. (1991) Somatotroph hypoplasia and dwarfism in transgenic mice expressing a non-phosphorylatable CREB mutant. *Nature (Lond.)* **350,** 622–624.

Suzuki, T., George, F. R., and Meisch, R. A. (1988) Differential establishment and maintenance of oral ethanol reinforced behaviour in Lewis and Fischer 344 inbred rat strains. *J. Pharm. Exp. Ther.* **245,** 164–170.

Terwilliger, R. Z., Beitner-Johnson, D., Sevarino, K. A., Crain, S. M., and Nestler, E. J. (1991) A general role for adaptations in G proteins and the cyclic AMP system in mediating the chronic actions of morphine and cocaine on neuronal function. *Brain Res.* **548,** 100–110.

Van Vliet, B. J., Van Rijswijk, A. L., Warden, G., Mulder, A. H., and Schoffelmeer, A. N. (1993) Adaptive changes in the number of G_s- and G_i- proteins underlie adenylyl cyclase sensitization in morphine-treated rat striatal neurons. *Eur. J. Pharmacol.* **245,** 23–29.

Walker, W. H., Fucci, L., and Habener J. F. (1995) Expression of the gene encoding transcription factor cyclic adenosine 3'5'-monophosphate (cAMP) response element-binding protein (CREB): Regulation by follicle stimulating hormone-induced cAMP signaling in primary rat sertoli cells. *Endocrinology* **136,** 3534–3545.

Wang, H., Watkins, D. C., Malbon, C. C. (1992) Antisense oligodeoxynucleotides to G$_s$ protein α-subunit sequence accelerate differentiation of fibroblasts to adipocytes. *Nature (Lond.)* **358,** 334–337.

Widnell, K. L., Russell, D. S., and Nestler, E. J. (1994) Regulation of expression of cAMP response element-binding protein in the locus coeruleus in vivo and in a locus coeruleus-like cell line in vitro. *Proc. Natl. Acad. Sci. USA* **91,** 10,947–10,951.

Widnell, K. L., Self, D. W., Lane, S. B., Russell, D. S., Vaidya, V., Miserendino, M., Rubin, C. S., Duman, R. S., and Nestler, E. J. (1996) Regulation of CREB expression: in vivo evidence for a functional role in morphine action in the nucleus accumbens. *J. Pharmacol. Exp. Ther.* **276,** 306–315.

Wise, R. A. (1990) The role of reward pathways in the development of drug dependence, in *Psychotropic Drugs of Abuse* (Balfour, D. J. K., ed.), Pergamon, Oxford, pp. 23–57.

Woolf, T. M., Jennings, C. G. B., Rebagliati, M., and Melton, D. A. (1990) The stability, toxicity and effectiveness of unmodified and phosphorothioate antisense oligodeoxynucleotides in *Xenopus* oocytes and embryos. *Nucleic Acids Res.* **18,** 1763–1769.

G Protein-Linked Opioid Receptors

Pushkar Sharma, Ivone Gomes,
and Shail K. Sharma

Introduction

Opiates, such as morphine and heroin, exert their pharmacological effects by binding to highly specific receptor proteins on the cell surface of neuronal and nonneuronal cells. In 1973, researchers in three different laboratories demonstrated the presence of receptors specific for opiates in preparations from brain and intestine (Pert and Snyder, 1973; Simon et al., 1973; Terenius, 1973). These studies satisfied the criteria first enunciated by Goldstein et al. (1971) for a narcotic receptor that is specific for the pharmacologically active D(–) enantiomer of the narcotics. The existence of specific opiate receptors in brain led many scientists to look for endogenous ligands in the brain for which opiate receptors exist. In 1975, Hughes et al. isolated and characterized two pentapeptides (Met- and Leu-enkephalins) from bovine brain that were able to mimic the actions of morphine in the pharmacological test systems. Now, it is known that a family of more than 20 endogenous opioid receptor ligands are generated from three precursor proteins, namely, proopiomelanocortin, proenkephalin, and prodynorphin (Evans et al., 1988). Based on the pharmacological potency and the selectivity of receptors for various opioid ligands, the receptors were classified into μ, δ, and κ receptors (Pasternak, 1986; Zukin et al., 1988). Recently, the δ opioid receptor (Evans et al., 1992), rat μ opioid receptor (Chen et al., 1993), and rat κ opioid receptor (Meng et al., 1993)

From: *Neuromethods, Vol. 31: G Protein Methods and Protocols*
Ed: R. K. Mishra, G. B. Baker, and A. A. Boulton Humana Press Inc.

were cloned. The δ opioid receptor cDNA when expressed in COS cells showed all the expected properties of the δ receptor (Keiffer et al., 1992). Indirect evidence for the coupling of opioid receptors to G proteins was first shown by Sharma et al. (1975a). In these studies, the GTP analog Gpp(NH)p decreased the agonist affinity for morphine receptors in NG108-15 cells. Koski and Klee (1981) further showed that opioid ligand binding to NG108-15 membranes resulted in the stimulation of low *Km* GTPase activity. Coupling of G protein to three classes of opioid receptors elicits cellular responses such as inhibition of adenylate cyclase (Sharma et al., 1975a), increased potassium conductance (North et al., 1987) and inactivation of calcium channels (Tsunoo et al., 1986). Despite cloning and expression of opioid receptors, it is still important to be able to solubilize active receptors from intact membranes. The purified receptor will be useful for structural studies and to map the domains of agonist/antagonist binding as well as the site of G_i binding.

Cultured Neuronal Cells as a Model System to Study Opiate Receptor

Most of the earlier studies on the identification of opiate binding were done with crude membrane preparations derived from a mixture of different cell types. In order to obtain a homogenous source, several clonal lines of neuroblastoma cells were screened for opiate binding. A neuroblastoma x glioma hybrid cell line (NG108-15), with well developed neural properties, was found to have high-affinity morphine receptors (Klee and Nirenberg, 1974). The relative binding affinities of narcotic analogs to NG108-15 cells were the same as those found with rat brain homogenates and compared well with the pharmacological potency of the narcotics reported for isolated intestinal segments. The availability of a relatively homogenous population of cells with and without morphine receptors, although still retaining the other neural properties, provided an opportunity to explore questions relating to the mechanism of action of morphine and the phenomena of tolerance and dependence.

Receptor-Mediated Signal Transduction

The chain of biochemical events following the combination of morphine with receptors was studied in cultured NG108-15 cells. In 1975, Sharma et al. first reported that morphine, on binding to the receptors, inhibits adenylyl cyclase activity. The inhibition was stereospecific and was reversed by the antagonist naloxone, thereby showing functional coupling between the morphine receptor and the enzyme. The relative affinities of narcotics for the opiate receptors agreed well with their effectiveness as inhibitors of adenylyl cyclase and with their pharmacological potency (Table 1). In these studies, complete reversal of morphine-induced inhibition of adenylyl cyclase in the presence of $1 \times 10^{-5}M$ Gpp(NH)p suggested for the first time that the morphine receptors are G protein-linked receptors (Fig. 1). We further designed experiments to test the hypotheses illustrated in Fig. 2 (Sharma et al., 1975b). It was shown that the addition of morphine to the cultured cells resulted in a rapid inhibition of adenylyl cyclase activity and a concomitant decrease in intracellular cAMP levels. Further incubation of cells with morphine revealed a second regulatory process manifested in a compensatory increase in adenylyl cyclase activity termed the late positive regulation. The later increase in adenylyl cyclase activity compensated for the inhibition of enzyme activity by morphine, and cAMP levels were restored to the normal value. Cells were now tolerant to morphine, and were also dependent on the narcotic, since withdrawal of the drug or the addition of a specific narcotic antagonist raised cAMP levels to abnormally high values and was followed by a gradual return to the normal level of adenylyl cyclase in NG108-15 cells. The dual regulation of adenylyl cyclase provided the first biochemical model in the delineation of tolerance and dependence to opiates, and continues to provide a valuable guideline to research activities in this area. The properties of

Table 1
Comparison of Narcotic Affinity for the Opiate Receptor
and Ability to Inhibit Adenylate Cyclase

Narcotic	Kd^a Narcotic receptor, nM	Ki^b Adenylate cyclase, nM
Etorphine	5	10
Levorphanol	200	200
Morphine	4000	2000
3-Allylyprodine	10,000	50,000
Dextrorphan	10,000	—
Naloxone	20	—

[a]Apparent dissociation constant of the narcotic–receptor complex.
[b]Narcotic concentration required for 50% of maximal inhibition of adenylate cyclase.
—, No inhibition
Data from Sharma et al. (1975a).

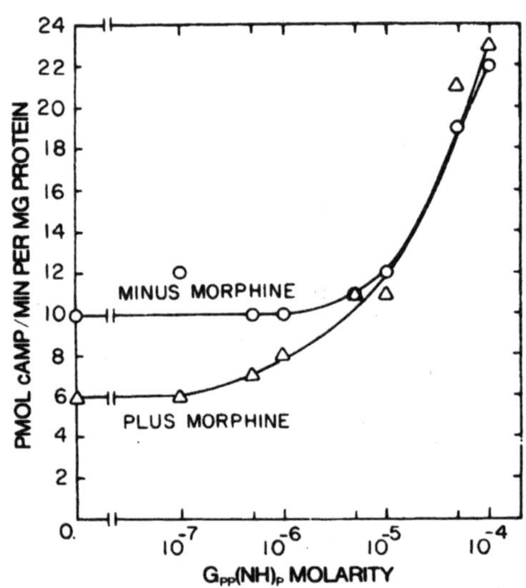

Fig. 1. Effect of Gpp(NH)p on adenylate cyclase activity of an NG108-15 homogenate, in the presence and absence of 10 μM morphine. Data from Sharma et al. (1975a).

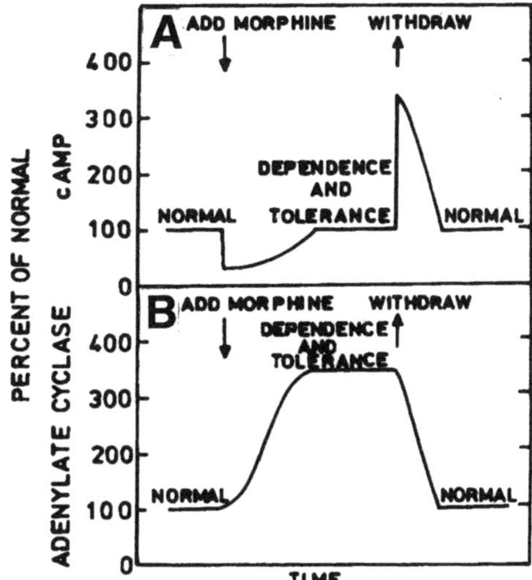

Fig. 2. A model of the role of adenylate cyclase regulation in the development of morphine tolerance and dependence. (**A**) Effects of morphine upon cAMP levels. (**B**) The effects of the opiate on adenylate cyclase activity as a function of time. Data from Sharma et al. (1975b).

adenylyl cyclase preparations from control and opiate-treated cells did not differ appreciably with respect to the concentration of effector agents, such as PGE_1, 2-chloroadenosine, morphine, naloxone, ATP, Mg^{2+}, or Mn^{2+} required for half-maximal effects (Sharma et al., 1977). However, Gpp(NH)p abolished all of the opiate-dependent compensatory increase in adenylyl cyclase activity (Fig. 3).

In NG108-15 cells, five different G proteins, which are pertussis toxin substrates namely G_{i2}, two isoforms of Go and two isoforms of G_{i3} (Roerig et al., 1992), are expressed. The nature of G proteins involved in transduction of δ-opioid receptor activities still remains obscure. The μ and κ opioid receptor types also transduce their signals via the G proteins family. Activation of δ-opioid receptors inhibits adenylyl cyclase (Sharma et al., 1975a,b; Blume et al., 1979), and voltage-dependent calcium flux (Porzig, 1990), and hyperpolarizes the cell by opening K^+ channels (North, 1989). Opioids also stimulate phosphoinositide-specific phospholipase C,

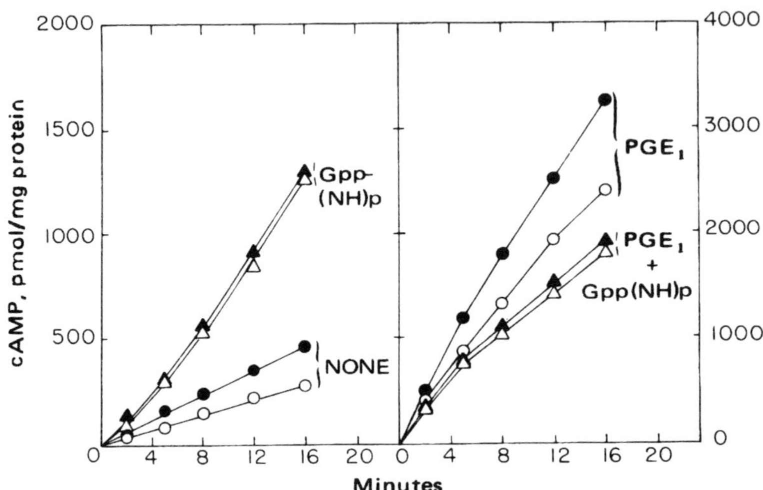

Fig. 3. Effect of Gpp(NH)p on the activity of basal (**left**) and PGE_1 stimulated (**right**) adenylate cyclase from NG108-15 cells (subculture 15) treated 3 days with or without 1 μM etorphine. Symbols: ○, no addition, control homogenate; ●, no addition, homogenate from cells treated with etorphine; △, 100 mM Gpp(NH)p, control homogenate; ▲, 100 μM Gpp(NH)p, homogenate from cells treated with etorphine. (Right) Each reaction mixture contained 10 μM PGE_1. Data from Sharma et al. (1977).

leading to an accumulation of inositol phosphates in NG108-15 cells (Jin et al., 1994). These findings support the inclusion of opioid receptors in the G protein-coupled receptor superfamily, which includes adrenergic, dopaminergic, muscarinic, and several peptide receptors.

Recently, a positive coupling of a subset of opioid receptors to basal adenylyl cyclase activity has been found in F-11 neuroblastoma-sensory neuron hybrid cells (Cruciani et al., 1993) and in rat olfactory bulb (Olianas and Onali, 1992). The availability of cell lines, such as NG108-15 with exclusively δ-opioid receptors (Sharma et al., 1975a). SK-N-BE, a neuroblastoma cell line with μ, δ1, and δ2 receptor subtypes (Polastron et al., 1994; Palazzi et al., 1996), 7315C-derived from a rat pituitary tumor with a homogenous population of μ-opioid receptors (Frey and Kebabian, 1984), human neuroblastoma SKNSH expressing both μ- and δ-opioid receptors (Yu et al.,

1986) and a neuroblastoma, Neuro 2A, stably expressing a cloned μ-opioid receptor (Chakrabarti et al., 1995) provides several cellular models to study opioid actions.

Selective Ligands for μ and δ Opioid Receptors

With the development of selective agonists and antagonists, it has become obvious that each type of opioid receptor has unique properties. Morphine is the prototype agonist of the μ-opioid receptor. DAMGO, an enkephalin analog; binds to μ-opioid receptors with about 100-fold greater affinity than to δ-opioid receptors. Opioid antagonists, naloxone and naltrexone, show modest μ-opioid receptor selectivity. The first useful μ-opioid receptor antagonist was the cyclic peptide CTP. Leu- and Met-enkephalin served as the prototype ligands during early studies of the opioid receptor. Several linear peptides, DADLE, DSLET and DTLET, are extensively used for the characterization of δ-opioid receptors DSTBULET, obtained by selective modification of DSLET, has over 100-fold selectivity for δ- relative to μ-opioid receptors and DPDPE is about 3000-fold more potent toward δ- relative to μ- or κ-opioid receptors. Selective δ-opioid receptors antagonists, ICI174,864 and TIPP and an irreversible δ-opioid receptor alkylating agent, DALCE have been described (Knapp et al., 1995).

Solubilization and Purification of Opioid Receptors

It has been difficult to solubilize the opioid receptor proteins from membranes, and in many cases, the solubilized receptor lacks appropriate binding stereospecificity. In the past, two basic approaches were used to dissociate the receptor from membranes: (1) Solubilization using nondenaturing detergents like CHAPS, Triton X-100 and digitonin, followed by molecular sieving and affinity chromatography; and (2) covalent labeling of the receptor in its native form embedded in membranes with a specific ligand, followed by solubilization and purification by affinity chromatography. Using the first approach, several investigators reported variable molecular weights for the purified protein, and in some cases, SDS-PAGE revealed more than one band of molecular weights 23, 35, and 43 (Bidlack et al., 1981), 65 (Gioannini et al., 1985), 58 (Cho et al., 1986), 43 and 65 (Simon et al., 1990), and 63 kDa (Ahmed et al., 1989). Simonds et al. (1985) labeled the NG 108-15 δ-opioid receptor with 3-methylfentanylisothiocynate (super-FIT) and reported a molecular weight of 58 kDa. In one study, a spurious μ-opioid receptor was purified, but later was shown to have considerable homology with the immunoglobulin superfamily, but not with the family of G protein-coupled receptors (Loh and Smith, 1990).

Although in recent years cloning of the opioid receptors has been achieved and the cloned DNA been expressed in appropriate cell lines, the pure receptor protein is still not available in sufficient quantities.

Gomathi and Sharma (1993) solubilized the NG108-15 δ-opioid receptor using CHAPS and purified it by affinity chromatography on 6-succinylmorphine coupled to Sepharose 4B columns. The protein was eluted with KCl and on SDS-PAGE followed by silver staining, which showed a single band of 58 kDa (Fig. 4, lane 1). The affinity-purified receptor was further subjected to WGA Sepharose chromatography and eluted with the competing sugar, N-acetyl-D-glucosamine. The WGA eluate showed a single band of 58 kDa and retained the capacity to bind (^{3}H)-DADLE (Fig. 4, lane 2). Using the same purification procedure, we could partially purify the receptor from bovine frontal cortex, which showed two bands of 58 and 30 kDa on SDS-PAGE as revealed by silver staining. This protein was used to raise a polyclonal antibody in mice. In an attempt to immunoprecipitate the antigen, surface proteins of NG108-15 cells were labelled with ^{125}I, and membranes prepared and solubilized using 10 mM CHAPS. The antigen was precipitated by incubating the solubilized protein (diluted 1:10) with the antisera, and the antigen–antibody complex, was collected on goat antimouse IgG + IgM coupled to microspheres. On SDS-PAGE and autoradiography, a single

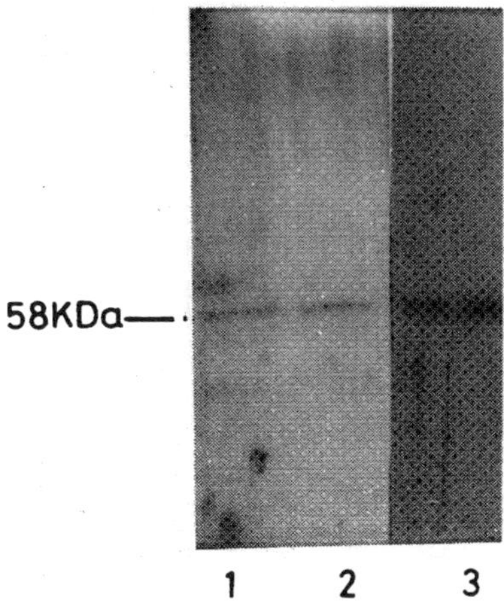

Fig. 4. Analysis of the receptor from NG108-15 cells by SDS-PAGE on 10% gels. Lanes 1 and 2 were stained with silver. Lane 1, 5 µg of affinity-purified receptor; Lane 2, 5 µg of WGA-Sepharose eluted receptor; Lane 3, autoradiograph pattern of ^{125}I-labeled antigen immunoprecipitated by IMS from NG-108-15 cells. Data from Gomathi and Sharma (1993).

band of 58 kDa was seen, which co-migrated with affinity-purified receptor (Fig. 4, lane 3).

Reconstitution of Receptor into Lipid Vesicles

Lipids are known to be required for optimal binding of ligand to opiate receptor. In several studies, the affinity-purified protein did not bind labeled opiates. However, on reconstitution with lipid, binding to opiates was significantly restored (Cho et al., 1983; Frey et al., 1989; Scheideler and Zukin 1990). Ueda et al. (1988) reconstituted the affinity-purified rat µ-opioid receptor with G_i, resulting in more than 200-fold higher binding affinity for DAGO, which was abolished in the presence of GTP. In our studies, reconstitution of the affinity-purified receptor from NG108-15 cells into lipid vesicles resulted in a ninefold enhancement of (^{3}H)DADLE binding (Table 2), suggesting that the membrane lipid environment is important for ligand binding to the δ-receptor. The 58-kDa band seen on SDS-PAGE was electroeluted (recovery

Table 2
[³H] DADLE and [³H] DPDPE Binding
to the Affinity-Purified Receptor from NG108-15 Cells[a]

	Specific binding, fmol/mg protein	
	[³H] DADLE	[³H] DPDPE
Affinity-purified receptor	1187	2226
Affinity-purified receptor reconstituted in lipid vesicles	10,304	ND
58-kDa band protein[b] reconstituted in lipid vesicles	4306	ND

[a][³H] DADLE and [³H] DPDPE, 5 nM, were used to study binding to affinity-purified receptor or affinity-purified receptor reconstituted into lipid vesicles (26 µg).

[b]SDS-PAGE of the affinity-purified receptor was carried out on 10% gels. A lane was cut and stained with silver. Gel from another lane corresponding to the 58-kDa band was excised, the protein electroeluted and reconstituted in lipid vesicles. Binding studies with [³H] DADLE were carried out. Control lipid vesicles without protein did not show any specific binding. Nonspecific binding was determined in the presence of DADLE, 10 µM and was 20–30% of total binding. Values shown are averages from two experiments performed in duplicate.

ND = not done.

Data from Gomathi and Sharma (1993).

was 48%) and reconstituted in lipid vesicles. Significant (³H) DADLE binding was observed after reconstitution into lipid vesicles, confirming the 58-kDa protein to be the δ-opioid receptor (Gomathi and Sharma, 1993).

Western Blot Analysis

Multiple antigenic peptides to the predicted C ([353]TRE-RVTACTPSDGPG[367]) and N-terminal ([3]LVPSARAELQSS-PLV[17]) of δ-opioid receptor were obtained from Research Genetics, USA. These peptides were used to immunize Balb/c mice, and the spleens used for fusion with a myeloma cell line, SP2/0-Ag-14. The hybridomas showing positive reactivity to the N- or C-terminal peptides were pooled separately. From the culture supernatants, the antibodies were purified by

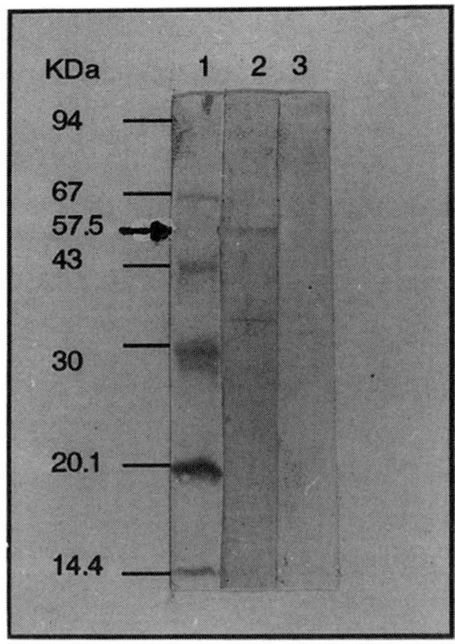

Fig. 5. Western blot analysis of affinity-purified protein from NG108-15 cells using antibodies to the predicted C-terminal region of the δ-opioid receptor. Lane 1, molecular weight standards; Lane 2, C-terminal antibodies; Lane 3, normal mouse IgG.

established procedures and used for Western blot studies. NG108-15 δ-receptor purified as described above was subjected to SDS-PAGE followed by Western blot analysis. The PVDF membranes were probed with antibodies to C353–367 and N3–17 terminal peptide or with an equivalent amount of normal mouse IgG followed by second antibody coupled to alkaline phosphatase. Figure 5 shows the presence of one major band of 58 kDa and a minor of 38 kDa. Using the C353–367/N3–17-terminal peptide antibodies, we also immunoprecipitated the δ-receptor from CHAPS-solubilized NG108-15 extracts. The immunoprecipitated protein showed specific binding with (^{3}H) DPDPE. The amino acid sequence of the 58-kDa band identified in the Western blot studies is under progress.

Preliminary work on the deglycosylation of the NG108-15 δ-receptor with β-galactosidase, neuraminidase, and PNGase followed by SDS-PAGE and Western blot showed a decrease in molecular weight of the purified receptor from 58 to 49 kDa (Gomes et al., 1997).

Acknowledgment

This work was supported by project No. BT/R&D /15/26/93 from the Department of Biotechnology, Government of India to S. K. S.

References

Ahmed, M. S., Zhou, D. H., Cavinato, A. G., and Maulik, D. (1989) Opioid binding properties of the purified kappa receptor from human placenta. *Life Sci.* **44,** 861–71.

Bidlack, J. M., Abood, L. G., Osei-Gyimah, P., and Archer, S. (1981) Purification of opiate receptor from rat brain. *Proc. Natl. Acad. Sci. USA* **78,** 636–639.

Blume, A. J., Lichtshtein, D., and Boone, G. (1979) Coupling of the opiate receptors to adenylate cyclase: requirement for Na^+ and GTP. *Proc. Natl. Acad. Sci. USA* **76,** 5626–5630.

Chakrabarti, S., Law, P.-Y., and Loh, H. H. (1995) Neuroblastoma Neuro 2A cells stably expressing a cloned μ-opioid receptor: a specific cellular model to study acute and chronic effects of morphine. *Mol. Brain Res.* **30,** 269–278.

Chen, Y., Mestek, A., Liu, J., Hurley, A., and Yu, L. (1993) Molecular cloning and functional expression of a mu opioid receptor from rat brain. *Mol. Pharmacol.* **44,** 8–12.

Cho, T. M., Ge, B. L., Yamato, C., Smith, A. P., and Loh, H. H. (1983) Isolation of opioid binding components by affinity chromatography and reconstitution of binding activities. *Proc. Natl. Acad. Sci. USA* **80,** 5176–5180.

Cho, T. N., Masegawa, J. I., Ge, B. L., and Loh, M. M. (1986) Purification to apparent homogeneity of a μ type opioid receptor from rat brain. *Proc. Natl. Acad. Sci. USA* **83,** 4138–4142.

Cruciani, R. A., Dvorkin, B., Morris, S. A., Crain, S. M., and Makman, M. H. (1993) Direct coupling of opioid receptors to both stimulatory and inhibitory guanine nucleotide-binding proteins in F11 neuroblastoma sensory neuron hybrid cells. *Proc. Natl. Acad. Sci. USA* **90,** 3019–3023.

Evans, C., Hammond, D. L., and Frederickson, R. C. A. (1988) The opioid peptides, in *The Opiate Receptors* (Pasternak, G. W., ed.), Humana, Clifton, NJ, pp. 23–74.

Evans, C. J., Keith, J. D. E., Morrison, H., Magendzo, K., and Edwards, R. H. (1992) Cloning of delta opioid receptor by functional expression. *Science* **258,** 1952–1955.

Frey, E. A. and Kebabian, J. W. (1984) A μ-opiate receptor in 7315c tumor tissue mediates inhibition of immunoreactive prolactin release and adenylate cyclase activity. *Endocrinology* **115,** 1799–1804.

Frey, E. A., Gosse, M. E., and Cote, T. E. (1989) Reconstitution of the solubilized m opioid receptor coupled to a GTP-binding protein. *Eur. J. Pharmacol.* **172,** 347–356.

Gioannini, T. L., Howard, A. D., Hiller, J. M., and Simon, E. J. (1985) Purification of an active opioid binding protein from bovine striatum. *J. Biol. Chem.* **260,** 15,117–15,121.

Goldstein, A., Lowney, L. I., and Pal, B. K. (1971) Stereospecific and nonspecific interactions of the morphine congener levorphanol in subcellular fractions of mouse brain. *Proc. Natl. Acad. Sci. USA* **68,** 1742–1747.

Gomathi, K. G., and Sharma, S. K. (1993) Purification and reconstitution of the δ opioid receptor. *FEBS Lett.* **330,** 146–150.

Gomes, I., Sharma, P., and Sharma, S. K. (1997) manuscript submitted.

Hughes, J., Smith, T. W., Kosterlitz, H. W., Fothergill, L. A., Morgan, B. A., and Morris, H. R. (1975) Identification of two related pentapeptides from brain with potent agonist activity. *Nature* **258,** 577–579.

Jin, W., Lee, N. M., Loh, H. H., and Thayer, S. A. (1994) Opioids mobilize calcium from inositol 1,4,6-triphosphate sensitive stores in NG108-15 cells. *J. Neurosci.* **14,** 1920–1929.

Kieffer, B. L., Befort, K., Gaveriaux-Ruff, C., and Hirth, C. G. (1992) The delta opioid receptor: Isolation of a cDNA by expression cloning and pharmacological characterization. *Proc. Natl. Acad. Sci. USA* **89,** 12,048–12,052.

Klee, W. A. and Nirenberg, M. (1974) A neuroblastoma × glioma hybrid cell line with morphine receptors. *Proc. Natl. Acad. Sci. USA* **71,** 3474–3477.

Knapp, R. J., Vaughn, L. K., and Yamamura, H. I. (1995) Selective ligands for μ and δ opioid receptors, in *The Pharmacology of Opioid Peptides,* (Tseng, L. F., ed.), Harwood Academic Publishers, Wisconsin, pp. 1–27.

Koski, G. and Klee, W. A. (1981) Opiates inhibit adenylate cyclase by stimulating GTP hydrolysis. *Proc. Natl. Acad. Sci. USA* **78,** 4185–4189.

Loh, H. H. and Smith, A. P. (1990) Molecular characterization of opioid receptors. *Annu. Rev. Pharmacol. Toxicol.* **30,** 123–147.

Meng, F., Xie, G. X., Thompson, R. C., Mansour, A., Goldstein, A., Watson, S. J., and Akil, H. (1993) Cloning and pharmacological characterization of rat κ-opioid receptor. *Proc. Natl. Acad. Sci. USA* **90,** 9954–9958.

North, R. A. (1989) Drug receptors and the inhibition of nerve cells. *Br. J. Pharmacol.* **98,** 13–28.

North, R. A., Williams, J. T., Suprenant, A., and Christie, M. J. (1987) μ and δ receptors belong to a family of receptors that are coupled to potassium channels. *Proc. Natl. Acad. Sci. USA* **84,** 5487–5491.

Olianas, M. C. and Onali, P. (1992) Characterization of opioid receptors mediating stimulation of adenylate cyclase activity in rat olfactory bulb. *Mol. Pharmacol.* **42,** 109–115.

Palazzi, E., Ceppi, E., Guglielmetti, F., Catozzi, L., Amoroso, D., and Groppetti, A. (1996) Biochemical evidence of functional interaction between μ and δ opioid receptors in SK-N-BE neuroblastoma cell line. *J. Neurochem.* **67,** 138–144.

Pasternak, G. W. (1986) Multiple morphine and enkephalin receptors: Biochemical and pharmacological aspects. *Ann. NY Acad. Sci.* **467,** 130–139.

Pert, C. B. and Snyder, S. H. (1973) Opiate receptor: Demonstration in nervous tissue. *Science* **179,** 1011–1014.

Polastron, J., Mur, M., Mazarguil, H., Puget, A., Meunier, J.-C., and Jauzac, P. (1994) SK-N-BE neuroblastoma cell line containing two sub types of δ opioid receptors. *J. Neurochem.* **62,** 898–906.

Porzig, M. (1990) Pharmacological modulation of voltage dependent calcium channels in intact cells. *Rev. Physiol. Biochem. Pharmacol.* **114,** 209–262.

Roerig, S. C., Loh, H. H., and Law, P. Y. (1992) Identification of three separate guanine-nucleotide binding proteins that interact with δ-opioid receptor in NG108-15 neuroblastoma x glioma hybrid cells. *Mol. Pharmacol.* **41,** 822–831.

Scheideler, M. A. and Zukin, R. S. (1990) Reconstitution of solubilized delta opioid receptor binding sites in lipid vesicles. *J. Biol. Chem.* **265,** 15,176–15,182.

Sharma, S. K., Nirenberg, M., and Klee, W. A. (1975a) Morphine receptors as regulators of adenylate cyclase activity. *Proc. Natl. Acad. Sci. USA* **72,** 590–594.

Sharma, S. K. Klee, W. A., and Nirenberg, M. (1975b) Dual regulation of adenylate cyclase accounts for narcotic dependence and tolerance. *Proc. Natl. Acad. Sci. USA* **72,** 3092–3096.

Sharma, S. K. Klee, W. A., and Nirenberg, M. (1977) Opiate dependent modulation of adenylate cyclase. *Proc. Natl. Acad. Sci. USA* **74,** 3365–3369.

Simon, E. J., Hiller, J. M., and Edelman, I. (1973) Stereospecific binding of the potent narcotic analgesic ^{3}H-etorphine to rat brain homogenate. *Proc. Natl. Acad. Sci. USA* **70,** 1947–1949.

Simon, J., Benyhe, J., Hepp, J., Varga, K., Medzihradsky, K., Borsodi, A., and Wollemann, M. (1990) Method of isolation of kappa-opioid binding sites by dynorphin affinity chromatography. *J. Neurosci. Res.* **25,** 549–555.

Simonds, W. F., Burke, T. R., Jr., Rice, K. C., Jacobson, A. E., and Klee, W. A. (1985) Purification of the opiate receptor of NG108-15 neuroblastoma-glioma hybrid cells. *Proc. Natl. Acad. Sci. USA* **82,** 4974–4978.

Terenius, L. (1973) Stereospecific interaction between narcotic analgesics and a synaptic plasma membrane fraction of rat cerebral cortex. *Acta. Pharmacol. Toxicol.* **32,** 317–320.

Tsunoo, A., Yoshii, M., and Narahashi, T. (1986) Block of calcium channels by enkephalin and somatostatin in neuroblastoma-glioma hybrid cells. *Proc. Natl. Acad. Sci. USA* **83,** 9832–9836.

Ueda, H., Harada, H., Nozaki, M., Katada, T., Ui, M., Satoh, M., and Takagi, H. (1988) Reconstitution of rat brain μ opioid receptors with purified guanine nucleotide-binding regulatory proteins, Gi and Go. *Proc. Natl. Acad. Sci. USA* **85,** 7013–7017.

Yu, V. C., Richards, M. L., and Sadee, W. (1986) A human neuroblastoma cell line expresses μ and δ opioid receptor sites. *J. Biol. Chem.* **261,** 1065–1070.

Zukin, R. Z., Eghlali, M., Olive, D., Unterwald, E. M., and Tempel, A. (1988) Characterization and visualization of rat and guinea pig brain κ opioid receptors: Evidence for κ_1 and κ_2 opioid receptors. *Proc. Natl. Acad. Sci. USA* **85,** 4061–4065.

G Protein-Coupled Melatonin Receptors

Lennard P. Niles

Introduction

Physiological Effects of Melatonin

Increasing evidence indicates that melatonin, the principal hormone secreted by the pineal gland, influences the function of diverse neuroendocrine and other systems in mammals. This indoleamine hormone is also involved in maintaining brain homeostasis, entraining biological rhythms and coordinating reproductive function to changes in photoperiod, particularly in seasonal breeders (Brown, and Niles, 1982; Tamarkin et al., 1985; Reiter, 1991). Other studies indicate a potentially important immunomodulatory role for melatonin (Maestroni, 1993) that binds with high affinity to T-lymphocytes (Gonzalez-Haba et al., 1995). Recent reports that melatonin is a potent free radical scavenger suggest that it may also play an important role as a protective antioxidant hormone (Hardeland et al., 1993; Reiter, 1996).

Circadian Rhythmicity

A variety of experimental approaches have established that the circadian clock or pacemaker in mammals and some other vertebrates resides in the suprachiasmatic nuclei (SCN) of the hypothalamus. SCN ablation causes a loss of rhythmicity in several biological and behavioral activities ranging from neuroendocrine to locomotor activity (Cassone, 1990). The marked diurnal rhythm in pineal melatonin production (and

From: *Neuromethods, Vol. 31: G Protein Methods and Protocols*
Ed: R. K. Mishra, G. B. Baker, and A. A. Boulton Humana Press Inc.

secretion) is regulated by the SCN pacemaker, which receives photic information via a direct retinohypothalamic tract, and transmits this information to the pineal via a multisynaptic pathway. Melatonin, in turn, is thought to influence circadian rhythmicity by acting not only on neuroendocrine targets (Brown and Niles, 1982), but also directly on receptor sites within the SCN (Krause and Dubocovich, 1991). Neurophysiological studies indicate that melatonin alters the electrical activity of SCN neurons in vitro (Shibata et al., 1989). Moreover, in vitro application of melatonin to SCN slices can induce phase shifts in the electrophysiologic firing of SCN neurons (McArthur et al., 1991), which supports a direct role for this hormone in synchronizing circadian rhythms.

Reproductive Function

There is extensive evidence that melatonin plays an important role in the regulation of reproductive activity in seasonal breeders (Reiter, 1991). High-affinity receptors in the pars tuberalis (PT) of the pituitary are thought to mediate the effects of melatonin on reproductive function in sheep and other seasonal breeders (Morgan et al., 1994). Although a similar role for melatonin is less established for other mammalian species, there is some evidence that this hormone may be involved in modulating human reproductive physiology. Melatonin levels in human follicular fluid exceed those in serum and exhibit marked circadian and circannual variations (Brzezinski et al., 1987; Ronnberg et al., 1990). Physiological concentrations of melatonin stimulate progesterone production by human granulosa cells (Webley and Luck, 1986). Similarly, a positive correlation between follicular fluid melatonin and progesterone concentrations and a negative correlation between melatonin and estradiol have been reported (Yie et al., 1995a). In addition, the presence of high-affinity binding sites for melatonin on human granulosa cells (Yie et al., 1995b) suggests that this hormone may influence steroidogenesis and human reproductive function by a direct action at the ovarian level.

Oncostatic Effects

Clinical studies have shown that circulating melatonin levels are elevated in patients with breast cancer and other malignancies (Touitou et al., 1985; Lissoni et al., 1986; Dogliotti et al., 1990). Since immune system-derived cytokines, such as interferon-γ, can enhance melatonin production (Withyakumnarnkul et al., 1990), the elevation of this hormone in cancer patients may represent one of the physiological mechanisms aimed at arresting tumor growth. The reported anticancer effects of melatonin may involve its enhancement of immune function (Maestroni et al., 1986; Maestroni, 1993). A direct action is also possible since in vitro studies show that physiological concentrations of this hormone can inhibit the proliferation of human breast cancer (MCF-7) cells (Hill and Blask, 1988). Intracellular targets for melatonin, such as calmodulin (Benítez-King and Antón-Tay, 1993; Benítez-King et al., 1993) and nuclear binding sites (Menendez-Pelaez and Reiter, 1993; Steinhilber et al., 1995), may be involved in its direct action on tumour cells. Recent evidence suggests that the antiproliferative effect of melatonin in breast cancer cells involves inhibition of estrogen receptor (ER) expression (Molis et al., 1994).

Physiological vs Pharmacological Actions

In several studies, excessively large doses or concentrations of melatonin have been used, either in vivo or in vitro, to examine what are reported to be the physiological effects of this hormone. However, there is evidence that pharmacological (micromolar) concentrations of melatonin can interact with benzodiazepine (BZ) receptors both in vitro (Marangos et al., 1981; Niles, 1991) and in vivo (Niles et al., 1987a; Tenn and Niles, 1995). This interaction involves central-type, G_i-coupled and other BZ receptor subtypes that mediate the pharmacological actions of melatonin, such as enhancement of GABAergic function (Coloma and Niles, 1988; Niles and Peace, 1990), inhibition of dopaminergic activity (Tenn and

Niles, 1995), and suppression of adenylyl cyclase (AC) activity (Niles and Hashemi, 1990a; Tenn et al., 1996). It has been proposed that the psychopharmacological actions of melatonin, such as its anticonvulsant, sedative, and anxiolytic effects (Rudeen et al., 1980; Sugden, 1983; Golombek et al., 1993), involve the allosteric enhancement of central GABA-ergic activity via BZ receptors (Niles, 1991). Evidence that the central BZ antagonist, flumazenil, blocks the anxiolytic (Golombek et al., 1993) and antidopaminergic (Tenn and Niles, 1995) effects of pharmacological doses of melatonin supports the foregoing.

Although these pharmacological effects of melatonin have important implications in terms of the therapeutic potential of this indoleamine and its analogs, it is important to differentiate the underlying sites and mechanisms involved from those responsible for the physiological effects of the endogenous hormone. This chapter will focus on the sites, pathways, and potential mechanisms that underlie the physiological actions of melatonin.

Melatonin Receptors

High-Affinity Binding Sites

The development of the radioiodinated melatonin receptor ligand 2-[^{125}I]iodomelatonin ([^{125}I]MEL), which exhibits selective high-affinity binding in the brain and elsewhere, has led to rapid progress in the characterization of melatonin receptors in recent years. Autoradiographic and homogenate binding studies have identified high- (picomolar) -affinity sites for melatonin in discrete areas of the vertebrate brain, such as the SCN, and also in peripheral areas, such as the pars tuberalis of the pituitary (Krause and Dubocovich, 1991). Melatonin binding to these high-affinity sites is modulated by guanine nucleotides and monovalent cations, indicating association of the high-affinity receptor with a G protein (Rivkees et al., 1989a; Morgan et al., 1989a; Niles, 1990; Laitinen and Saavedra, 1990a). In keeping with the above, agonist competition and saturation binding experiments have indicated that the high-affinity receptor exists in two interconvertible affinity states, with abolishment of the high-affinity state in the presence of guanine nucleotides and/or monovalent cations (Ying et al., 1992).

High-affinity binding sites for melatonin have been detected in very low density on human malignant melanoma (M-6) cells (Ying et al., 1993). The pharmacological characteristics and guanine nucleotide sensitivity of [^{125}I]MEL binding sites on M-6 cell membranes, together with the ability of melatonin to inhibit forskolin-stimulated cyclic AMP (cAMP) production in M-6 cells, suggest that these cells express G_i-coupled receptors for this hormone, which are similar to those in the hypothalamus/SCN, PT and other areas. A similar functional high-affinity receptor has been found in a murine retina tumor (RT2-2) cell line (Mahle et al., 1993). Putative high–affinity receptors have also been observed in human fetal brain membranes (Yuan et al., 1991), and in various peripheral tissues, such as the spleen, adrenal gland, heart, and kidney of various vertebrates (Hu et al., 1991; Persengiev, 1992; Pang et al., 1993; Song et al., 1993).

In addition to the melatonin receptor described above, a second high-affinity receptor has been found in rat (circle of Willis) cerebral and caudal arteries (Viswanathan et al., 1990; Seltzer et al., 1992). Although this receptor exhibits a pharmacological profile similar to that of the G protein-coupled receptor present in the SCN and elsewhere, it appears to be coupled to a different signaling and functional pathway. Unlike the high-affinity G_i-coupled melatonin receptor, in the hypothalamus/SCN and pars tuberalis, which can suppress AC activity as discussed later, the arterial receptor is reportedly not linked to modulation of this enzyme, but mediates potentiation of norepinephrine-induced contraction by melatonin (Viswanathan et al., 1990). It now appears that the putative receptor on blood vessels in the rat circle of Willis is linked to a G protein (Capsoni et al., 1993), but its identity and transduction pathway are presently unknown. Interestingly, in contrast to the potentiating contractile effect of melatonin seen in rat cerebral and caudal arteries, this hormone has also been reported to produce a vasorelaxant effect in the rat aorta and sheep pulmonary circulation (Weekley 1991, 1993). The foregoing suggests that the diverse actions of melatonin are mediated by multiple receptor subtypes and/or transduction and signaling pathways in the mammalian brain and other tissues.

Low-Affinity Binding Sites

A binding site with relatively low (nanomolar) affinity for melatonin is present in the brain of the hamster (Niles, et al., 1987b; Duncan et al., 1988; Pickering and Niles, 1990). This site is also present in peripheral organs, including the hamster spleen and testes (Niles, 1989; Pickering and Niles, 1989) and in a syrian hamster melanoma (RPMI-1846) cell line (Pickering and Niles, 1992). In contrast to the restricted distribution and low density of the picomolar-affinity receptor described above, the putative nanomolar-affinity receptor is present in significantly greater density (Niles, 1990), and it shows a widespread distribution throughout the central nervous system of the Syrian hamster (Duncan et al., 1988; Pickering and Niles, 1990). Moreover, unlike the picomolar receptor, nano-molar-affinity sites in the hamster brain and on RPMI-1846 cells appear to be insensitive to GTP (Duncan et al., 1988; Pickering and Niles, 1992). Nonetheless, this apparent lack of sensitivity to GTP belies an association with a G protein, as discussed later.

Regulation of Melatonin Receptors

Downregulation

Several studies have shown that exposure of cells to their receptor agonists results in desensitization, so that there is a reduced response to subsequent stimulation (Hausdorff et al., 1990). This attenuation of subsequent responses to a hormone or drug may be classified as either homologous (agonist-specific) or heterologous (agonist-nonspecific) desensitization, and it may involve various mechanisms, including receptor phosphorylation, sequestration, and downregulation (Sibley and Lefkowitz, 1985; Sibley et al., 1988). Phosphorylation with uncoupling of signal transduction or agonist-induced sequestration of receptors within the target cell can occur rapidly within a time frame of minutes. In contrast, downregulation, which results in a decrease in the number of receptors in the cell, occurs after exposure to an agonist for one or more hours (Lefkowitz et al., 1990).

The marked diurnal rhythm in circulating levels of melatonin provides an excellent physiological system for studies of melatonin receptor regulation, as indicated by recent reports. In a quantitative autoradiographic study, a diurnal variation in melatonin receptor density was observed in the rat SCN (Laitinen et al., 1989). High-affinity [^{125}I]MEL binding was highest around the dark–light transition and lowest at the end of the light phase (Laitinen et al., 1989). These findings are unusual in that high-affinity binding increased during the peak of the melatonin rhythm, suggesting that melatonin upregulates its high-affinity receptor, in contrast to the agonist–induced receptor downregulation typically seen with G protein-coupled receptors (Lefkowitz et al., 1990). In a second autoradiographic study, these authors observed no changes in high-affinity binding in the SCN following either depletion of circulating melatonin (by light exposure, pinealectomy, or superior cervical ganglionectomy) or chronic melatonin injection, leading to their suggestion that melatonin does not regulate its own signaling, at least at the level of agonist–receptor interaction (Laitinen et al., 1992).

Another study, utilizing hypothalamic membranes containing the SCN, also indicated a diurnal rhythm in the density of high-affinity receptors for melatonin. However, in marked contrast to the above findings, this study detected an inverse correlation between the density of high-affinity sites and the concentration of the natural agonist, melatonin (Tenn and Niles, 1993). Thus, high-affinity binding was highest late in the light phase, following prolonged depletion of the agonist, and lowest during darkness when exposure to elevated melatonin concentrations presumably downregulated the high-affinity receptor (Tenn and Niles, 1993). In keeping with these diurnal findings, a significant increase in high-affinity binding has been observed in the median eminence (ME)/PT (ME/PT) of rats killed at the end of the light phase compared with animals killed in the morning (Vanecek et al., 1990). Similarly, suppression or depletion of circulating melatonin levels by exposure to constant light or pinealectomy caused a significant increase in the density of high-affinity sites in the PT of the rat and hamster (Gauer et al., 1992).

Conversely, a single injection of melatonin reversed the effect of constant light or pinealectomy on high-affinity binding in the rat PT and SCN (Gauer et al., 1993). Moreover, preincubation of cultured ovine PT cells in the presence of melatonin (100 pM or 1M) for 24 h resulted in a significant decrease in [^{125}I]MEL binding in crude PT membranes (Hazlerigg et al., 1993). Taken together, most reports suggest that melatonin is involved in the regulation of its high-affinity receptor. Interestingly, diurnal studies have indicated that only the high–affinity state of the rat SCN receptor for melatonin can be detected during the light period, whereas both high- and low-affinity states are present during darkness (Tenn and Niles, 1993). These authors have suggested that the G protein (G_i) that is essential for high-affinity binding may itself exhibit a diurnal rhythm with peak levels occurring during the light phase, when the receptor is maintained in a high-affinity state. This intriguing possibility is supported by evidence that G_i and other G proteins can be downregulated following exposure of their coupled receptors to agonists (Green et al., 1990; Milligan and Green, 1991). Further support comes from the fact that melatonin levels in the SCN would be depleted during the day, thus favoring G_i elevation. However, it should be borne in mind that a variety of other agents, such as neurotransmitters and hormones, may also influence G_i levels via their receptors in the SCN, and thus influence melatonin receptor function. In keeping with this view, there is a significant decrease in high-affinity binding in the SCN, late in the light phase, before melatonin levels increase (Tenn and Niles, 1993).

Desensitization

As discussed above, receptor downregulation results in desensitization or attenuation of receptor mediated responses to agonists in many biological systems. In a recent study, it was observed that although a prolonged (16-h) pretreatment of PT cells with melatonin caused a significant decrease in GTP-sensitive [^{125}I]MEL binding in cell membranes, there was no attenuation in the maximal inhibition by this hormone of forskolin-stimulated production of cAMP in pretreated cells (Hazlerigg et al., 1993). There was a 10-fold increase in the concentration (IC$_{50}$) of melatonin required to produce a half-maximal inhibition of enzyme activity, in pretreated cells, which may be indicative of functional desensitization. However, since melatonin's inhibitory potency is reduced in control PT cells as the concentration of forskolin is increased (Morgan et al., 1991), but pretreatment with this hormone sensitizes these cells to forskolin, the authors suggest that the increase in IC$_{50}$ for melatonin may be owing to the enhanced response to forskolin in sensitized PT cells (Hazlerigg et al., 1993). Additional studies are required to clarify whether melatonin-induced downregulation of its G$_i$-coupled receptor results in desensitization to subsequent signaling by this hormone.

Molecular Characteristics of Melatonin Receptors

Pharmacology and Function

Recently, a cDNA encoding a high-affinity melatonin receptor was isolated from *Xenopus laevis* (frog) dermal melanophores via an expression cloning strategy (Ebisawa et al., 1994). Subsequently, a PCR strategy, using degenerate primers based on the sequence of the frog receptor, was used to clone a melatonin receptor designated MEL$_{1a}$ from sheep PT and human hypothalamic SCN (Reppert et al., 1994).

Competition binding assays, using COS-7 cells transfected with the frog, sheep, or human melatonin receptor cDNA, indicated a rank order of inhibition that is characteristic of the high–affinity receptor. Melatonin and its halogenated analogs (2-iodomelatonin and 6-chloromelatonin) were significantly more potent than the melatonin precursor, *N*-acetylserotonin, whereas serotonin was ineffective. In functional studies using CHO cells transfected with the *Xenopus* melatonin receptor clone (Ebisawa et al., 1994) or NIH 3T3 cells transfected with the sheep cDNA (Reppert et al., 1994), melatonin inhibited cAMP production. Moreover, pretreatment of transfected NIH 3T3 cells with pertussis toxin blocked the inhibitory effect of melatonin on forskolin-stimulated cAMP accumulation (Reppert et al., 1994). The pharmacological and functional characteristics of the cloned MEL_{1a} receptor from human SCN and sheep PT indicate that it is similar to the high-affinity G_i-coupled melatonin receptor described above.

More recently, a second human melatonin receptor designated MEL_{1b} has been cloned (Reppert et al., 1995a). Interestingly, binding and functional studies with transfected cells indicate that this high-affinity receptor exhibits a pharmacological profile that is similar to that of the MEL_{1a} receptor. In addition, like the latter, the MEL_{1b} receptor is also coupled to inhibition of AC. However, in contrast to the MEL_{1a} receptor, the MEL_{1b} receptor does not appear to be present in either the SCN or the PT, but it is enriched in the retina and also present in lower density in the human hippocampus (Reppert et al., 1995a). In addition to these two subtypes, a new melatonin receptor subtype, designated the MEL_{1c} receptor, has now been identified in the chick brain (Reppert et al., 1995b). At present, there is no evidence that the MEL_{1c} receptor is expressed in mammals.

Molecular Structure

Sequences of the frog, sheep, and two human cDNAs revealed that they encode proteins of 420, 366, 350, and 360 amino acids, with predicted molecular weights of 47.4, 40.4, 39.4 and 40.2 kDa, respectively (Ebisawa et al., 1994; Reppert et al., 1994, 1995a). These molecular weights are similar to that previously determined from radiation inactivation studies, which indicated a molecular mass of 44 ± 9 kDa for the chick retinal melatonin receptor (Pickering et al., 1990).

In keeping with the foregoing, hydropathy analysis of the *Xenopus* receptor revealed seven hydrophobic transmembrane segments, which is a well-known characteristic of G protein-coupled receptors. The amino-terminus contains one consensus site for N-linked glycosylation in the frog receptor and two such sites in each mammalian receptor. Several consensus sites for protein phosphorylation are present within the carboxyl-terminal tail, which is 119 amino acids long in the frog receptor (Ebisawa et al., 1994). The coding regions of the sheep PT and human SCN receptors exhibit an overall homology of 60% with that of the frog receptor and a 77% homology to the frog receptor within the transmembrane domains. An overall homology of 80% is present between the sheep and human receptors, and this increases to 87% for the transmembrane domains. Partial cDNAs from the hamster and rat brain were found to be 94% identical and to have an 86% identity to the corresponding region of the sheep PT and human SCN clones (Reppert et al., 1994), indicating that they belong to the MEL_{1a} subtype. In contrast, the human MEL_{1a} and MEL_{1b} receptors exhibit only a 60% homology at the amino acid level (Reppert et al., 1995a), in keeping with their expression by different genes. Interestingly, both of these receptors (and also the chick MEL_{1c} receptor) contain an intron in the same position in the first cytoplasmic loop, suggesting that spliced variants may occur.

Signal Transduction by G Proteins

Multiplicity of G Proteins

GTP binding proteins (G proteins) are members of a large family of proteins that bind guanine nucleotides and play a major role as signal transducers for a variety of biologically active agents, including hormones, neurotransmitters, and growth factors (Spiegel, 1992). G proteins are heterotrimeric in structure, consisting of three distinct α- (39–52 kDa), β- (35–36 kDa), and γ- (7–10 kDa) subunits (Kaziro et al., 1991). Molecular studies have identified at least 23 distinct G protein α-subunits which are encoded by 17 different genes (Gudermann et al., 1996). On the basis of structural relationships, these subunits have been assigned to four major subfamilies designated as: $G\alpha_s$, $G\alpha_i$, $G\alpha_q$, and $G\alpha_{12}$ (Simon et al., 1991). This complexity is compounded by the existence of several distinct β- and γ-subunits (Watson et al., 1994; Ray et al., 1995), which, together with α-subunits, make several hundred heterotrimeric combinations possible (Hepler and Gilman, 1992).

Activation of G Proteins

G protein activation has been extensively studied for adrenergic receptors coupled to either stimulation or inhibition of AC activity. However, the general mechanism of activation and signal transduction is applicable to all G protein-coupled receptors. In the inactive state, the α-subunit is bound with GDP and associated with β- and γ-subunits. When bound to a G protein, e.g., G_s, which stimulates AC activity, receptors exhibit higher affinity for agonists. Binding of an agonist causes a conformational change in the receptor, leading to an increase in the rate at which GDP dissociates and GTP binds to the G protein α-subunit. The binding of GTP to the α-subunit results in dissociation of the G protein from the receptor, which reverts to a low-affinity conformation, whereas the GTP-charged α-subunit separates from the $\beta\gamma$ complex and goes on to activate appropriate effectors, such as AC, phospholipase C (PLC), cGMP phosphodiesterase, and ion channels (Bourne et al., 1990; Simon et al., 1991). The activation cycle is completed by hydrolysis of bound GTP by the intrinsic GTPase activity of the α-subunit. The GDP-bound α-subunit can then reassociate with $\beta\gamma$-subunits to revert to an inactive state before re-entering the GTPase cycle (Fig. 1).

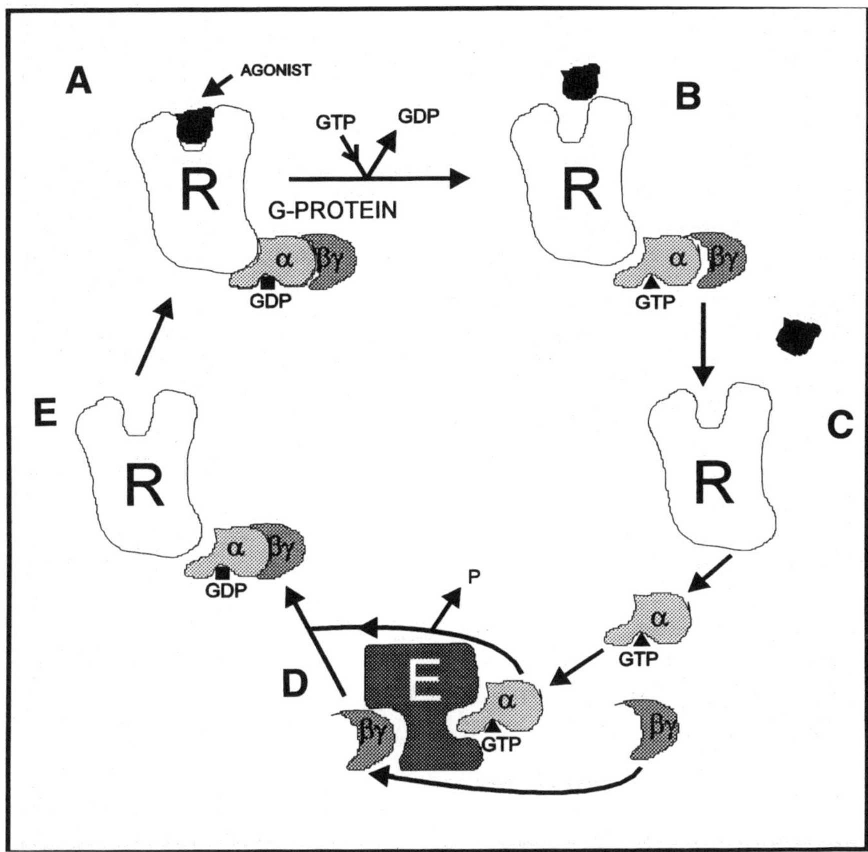

Fig. 1. G protein activation and signal transduction. (**A**) An agonist-induced conformational change in the receptor promotes dissociation of GDP from the G protein subunit. (**B**) Binding of GTP to the α-subunit results in dissociation of the G protein from the receptor, which assumes a low-affinity state with decreased affinity for agonist. GTP also reduces the affinity of α for βγ resulting in the disassociation of these subunits. (**C**) The GTP-charged α-subunit directly regulates effector function. The βγ dimer may also directly influence effector activity. (**D**) The intrinsic GTPase activity of the α-subunit terminates its effect by hydrolyzing GTP to GDP, resulting in its disassociation from the effector. (**E**) The GDP bound α-subunit reassociates with βγ to return to the basal state before re-entering the cycle.

Roles of G Protein Subunits

Although the heterotrimer is necessary for G protein coupling to receptors, the α-subunit is thought to confer specificity in coupling and also, in most cases, in effector regulation. In the case of AC activation, it is accepted that this enzyme is stimulated by the GTP-bound $G\alpha_s$-subunit, until its intrinsic GTPase shuts off the signal (Spiegel, 1992). However, there is evidence that βγ-subunits can modulate the activities of some forms of AC. For example, both type II and type IV enzymes, which are present in the brain and peripheral tissues (Gao and Gilman, 1991), can be stimulated by βγ-subunits in the presence of activated $G\alpha_s$ (Tang and Gilman, 1991). Furthermore, βγ-subunits derived from inhibitory G proteins, which are sensitive to pertussis toxin, can stimulate AC II in transfected cells (Federman et al., 1992), suggesting multifunctional roles for these signal transducers.

Inhibition of AC activity is mediated by an inhibitory G protein (G_i), but there has been considerable controversy regarding the mechanism of enzyme suppression. One model of AC inhibition, referred to as the "subunit exchange model", proposes that free βγ-subunits, released by activation of G_i, combine with $G\alpha_s$ to form inactive heterotrimers and thus decrease enzyme activity (Reithmann et al., 1991). Other investigators have proposed that in addition to the above indirect mechanism, the activated α-subunit of G_i can act directly, presumably on the catalytic component of the enzyme, to inhibit AC activity. This view is supported by various findings, including evidence that inhibitory hormones can inhibit AC in the cyc⁻ mutant of S49 lymphoma cells, which lack the $G\alpha_s$-subunit. However, recent evidence that βγ dimers can act directly to inhibit calmodulin-activated (type I) AC in the presence of activated $G\alpha_s$ (Tang et al., 1991; Taussig et al., 1993), suggests that agonist-induced responses depend on the type(s) of AC and G proteins present in target cells. Although the foregoing has focused on AC systems, it should be noted that βγ-subunits may also regulate the activity of other effectors (Iniguez-Lluhi et al., 1993).

Transmembrane Signaling by Melatonin

Modulation of AC Activity

cAMP

AC is a membrane-bound enzyme with a binding site for ATP on its cytoplasmic surface. The substrate for this enzyme is a complex of ATP and Mg^{2+}, but free levels of this divalent cation, in excess of ATP are necessary for AC activity. Evidence from biochemical and molecular studies suggests the existence of several isoforms of AC in the brain and other tissues (Krupinski et al., 1989; Tang and Gilman, 1991). Numerous hormones, neurotransmitters, and other agents can directly modulate AC activity via stimulatory (G_s-coupled) or inhibitory (G_i-coupled) receptors to alter the production of adenosine 3′, 5′-monophosphate or cAMP. Production of cAMP by AC is owing to the creation of a cyclic phosphodiester bond with the α-phosphate group of ATP, whereas release of pyrophosphate supplies energy for the reaction (Nestler and Duman, 1994). This cyclic nucleotide, discovered in the late 1950s by Sutherland and coworkers (Rall et al., 1957; Sutherland and Rall, 1958), is now known to be a ubiquitous second messenger involved in wide-ranging intracellular signaling. The effects of cyclic AMP (cAMP) are largely mediated by a cAMP-dependent protein kinase A (PKA), which catalyzes the phosphorylation of various cellular proteins (Walsh et al., 1968). In the absence of cAMP, PKA is maintained in an inactive form by a regulatory (R) subunit. PKA is activated when cAMP binds to the R subunit to induce a conformational change that results in dissociation of the catalytic (C) subunit (Taylor, 1989).

Effects of Melatonin

As noted earlier, numerous studies have indicated that melatonin influences a variety of biological processes. However, for several years, descriptions of melatonin's effects far exceeded the information available about the target sites and mechanisms involved. Considerable progress has been made in recent years in the characterization and localization of binding sites for melatonin. Similarly, progress has been made in understanding how the melatonin signal is transduced and the cellular effects of this signaling, particularly with regard to high-affinity receptors that mediate modulation of AC activity. In an early study of the cellular effects of melatonin, Vacas et al. (1981) observed that melatonin and related analogs inhibited basal production of cAMP in rat medial basal hypothalamus. This effect was observed at indoleamine concentrations ranging from 10^{-8}–$10^{-5}M$, but there was no clear concentration–effect relationship. In another study, melatonin and its biologically active analogs, 5-methoxytryptophol and 6-chloromelatonin, in concentrations of 10^{-7}–$10^{-5}M$, inhibited β-adrenoceptor-stimulated cAMP accumulation in cultured rat astroglial cells (Vacas et al., 1984a). A subsequent examination of the effects of melatonin on AC activity revealed that in the presence of GTP, nanomolar concentrations of this hormone could suppress enzyme function in the rat hypothalamus and retina (Niles, 1985). Evidence in support of these early reports that melatonin could inhibit AC activity and/or cAMP production came from the observation that this hormone could reverse the cAMP-dependent pigment-dispersing effect of forskolin, in melanophore-containing meningeal explants from *Xenopus laevis* tadpoles (White et al., 1987). Moreover, prior treatment of meningeal explants with pertussis toxin blocked the pigment-aggregating effect of melatonin, suggesting involvement of an inhibitory regulatory protein (White et al., 1987).

Recent advances in the localization and characterization of binding sites for melatonin have rekindled an interest in studies of melatonin signaling. As noted above, picomolar-affinity sites for melatonin were found to be sensitive to guanine nucleotides and monovalent cations, indicating their coupling to a G protein and potential functional significance. Not surprisingly, therefore, several recent studies of melatonin signaling have focused on central or peripheral areas containing these high-affinity sites. One such study indicated that melatonin (10 n*M*) suppresses basal cAMP levels in neonatal rat anterior pituitary. Treatment with luteinizing hormone-releasing hormone (LH-RH) increased the production of cAMP in this tissue, but in the presence of LH-RH, melatonin caused a dose-dependent inhibition of cAMP accumulation, with half-maximal inhibition at a concentration of about 250 p*M* (Vanecek and Vollrath, 1989). In addition, melatonin inhibited forskolin–stimulated cAMP production in neonatal rat pars tuberalis (Vanecek and Vollrath, 1989). In accordance with the foregoing, melatonin was found to inhibit forskolin-stimulated cAMP accumulation in primary cultures of ovine PT cells (Morgan et al., 1989b), in ME/PT explants from Djungarian hamsters (Carlson et al., 1989), and in human melanoma cells (Ying et al., 1993). However, no effect was observed on basal cAMP levels in ovine PT cells or on forskolin-stimulated AC activity in ovine PT homogenates (Morgan et al., 1989b).

An examination of the effects of melatonin on forskolin-stimulated AC activity in Syrian hamster hypothalamic homogenates indicated that hormonal concentrations of 10 p*M* to 1 n*M* suppressed enzyme activity, whereas higher concentrations were ineffective (Niles and Hashemi, 1990b). This apparently biphasic response may be explained by the fact that low-affinity binding sites for melatonin, which do not mediate inhibition of AC activity, are present in far greater density than the picomolar-affinity sites, which are coupled to inhibition of AC activity in the Syrian hamster hypothalamus (Niles, 1990). Consequently, when this tissue is exposed to hormone concentrations above ~1 n*M*, activation of nanomolar-affinity sites may negate picomolar signaling by stimulating AC activity or blocking inhibitory signaling via another system, such the PLC pathway (Niles and Hashemi, 1990b). In keeping with this hypothesis, activation of the PLC pathway has been shown to block inhibition of AC in various tissues as a result of protein kinase C (PKC)-induced phosphorylation of G_i (Katada et al., 1985; Summers et al., 1988). One might question this hypothesis, since the bimodal effects of melatonin were observed in hypothalamic membranes (Niles and Hashemi, 1990b), whereas inactivated PKC is primarily localized in soluble fractions of various tissues (Jakobs et al., 1985; Nishizuka, 1986). However, it should be noted that in contrast to several peripheral tissues, there is an appreciable localization of PKC in particulate fractions from the brain (Kikkawa et al., 1982; Girard et al., 1986), and direct phorbol ester-induced activation of this enzyme, with blockade of the inhibitory pathway of adenylyl cyclase, has been demonstrated in rat brain membranes (Olianas and Onali, 1990). Moreover, recent evidence that nanomolar-affinity sites are indeed coupled to PLC activation (Eison and Mullins, 1993) supports the foregoing possibility of PKC-mediated crosstalk between the AC and PLC pathways.

In addition to the mammalian tissues described above, high-affinity receptors in the chick brain and retina also mediate inhibition of AC activity (Niles et al., 1991). It should be noted that this effect was observed with synaptosomal fractions from the chick forebrain, but not with whole-tissue homogenates. This finding suggests that the melatonin receptors involved are present on brain neurons, in keeping with the autoradiographic localization of high-affinity sites in several chick brain nuclei (Rivkees et al., 1989b). In contrast to brain synaptosomal fractions, retinal synaptosomal membranes were unresponsive to melatonin. This was apparently owing to interference by melanin, a nonspecific, high-capacity binder of melatonin (Laitinen and Saavedra, 1990b), since removal of this substance allowed detection of the inhibitory effect of melatonin on AC activity (Niles et al., 1991). Recently, melatonin was found to inhibit basal levels of cAMP accumulation in the golden hamster retina (Faillace et al., 1994). This effect appeared to be time-dependent with inhibition detected in retinas taken from hamsters killed at 12:00 PM and 12:00 AM, but not in those obtained at 4:00 AM, when endogenous levels of melatonin were elevated (Faillace et al., 1994). Assuming that a G_i-coupled receptor is involved, one might speculate that melatonin-induced downregulation of this receptor (as discussed earlier) resulted in desensitization of the inhibitory signaling pathway during the dark phase of the diurnal cycle.

G Proteins

Studies utilizing pertussis toxin, which blocks G_i function by catalyzing the ADP ribosylation of its α-subunit (Murayama and Ui, 1983; Katada et al., 1985), confirm that this G protein transduces the melatonin signal in vertebrates, such as the hamster (Carlson et al., 1989) and chick (Niles et al., 1991). In addition to pertussis toxin, other G_i-uncoupling agents, such as the alkylating compound *N*-ethylmaleimide (NEM) (Ying et al., 1992) and heparin (Niles et al., 1994), can block melatonin signaling in chick brain membranes, further supporting G_i involvement. However, on the basis of evidence that a combination of GTP and a maximally effective concentration of pertussis toxin produces an additive inhibition of [^{125}I]MEL binding in ovine PT cell membranes, it has been suggested that both pertussis toxin-sensitive and insensitive G proteins are involved in transducing the melatonin signal in ovine PT cells (Morgan et al., 1991).

There is certainly evidence that a single receptor may be coupled to more than one G protein. For example, studies in smooth muscle cells have suggested that a single type of α1-adrenoceptor may be linked to two different signal transduction systems. One involves activation of PLC by a pertussis toxin-insensitive G protein, whereas a second system involves modulation of calcium channel function via a pertussis toxin-sensitive G protein (Oriowo and Ruffolo, 1992; Oriowo et al., 1992). Other receptors exhibiting interaction with more than one G protein include the thyrotropin receptor (Van Sande et al., 1990; Kosugi et al., 1992) and the α2-adrenoceptor (Eason et al., 1992). It is also possible that closely related, but distinct receptor subtypes can activate separate G proteins and effectors (Camps et al., 1992). At present, it is not known which, if any, of the above possibilities pertains to the high-affinity melatonin receptor in the ovine PT or elsewhere. However, recent transfection studies indicating that the *Xenopus* melatonin receptor can inhibit AC activity via either pertussis toxin-sensitive G_i-like proteins or the α-subunit of the insensitive G_z protein suggest that the latter may be involved in melatonin signaling in the ovine PT (Yung et al., 1995). Another unanswered question concerns the identity of the G protein subtype (G_{i1}, G_{i2}, or G_{i3}) that transduces inhibitory signaling by melatonin. Nonetheless, there is now convincing evidence that the melatonin receptor is coupled to a pertussis toxin-sensitive G protein that inhibits AC activity, leading to a decrease in cAMP accumulation in target tissues. Presumably, activation of this pathway by melatonin results in a decrease in phosphorylation via a reduction in the activity of cAMP-dependent PKA with physiological consequences dependent on the target cells involved, as illustrated in Fig. 2.

Modulation of PLC Activity

IP$_3$ and Diacylglycerol

Signaling by some hormones, neurotransmitters, and growth factors involves activation of an inositol phospholipid-specific phosphoinositidase (PLC). Biochemical and molecular studies have indicated the presence of multiple forms of PLC in neural and other tissues (Boyer et al., 1989). It is known that at least 16 isozymes of PLC exist, which are designated β, γ, and δ (Meldrum et al., 1991; Rhee and Choi, 1992). PLC activity is present in both cytosol and membrane fractions, but the relative intracellular distribution varies for different isozymes (Lee et al., 1987). Studies of the regional distribution of various isozymes in the CNS indicate that PLC-β is localized in neurons primarily in the striatum, hippocampus, and certain thalamic nuclei, whereas PLC-γ appears to be uniformly present in neurons in all brain regions (Gerfen et al., 1988). In contrast to the neuronal presence of these two isozymes, PLC-δ immunoreactivity is localized in astroglia (Choi et al., 1989).

Hydrolysis of phosphatidylinositol 4,5-bisphosphate (PIP$_2$) by PLC generates two second messengers, inositol

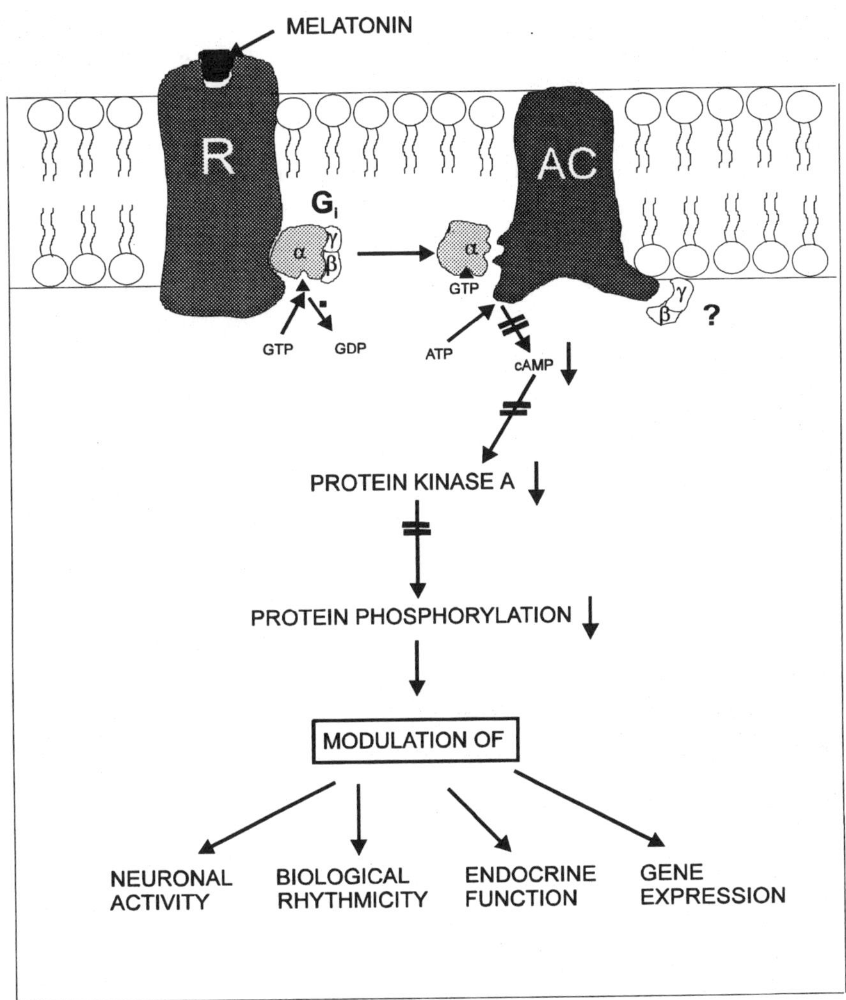

Fig. 2. Possible cellular consequences of G_i activation by melatonin. As illustrated, melatonin binding to its receptor (R) induces an exchange of GTP for GDP on the α-subunit of G_i. The GTP-charged α-subunit dissociates from the receptor and from the βγ dimer and suppresses the activity of the effector, AC, resulting in a decrease in cAMP production. This action initiates a cascade of effects, such as decreased activation of PKA and decreased phosphorylation of its target proteins, leading to changes in cellular and physiological activity. Horizontal bars indicate steps at which cellular activity is decreased.

1,4,5,-trisphosphate (IP_3) and diacylglycerol (DAG) (Berridge, 1987). The hydrophilic messenger, IP_3, is released into the cytoplasm and binds to specific receptors on nonmitochondrial Ca^{2+} storage sites to open channels for this cation, thus increasing cytoplasmic-free Ca^{2+}. IP_3 can also act on calcium channels in the plasma membrane to stimulate the influx of extracellular Ca^{2+}, further increasing the concentration of intracellular calcium, i.e., $[Ca^{2+}]_i$ (Berridge and Irvine, 1989). DAG and calcium activate PKC and other calcium–dependent kinases to alter the phosphorylation and function of target proteins. Growth factors utilize receptor autophosphorylation in their interaction with the γ-isozymes of PLC, whereas hormones, neurotransmitters, and various drugs typically alter phosphoinositide turnover via G protein activation of β-isozymes of PLC (Boyer et al., 1989; Rhee and Choi, 1992).

G Proteins

Early evidence that guanine nucleotides could activate PLC indicated the involvement of G proteins in the regulation of this enzyme (Gonzales and Crews, 1985). Guanine nucleotides can also modulate the binding of hormones or drugs, such as α-adrenergic amines, thyrotropin-releasing hormone (TRH), and muscarinic agonists, which stimulate phosphoinositide hydrolysis (Litosch and Fain, 1986; Cockcroft, 1987; Fain et al., 1988). In some cells, the effects of agonists on phosphoinositide breakdown are blocked by pertussis toxin, but this toxin is ineffective in other systems, suggesting the involvement of both pertussis toxin-sensitive and insensitive G proteins in this signaling pathway (Fain et al., 1988).

A novel G protein, G_q, which is not sensitive to pertussis toxin, has been linked to activation of PLC-β (Smrcka et al., 1991; Taylor et al., 1991). The G_q family includes G_{11} and G_{14-16}, whose α-subunits can also stimulate PLC-β and are also insensitive to pertussis toxin (Strathmann and Simon, 1990; Simon et al., 1991). However, G_q and G_{11} are thought to be the signal transducers for receptors linked to PI hydrolysis (Gutowski et al., 1991; Wu et al., 1992). In keeping with the apparent ability of some effectors to act as GTPase-activating proteins or GAPs, PLC-β1 stimulates the hydrolysis of G_q-bound GTP at least 50-fold (Berstein et al., 1992). Moreover, the GAP activity of PLC-β1 is specific for G_q, since other G proteins are not affected. Although PLC-β enzyme activity has been detected in both particulate and cytosol fractions (Lee et al., 1987), it is presumed that the membrane-bound enzyme is regulated by agents acting via G protein-coupled receptors. In addition to the α-subunits of the G_q family, $\beta\gamma$-subunits have been implicated in PLC regulation. The possibility that these $\beta\gamma$-subunits are derived from G_i or G_o may account for the sensitivity of some PLC-coupled G proteins to pertussis toxin (Exton, 1994).

Effects of Melatonin

As discussed earlier, there is now considerable evidence that the picomolar sites in areas, such as the mammalian SCN and PT and chick retina and forebrain, are negatively coupled to AC. In contrast, until recently, relatively little was known about the signaling pathway(s) for nanomolar sites, although they had been reported to be insensitive to guanine nucleotides (Duncan et al., 1988) and not coupled to AC (Pickering and Niles, 1992). In one study, micromolar concentrations of melatonin, its analogs, 6-chloromelatonin and 2-iodomelatonin, and its precursor, *N*-acetylserotonin (NAS) were found to increase phosphatidylinositol metabolism in chicken brain slices (Popova and Dubocovich, 1995). Other studies, using a Syrian hamster RPMI 1846 melanoma cell line, which expresses nanomolar-affinity sites for melatonin (Pickering and Niles, 1992), have also indicated that these sites are coupled to phosphoinositide hydrolysis (Eison and Mullins, 1993). The stimulatory effect of agonists was blocked by the putative antagonist *N*-acetyltryptamine and by prazosin (Eison and Mullins, 1993), which has been proposed to be a potential antagonist at binding sites for melatonin (Niles et al., 1987b; Pickering and Niles, 1989, 1990, 1992).

A comparison of the concentration-dependent stimulation of phosphoinositide hydrolysis by melatonin and related compounds indicated that NAS was the most efficacious of the indoleamines studied (Eison and Mullins, 1993). This is not surprising, since several binding studies have found that NAS exhibits either similar or greater affinity than melatonin for the nanomolar-affinity sites labeled by [^{125}I]MEL (Duncan et al., 1988; Pickering and Niles, 1990). Indeed, it has been suggested that these nanomolar-affinity sites may in fact be present on receptors for NAS and not on melatonin receptors, as generally believed (Pickering and Niles, 1992). This view is supported by the foregoing and also by the fact that circulating levels of NAS are in the nanomolar range (Pang et al., 1980), which is significantly greater than the picomolar concentrations of circulating melatonin, and certainly more compatible with the low (nanomolar) affinity of the putative receptor sites coupled to phosphoinositide hydrolysis. However, it is also possible that melatonin may be present in sufficiently high concentrations in extrapineal tissues, such as the retina and gut (Reiter, 1991), to allow its paracrine interaction with similar low-affinity sites.

Modulation of Calcium Influx

Calcium

There is extensive evidence of the involvement of calcium in a variety of cell functions, including stimulus–secretion coupling, gene expression, cell differentiation, growth, and division (Rasmussen and Barrett, 1984; Davis, 1992; Bading et al., 1993). As described above, some agents utilize the second messenger IP$_3$, generated from phosphoinositide hydrolysis, to increase [Ca^{2+}]$_i$ via mobilization of intracellular stores and enhancement of Ca^{2+} influx. Other agents can directly increase or decrease [Ca^{2+}]$_i$, the second messenger in this case, via G-proteins that regulate Ca^{2+} channel activity (Birnbaumer et al., 1991). In neuronal tissue, Ca^{2+} channels can be regulated by either synaptic activity or impulse activity, in the case of voltage-gated channels. The changes in [Ca^{2+}]$_i$, resulting from direct channel regulation or via indirect pathways, such as phosphoinositide hydrolysis, influence the activity of intracellular proteins, such as calmodulin and protein kinases, with diverse effects on cellular physiology and gene expression (Dash et al., 1991; Gnegy, 1993).

G Proteins

In recent years, there has been increasing evidence that G proteins can directly regulate the activity of ionic channels (Birnbaumer et al., 1990). It is now known that $G\alpha_s$ in addition to its well known role in stimulating adenylyl cyclase activity, can activate dihydropyridine-sensitive Ca^{2+} channels (Brown and Birnbaumer, 1988). Moreover, all splice variants of $G\alpha_s$ can activate both AC and dihydropyridine-sensitive Ca^{2+} channels (Mattera et al., 1989). The pertussis toxin-sensitive G-proteins have also been implicated in ion channel regulation, as shown by the stimulation of atrial K^+ channels by the α-subunits of G_{i1}, G_{i2}, and G_{i3} (Yatani et al., 1988). Other studies indicate that G_o inhibits neuronal calcium channels (Hescheler et al., 1987), but activates neuronal K^+ channels (Van Dongen et al., 1988).

Effects of Melatonin

Nanomolar or higher concentrations of melatonin can inhibit Ca^{2+} uptake in the rat hypothalamus (Zisapel and Laudon, 1983) and in brain synaptosomes (Vacas et al., 1984b). Melatonin concentrations ranging from 10 nM to 1 μM have also been reported to either stimulate or inhibit depolarization-induced $^{45}Ca^{2+}$ influx in hypothalamic synaptosomes, depending on the time of preincubation with the hormone (Rosenstein et al., 1991). In another approach using neonatal rat pituitary cells, it was observed that melatonin inhibited the GnRH-induced increase in intracellular calcium ($[Ca^{2+}]_i$) and LH release in a dose-dependent manner. Pretreatment with pertussis toxin antagonized the effects of melatonin on $[Ca^{2+}]_i$, suggesting involvement of a pertussis toxin-sensitive G-protein (Vanecek and Klein, 1992). Since GnRH stimulates LH release by increasing $[Ca^{2+}]_i$ (Huckle and Conn, 1988; Catt and Stojilkovic, 1989), it was also suggested that blockade of this action by melatonin may explain its ability to inhibit

GnRH-induced LH release from neonatal rat gonadotrophs (Vanecek and Klein, 1992), as observed earlier by Martin et al. (1977).

Conclusions and Future Perspectives

Multiple Signaling Pathways for Melatonin

Although considerable progress has been made in recent years in studies of the receptors and signaling pathways employed by melatonin, several questions regarding the cellular and molecular mechanisms underlying the effects of this hormone await clarification (*see* Table 1). In addition, the functional and behavioral correlates of these cellular effects need to be addressed.

Various studies have now confirmed that a picomolar-affinity receptor for melatonin, in the hypothalamus and other tissues, is coupled to the inhibition of cAMP accumulation, via a pertussis toxin-sensitive G protein. As discussed in an earlier section, the G_i subtype(s) involved in melatonin signaling is not known. Studies using antibodies selective for the α-subunits of the various G proteins have identified G_{i2} as the transducer of receptor-mediated inhibition of AC activity in δ-opioid (McKenzie and Milligan, 1990) and α2-adrenergic pathways (Simonds et al., 1989; McClue et al., 1992). Based on the apparent predominance of G_{i2} as an inhibitory transducer, one might hypothesize that it mediates the suppressive effect of melatonin in the AC pathway. Nonetheless, studies are required to resolve this issue and also to clarify the potential roles of $G\alpha_i$ and $G\beta\gamma_i$ in melatonin signaling. Although the suppression of AC activity and cAMP is now established as a major signaling pathway for melatonin, there is also evidence that it may stimulate or enhance cAMP production. For example, picomolar concentrations of melatonin were found to potentiate the stimulatory effect of physiological amounts of vasoactive intestinal peptide (VIP) on cAMP production in human lymphocytes (Lopez-Gonzalez et al., 1992). These investigators did not attempt to clarify the sites or mecha-

Table 1
G Protein-Coupled Melatonin Receptors and Signaling Pathways[a]

Receptor	Localization species	G protein	Effector	Cellular response
High affinity (picomolar)	Hypothalamus/ SCN (human, rat, hamster)	$G\alpha_i$	AC	cAMP↓
High affinity	Retina and brain (human)	$G\alpha_i$	AC	cAMP↓
High affinity	Anterior pituitary (neonatal rat)	$G\alpha_i$	AC	cAMP↓
High affinity	PT (sheep)	G_i/G_z?	AC	cAMP↓
High affinity	Brain and retina (chick)	$G\alpha_i$	AC	cAMP↓
High affinity	Dermal melanophores (*Xenopus*)	$G\alpha_i$	AC	cAMP↓
High affinity	Human lymphocytes	$G\beta\gamma_i$?	AC	Potentiation of VIP-induced cAMP production
High affinity	Anterior pituitary (neonatal rat)	G_i/G_o?	Calcium channel?	Inhibition of calcium influx
High affinity	Retina (rabbit and chick)	G_i/G_o?	Calcium channel?	Inhibition of dopamine release
High affinity	Circle of Willis and caudal artery (rat)	G_o/G_s?	Ion channel?	Potentiation of NE-induced contraction
Low affinity (nanomolar)	RPMI 1846 cells (hamster) and brain slices (chicken)	G_q/G_{11}	PLC-β	Phosphoinositide hydrolysis

[a]Although both high- and low-affinity binding sites have been found in several other species and tissues, the above list is restricted to those areas where effector or cellular responses have been detected in native tissues or transfected cells. *See text* for details.

nisms involved in this action, but suggest that it may be related to the modulatory effect of melatonin on immune function. The effectiveness of a 10-pM concentration of melatonin suggests the involvement of a high-affinity receptor, which is consistent with recent evidence that high-affinity binding sites for melatonin are present on human lymphocytes (Gonzalez-Haba et al., 1995). However, since almost all of the known high-affinity melatonin receptor subtypes are coupled to G_i, it is not clear how the above potentiating effect is induced. An intriguing possibility is that this effect of melatonin may be mediated by $\beta\gamma$-subunits released from the G protein (G_i?) coupled to its putative lymphocyte receptor. It should be recalled that $\beta\gamma$-subunits can stimulate type II and type IV AC in the presence of activated $G\alpha_s$ (Tang and Gilman, 1991). Of course, there are other mechanisms that may be involved in the effects of melatonin on lymphocyte function, particularly in view of the apparent absence of a phosphodiesterase inhibitor in the above study. For example, melatonin-induced activation of a guanylate cyclase pathway would enhance intracellular levels of cyclic guanosine monophosphate (cGMP) (Vesely, 1981), which can inhibit a cAMP phosphodiesterase, leading to enhanced accumulation of cAMP in some tissues (Goy, 1991). It is also possible that melatonin may bind to calmodulin, thus blocking activation of calcium/calmodulin-dependent phosphodiesterase (Benítez-King et al., 1993).

Modulation of Circadian Rhythmicity

In assessing the functional correlates of AC inhibition by melatonin, it is useful to note the relatively restricted distribution of the G_i-coupled receptors for this hormone in the mammalian brain. The fact that the SCN are primary targets for melatonin in various vertebrate species, coupled with their role in regulating circadian rhythmicity (Moore, 1983; Meijer and Rietveld, 1989) suggests that high-affinity sites in these hypothalamic nuclei may mediate the modulatory effects of melatonin on biological rhythms (Cassone, 1990). Since melatonin signaling in the hypothalamus/SCN clearly involves the AC pathway, it is important to consider the potential role of this pathway in circadian function.

Earlier studies have shown that forskolin-induced activation of AC can produce either delayed or advanced phase shifts in the circadian rhythm of spontaneous nerve impulses in the eye of *Aplysia,* depending on the time of drug treatment (Eskin and Takahashi, 1983). Similar effects were produced by serotonin, which also stimulates AC activity causing an increase in cAMP production. Moreover, the cAMP analog 8-benzylthio-cyclic AMP mimics the effect of serotonin, whereas phosphodiesterase inhibitors potentiate this effect (Eskin et al., 1982), indicating an important role for cAMP in the regulation of circadian function.

There is considerable evidence that the SCN are involved in regulating the well-documented diurnal rhythm in melatonin production by the pineal gland (Brown and Niles, 1982; Reiter, 1991). Melatonin, in turn, may modulate SCN activity by acting directly on this circadian oscillator (McArthur et al., 1991). Interestingly, when applied to SCN-containing hypothalamic slices in vitro, melatonin (at an appropriately low concentration of 1 nM) induced a significant advance in the electrical activity rhythm in the SCN, only when added late in the subjective day or early subjective night (McArthur et al., 1991). It is noteworthy that the phase-shifting effect of melatonin was most pronounced late in the day, before the day–night transition, when the high-affinity G_i-coupled receptors for this hormone are at their maximal density and sensitivity in the SCN (Tenn and Niles, 1993). Since these receptors mediate the inhibitory effect of melatonin on cAMP production, it seems reasonable to assume that a decrease in the intracellular concentration of this second messenger is involved in the above phase-shifting effect. In accordance with the foregoing, cAMP and its analogs have been shown to reset the SCN circadian clock in vitro (Gillette and Prosser, 1988; Prosser and Gillette, 1989).

Another second messenger, cGMP, which can be modulated by melatonin (Vesley, 1981; Vacas et al., 1981), has also been implicated in the regulation of the circadian pacemaker. It was found that cGMP analogs can reset the phase of SCN oscillation in vitro only when applied during the subjective night of the circadian cycle (Prosser et al., 1989). A comparison of the in vitro effects of cAMP and cGMP analogs indicated a difference of about 12 h in the periods of SCN sensitivity to these second messengers, which appear to utilize distinct biochemical pathways in modulating the SCN clock (Prosser et al., 1989). Since melatonin stimulates cGMP production and its circulating levels are elevated at night, when the cGMP pathway is involved in regulating pacemaker activity, it appears that this pathway may also be involved in mediating the modulatory effects of melatonin on circadian physiology.

Modulation of Electrophysiological Activity

Several studies have established that melatonin can alter the electrical activity of neurons in the SCN and other brain areas. The predominant effect in the rat SCN appears to be inhibitory (Shibata et al., 1989; Stehle et al., 1989). A similar effect has been observed in the rat amygdala, whereas stimulatory and/or inhibitory effects have been detected in areas, including the somatosensory cortex, caudate nucleus, medial thalamus, and mesencephalic reticular formation (Naranjo-Rodríguez et al., 1991; Castillo-Romero et al., 1993). In another iontophoretic study, melatonin (1 μM), and its analog, 2-iodomelatonin (100 nM), inhibited the spontaneous firing of single neurons in the rabbit parietal cortex (Stankov et al., 1992). Interestingly, both of these indoleamines also potentiated the inhibitory effect of γ-aminobutyric acid (GABA). This is reminiscent of the allosteric enhancement of GABAergic activity by large IP doses of melatonin and its analogs, which can interact with BZ sites on the BZ/GABA receptor complex (Niles, 1991; Tenn and Niles, 1995). Although high-affinity G_i-coupled receptors, which mediate melatonin-induced suppression of cAMP production, are relatively enriched in this tissue (Stankov et al., 1992), it is not clear whether these receptors are involved in the inhibitory action of melatonin on neuronal firing in the rabbit cortex. However, in view of the regulatory effects of cAMP, cGMP, and calcium on neuronal activity (Huganir, 1987; Nestler and Duman, 1994), it is possible that modulation of the intracellular levels of one or more of these second messengers by melatonin is involved.

Central administration of selective antagonists or receptor–G protein uncoupling agents, such as pertussis toxin, prior to electrophysiological studies, may clarify whether G_i-coupled melatonin receptors are involved in the modulatory effects of this hormone in areas like the rodent SCN and rabbit cortex, where at least one subtype of this receptor is known to exist. In brain areas where the G_i-coupled receptor is not present, other putative melatonin receptors, such as the high-affinity receptor present on cerebral and caudal arteries (Viswanathan et al., 1990), or the nanomolar-affinity receptor, which is coupled to phosphoinositide metabolism (Eison and Mullins, 1993), may be involved in mediating the electrophysiological effects of melatonin. Since the high-affinity arterial sites appear to be coupled to a G protein, but not to inhibition AC (Viswanathan et al., 1990; Capsoni et al., 1993), the transduction systems involved may be clarified by selective blockade of potential G protein candidates, such as G_o and G_s, which are involved in modulating ion channel activity (Birnbaumer et al., 1990). In this regard, experimental approaches utilizing antisense oligonucleotides designed to block expression of the α-subunits of specific G proteins selectively (Albert and Morris, 1994; Goetzl et al., 1994) should be beneficial.

As selective antagonists for various melatonin receptor subtypes become available, they will help to clarify the cellular and functional roles of these subtypes. Although several potential melatonin agonists have been described, there is limited information regarding antagonists for high-affinity melatonin receptors (Yous et al., 1992; Spadoni et al., 1993). The putative antagonist, luzindole, blocks the inhibitory effect of melatonin on the calcium-dependent release of dopamine in the rabbit retina (Dubocovich, 1988). However, the finding that luzindole does not antagonize the suppressive effect of melatonin on cAMP production in ovine PT cells (Howell and Morgan, 1991) suggests that it may not be useful as an antagonist at G_i-coupled receptors for melatonin. This remains to be confirmed in various species and tissues where melatonin-induced inhibition of AC and/or cAMP production has been demonstrated. Studies with a series of 2-substituted 5-methoxy-*N*-acyltryptamines have revealed that 2-phenylmelatonin acts as an antagonist in the Syrian hamster gonadal regression model, while exhibiting mixed activity on neuronal activity in the rabbit parietal cortex (Spadoni et al., 1993). In contrast, 2-phenylmelatonin has been found to act as an agonist in the *Xenopus melanophore* assay (Garratt et al., 1995). Future studies with this putative melatonin antagonist should clarify whether the above discrepancies are related to receptor subtype differences in various target tissues. Evidence from binding and functional studies that the nanomolar-affinity binding sites for melatonin are sensitive to prazosin (Pickering and Niles 1989, 1992; Eison and Mullins, 1993) indicates the potential usefulness of this putative antagonist for investigating the possible involvement of these low-affinity sites in the neuronal and other effects of melatonin.

Modulation of Endocrine Function

The SCN have been implicated in the regulation of reproductive function by melatonin, but this view is controversial, since lesions of these nuclei either block (Rusak and Morin, 1976) or do not block (Bittman et al., 1979; Maywood et al., 1990) melatonin-induced gonadal regression in hamsters. Recent microdialysis studies, utilizing infusion of picogram doses of melatonin in Siberian hamsters, suggest involvement of SCN sites and also sites in the paraventricular and reuniens nuclei of the thalamus, in the reproductive effects of melatonin in this species (Badura and Goldman, 1993). The presence of high-affinity binding sites for melatonin in the PT of the rat and seasonal breeders like the hamster and sheep, together with evidence that the functional activity of this tissue is influenced by photoperiod in Djungarian hamsters (Wittkowski et al., 1988), suggests that these sites may mediate the effects of melatonin on endocrine function. Other potential sites include the preoptic area and ventromedial nucleus in the hypothalamus and the anterior paraventricular nucleus of the thalamus, which contain high-affinity sites for melatonin in the Syrian hamster (Williams et al., 1989). Although these sites exhibit binding characteristics that are similar to those in the SCN and PT, their signaling pathways are not known. Further studies are required to clarify which of the above-mentioned sites mediate the reproductive and other endocrine effects of melatonin. In this regard, it would be worthwhile also to consider the potential role of peripheral sites on organs, such as the testis, ovary, adrenal, and thyroid, where melatonin may exert direct effects on endocrine activity (Vesely, 1981; Persengiev, 1992; Yie et al., 1995b).

As discussed earlier, studies with neonatal rat pituitary cells indicate that melatonin-induced inhibition of Ca^{2+} influx may explain its ability to antagonize GnRH stimulation of LH release in these cells (Vanecek and Klein, 1992). Based on their observation that pertussis toxin blocks the action of melatonin, the above authors suggest that melatonin controls $[Ca^{2+}]$ by an action involving an inhibitory GTP-binding regulatory protein. Since the effect of melatonin depends on extracellular Ca^{2+}, it was also suggested that this hormone acts via a G protein to inhibit Ca^{2+} influx or activate outward K^+ conductance, with the resulting hyperpolarization leading to blockade of voltage-sensitive Ca^{2+} channels (Vanecek and Klein, 1992). These are quite plausible mechanisms and raise the question of which receptor is involved. The presence of a high-affinity G_i-coupled receptor, which mediates inhibition of cAMP production by melatonin in neonatal anterior pituitary cells (Vanecek and Vollrath, 1990a), suggests its involvement in Ca^{2+} modulation. There is evidence that a single receptor may be coupled to more than one effector system, via either one or more G proteins (Roerig et al., 1992; Milligan, 1993; Jouneaux et al., 1993). G_i has been implicated in the modulation of both Ca^{2+} and K^+ channels (Freissmuth et al., 1989). Therefore, it is possible that melatonin signaling, via its G_i-coupled receptor, results in modulation of both cAMP synthesis and intracellular Ca^{2+}. However, it should be noted that the targets for pertussis toxin include not only the three known G_i subtypes, but also G_o. Moreover, G_o is more potent than G_i in regulating Ca^{2+} and K^+ channels (Hescheler et al., 1987; Van Dongen et al., 1988). Therefore, its possible involvement in the inhibitory effect of melatonin on Ca^{2+} influx cannot be discounted at present.

Another possibility involves the PLC pathway, which has been implicated in GnRH-induced secretion of LH (Andrews and Conn, 1986; Chang et al., 1987). PLC activation results in calcium release from intracellular stores by IP_3, whereas DAG stimulation of PKC can lead to Ca^{2+} channel activation (Berridge and Irvine, 1989). Therefore, it is possible that melatonin may inhibit the GnRH-induced increase in $[Ca^{2+}]_i$ by inhibiting PLC activity. The ability of melatonin to inhibit GnRH-stimulated incorporation of [³H]glycerol into [³H]diacylglycerol (DAG) via a pertussis toxin-sensitive mechanism, in the anterior pituitary of immature rats (Vanecek and Vollrath, 1990b), supports the foregoing. Other G_i-coupled receptors, such as the serotonin 5-HT1A receptor (Claustre et al., 1988) and the adenosine-A1 receptor (Linden and Delahunty, 1989), can block PLC-induced phosphoinositide metabolism, suggesting the possible involvement of G_i in this action.

The ability of melatonin to inhibit Ca^{2+} influx in some tissues may be related to its potent inhibition of calcium-dependent dopamine release in retinas from rabbit and chick (Dubocovich 1983; Dubocovich and Takahashi, 1987). It is also possible that the cAMP pathway may be involved, since melatonin can suppress AC activity in the chick retina (Niles et al., 1991; Iuvone and Gan, 1995). However, the calcium dependency of dopamine release in the retina suggests that the action of melatonin involves blockade of calcium channel activity and/or antagonism of calmodulin.

Modulation of Gene Expression

The molecular characterization of several G protein-coupled receptors indicates certain basic structural similarities, including the presence of seven hydrophobic transmembrane segments separated by hydrophilic segments, leading to their designation as R7G receptors (Strosberg, 1991). Another characteristic of these R7G receptors is their linkage to intracellular biochemical pathways, which exhibit extensive interaction or crosstalk (Houslay, 1991). This crossregulation has been extensively studied in adrenoreceptor-linked AC systems, where activation of G_s-coupled receptors can lead to alterations in transcription of target receptors and other receptors on the same cell. For example, long-term (more than 4 h) exposure to β-adrenergic agonists causes a decrease in β-adrenergic receptor mRNA levels (Bouvier et al., 1989; Guest et al., 1990). This long-term activation of β-adrenergic receptors linked to stimulation of cAMP synthesis results in desensitization of this pathway, while also enhancing signaling via the inhibitory α2-adrenergic receptor pathway. It appears that sensitization of the inhibitory pathway results from an increase in $G\alpha_{i2}$ mRNA expression (Hadcock et al., 1990) and also upregulation of α2-adrenergic receptor gene expression (Sakaue and Hoffman, 1991). Conversely, prolonged activation of the inhibitory adenylyl cyclase pathway causes an increase in β2-adrenergic receptor expression (Hadcock et al., 1991).

Several studies indicate that cAMP plays an important role in the interaction between its stimulatory and inhibitory pathways. The ability of stimulatory agents to reduce receptor expression clearly involves cAMP, since S49 mouse lymphoma cell variants, which do no synthesize this second messenger, do not downregulate receptor mRNA. Moreover, kin⁻ cells, which lack cAMP-dependent protein kinase activity, synthesize cAMP in response to β-adrenergic agonists or forskolin, but do not show a decrease in receptor mRNA, indicating the importance of this kinase for agonist-induced downregulation of stimulatory receptors (Hadcock and Malbon, 1993). Stimulatory agonists downregulate β-adrenergic receptors via a PKA-dependent destabilization of receptor mRNA (Hadcock and Malbon, 1988; Hadcock et al., 1989). Therefore, it has been suggested that inhibitory agents may enhance β-adrenergic receptor expression by suppressing cAMP levels and, consequently, PKA activity, resulting in an increase in receptor mRNA stability (Hadcock and Malbon, 1993). Since the high-affinity receptor for melatonin, in tissues such as the SCN and PT, is coupled to a G_i-protein, it is very likely that suppression of cAMP and PKA activity by this hormone allows it to influence the expression of various receptors and G-proteins, particularly during its period of peak levels at night.

In addition to the foregoing, melatonin may also modulate gene expression in influencing circadian function. The major role of light in entraining circadian rhythmicity, via the retinohypothalamic projection from the retina to the SCN, is well documented (Meijer and Rietveld, 1989). It is also now known that light is a potent inducer of the expression of immediate early genes (IEGs), like c-*fos* and *jun*-B, in the SCN, and it also increases the activity of the transcription factor AP-1 (Kornhauser et al., 1992), which may be formed from the dimerization of the c-*fos* product, Fos, with c-Jun or other members of the Jun family, or between members of this family (Morgan and Curran, 1991). Evidence that the stimulatory effect of light on the expression of c-*fos* and *jun*-B, and also on AP-1 activity in the SCN is regulated by the circadian clock suggests the involvement of these and perhaps other IEGs and transcription factors in the regulation of circadian function (Takahashi and Kornhauser, 1993). A similar regulatory role for melatonin is suggested by its ability to induce Fos expression in the rat SCN (Kilduff et al., 1992). It remains to be determined whether this effect of melatonin in the SCN is linked to increased AP-1 activity with consequent transcriptional activation and changes in circadian physiology.

Various elements within the 5'-flanking region of the c-*fos* gene can mediate induction by second messengers and other agents (Fisch et al., 1989; Ryan et al., 1989). One such nuclear target, the cAMP response element (CRE), can mediate the stimulatory effect of both cAMP and calcium acting via a CRE binding protein (CREB), which is activated following phosphorylation by either a cAMP-dependent or Ca^{2+}/calmodulin-dependent protein kinase (Dash et al., 1991). This phosphorylation may be affected by melatonin acting via either or both of these pathways, as already discussed. Therefore, it is possible that the effects of melatonin on circadian rhythmicity involve changes in CREB and AP-1 activity, with consequent modulation of gene expression in the SCN. The potential involvement of cGMP in this action of melatonin (Prosser et al., 1989) may be related to its ability to modulate cAMP levels indirectly via either activation or inhibition of specific cAMP phosphodiesterases, and to alter the activity of various phosphoproteins or transcription factors, which are targets for cGMP-dependent kinases (Goy, 1991).

Clinical Implications

Several studies have suggested an association between endocrine abnormalities and psychiatric disorders (Holsboer, 1995). Glucocorticoids, such as cortisol, gonadal steroids, such as estrogen, and thyroid hormones can interact with specific target sites in the CNS to influence neuronal function and behavior. There is also considerable evidence that the pineal hormone, melatonin, can modulate neuroendocrine function and biological rhythmicity, as discussed earlier. Therefore, it is hardly surprising that melatonin has been implicated in mental illness either as a causative agent or biological marker of adrenergic activity (Brown and Niles, 1982; Lewy, 1983; Brown et al., 1985). However, the mechanisms underlying the involvement of melatonin in psychiatric disorders await clarification.

It is now apparent that hormones can exert modulatory effects on G protein function in various target tissues, including the brain (Malbon et al., 1988; Manji, 1992). Since the high-affinity melatonin receptors in the human SCN (hypothalamus) and hippocampus are coupled to G proteins, their function may be modulated by various hormones or neurotransmitters, which can alter G protein expression. This view is supported by evidence that although the marked circadian rhythm in high-affinity G protein-coupled melatonin receptor density is inversely related to the circulating levels of this hormone, other factors appear also to be involved in regulating high-affinity binding in the SCN (Tenn and Niles, 1993). The major glucocorticoid in rodents, corticosterone, has been shown to inhibit the expression of $G\alpha_i$ while enhancing that of $G\alpha_s$ in the rat brain (Saito et al., 1989). Presumably, the major glucocorticoid in humans, cortisol, may exert similar effects on these G proteins, with consequent changes in the function of associated receptors. Therefore, glucocorticoid excess, as occurs in Cushing's syndrome, may adversely affect G protein function with consequent changes in brain physiology and behavior (Dubrovsky, 1993). Similarly, abnormalities in other endocrine systems, such as the hypothalamic–pituitary–thyroid or gonadal axes, may also influence the activity of G protein-coupled receptors (Levine et al., 1990; Maus et al., 1990), including those of melatonin. As discussed earlier, melatonin plays a role in modulating circadian rhythmicity. Therefore, drug-, hormone-, or neurotransmitter–induced changes in the function of its G protein-coupled receptors in the SCN, the locus of the circadian clock, may affect biological rhythms. Since disturbances in endocrine and other biological rhythms are often associated with affective disorders (Wirz-Justice, 1995; Hallonquist et al., 1986), abnormalities in melatonin signaling may contribute to depression and related illnesses. Clearly, studies are required to clarify the interplay in the SCN between melatonin and various hormones and neurotransmitters, such as serotonin, which also plays an important role in circadian rhythmicity and affective disorders (Maes and Meltzer, 1995). It would also be beneficial to examine the effects of various antidepressants, including lithium, on melatonin receptor expression and function. Lithium has been implicated in the modulation of multiple G protein-coupled pathways in the CNS, leading to the suggestion that its effectiveness as a mood stabilizer involves direct effects on G protein activity (Manji, 1992). Indeed, lithium has been found to alter the affinity of the melatonin receptor in the rat and chick brain (Laitinen and Saavedra, 1990a; Niles et al., 1994). Thus, similar modulatory effects on the activity of this receptor in the human brain may be involved in the circadian phase-shifting and psychotropic effects of this agent.

Over the past several months, there has been a barrage of mass media reports about the potential benefits of melatonin for the treatment of a variety of health-related problems, including Alzheimer's disease, aging, autism, depression, and heart disease (Turek, 1996). Although there is limited experimental or clinical support for the numerous benefits attributed to melatonin, there is increasing evidence that it can alleviate sleep disorders. Studies utilizing various pharmacological doses of melatonin have demonstrated that it produces sedative effects in humans (MacFarlane et al., 1991; Dollins et al., 1993, 1994). The suggestion that the nocturnal increase in serum melatonin may provide a physiological signal for sleep onset (Dollins et al., 1994) is supported by the presence of decreased malatonin levels in long-term insomniacs (Hajak et al., 1995). Moreover, consistent with the age-related decline in circulating melatonin levels (Waldhouser et al., 1988), controlled-release melatonin replacement therapy has been found to improve sleep quality in elderly insomniacs significantly (Garfinkel et al., 1995). An important question concerns the mechanism(s) underlying the sedative action of melatonin. The effectiveness of relatively low melatonin doses, which produce normal nocturnal levels (Dollins et al., 1994), suggests that physiological effects mediated by high-affinity binding sites in the CNS are involved. Moreover, multiple mechanisms may be involved in the sleep-inducing action of melatonin. It has been suggested that long-term treatment for at least 3 wk may be necessary to resynchronize the sleep/wake cycle with day/night cycles in elderly insomniacs, whose circulating melatonin levels and receptor densities in the SCN circadian clock are presumably decreased (Garfinkel et al., 1995). Conversely, the acute sleep-inducing effects observed in young healthy subjects (Dollins et al., 1994) could involve not only G protein-coupled receptors in the SCN, hippocampus, and perhaps elsewhere in the brain, but also the ubiquitous intracellular calmodulin binding sites, which exhibit high affinity for this lipophilic hormone (Benítez-King et al., 1993). Since melatonin is a potent calmodulin antagonist, it may acutely suppress cellular activity by impeding the function of this important mediator of calcium signaling in various brain areas, including the ascending reticular activating system, with consequent tranquilization and sleep induction. It is likely that these mechanisms are also involved in the circadian phase-shifting effects of melatonin in humans and its ability to alleviate jet lag (Arendt et al., 1995).

In recent years, there have been major advances in knowledge of how signals, initiated by the binding of diverse hormones, neurotransmitters, and other agents to G protein-coupled receptors are specifically recognized and transduced at the cellular level. These advances have usually been spearheaded by the application of molecular approaches to characterize and unequivocally identify distinct receptor subtypes for various neurotransmitters and hormones. Notable examples include the cloned subtypes of the adrenergic, serotonergic, muscarinic, and dopaminergic receptors (Strosberg, 1991; Kobilka, 1992; Seeman and Van Tol, 1994). Site-directed mutagenesis together with truncation and deletion studies have led to a better understanding of the structural domains involved in receptor binding, G protein coupling, and receptor desensitization (Houslay, 1992). Other strategies, such as the antisense knockout of specific receptor subtypes, G proteins, or other cellular targets, provide novel and selective tools for the elucidation of signaling pathways and receptor function (Albert and Morris, 1994; Wahlestedt, 1994).

With the recent cloning of high-affinity G protein-coupled melatonin receptors from various vertebrate species (Ebisawa et al., 1994, Reppert et al., 1994, 1995a,b; Liu et al., 1995), it is now possible to examine the modulatory effects of diverse hormones and other agents on the expression and function of these receptors. Conversely, it is also likely that melatonin will be found to modulate the expression of its own receptors and associated G proteins, as well as those for various neurotransmitters, hormones, and other biological messengers, via the crossregulatory mechanisms often utilized by G protein-coupled receptors (Hadcock and Malbon, 1993; Milligan, 1993). Ultimately, the true benefits of this anticipated knowledge will derive from its integration into an overall understanding of the multiple physiological actions of melatonin and its potential as an important therapeutic agent.

References

Albert, P. R. and Morris, S. J. (1994) Antisense knockouts: molecular scalpels for the dissection of signal transduction. *Trends Pharmacol. Sci.* **15,** 250–254.

Andrews, W. V. and Conn, P. M. (1986) Gonadotropin-releasing hormone stimulates mass changes in phosphoinositides and diacylglycerol accumulation in purified gonadotrope cell cultures. *Endocrinology* **118,** 1148–1158.

Arendt, J., Deacon, S., English, J., Hampton, S., and Morgan, L. (1995) Melatonin and adjustment to phase shift. *J. Sleep Res.* **4 (Suppl. 2),** 74–79.

Bading, H., Ginty, D. D., and Greenberg, M. E. (1993) Regultion of gene expression in hippocampal neurons by distinct calcium signalling pathways. *Science* **260,** 181–186.

Badura, L. L. and Goldman, B. D. (1993) Central sites mediating reproductive responses to melatonin in juvenile male Siberian hamsters. *Brain Res.* **598,** 98–106.

Benítez-King, G. and Antón-Tay, F. (1993) Calmodulin mediates melatonin cytoskeletal effects. *Experientia* **49,** 635–641.

Benítez-King, G., Huerto-Delgadillo, L., and Antón-Tay, F. (1993) Binding of ^{3}H-melatonin to calmodulin. *Life Sci.* **53,** 201–207.

Berridge, M. J. (1987) Inositol trisphosphate and diacylglycerol:two interacting second messengers. *Annu. Rev. Biochem.* **56,** 159–193.

Berridge, M. J. and Irvine, R. F. (1989) Inositol phosphates and cell signalling. *Nature* **341,** 197–205.

Berstein, G., Blank, J. L., Jhon, D.-Y., Exton, J. H., Rhee, S. G., and Ross, E. M. (1992) Phospholipase C-β1 is a GTPase-activating protein for $G_{q/11}$, its physiologic regulator. *Cell* **70,** 411–418.

Birnbaumer, L., Mattera, R., Yatani, A., Van Dongen, A., Graf, R., Sanford, J., Codina, J., and Brown, A. M. (1990) Roles of G-proteins in coupling of receptors to ionic channels, in *Transmembrane Signalling, Intracellular Messengers and Implications for Drug Development* (Nahorski, S. R., ed.), John Wiley, New York, pp. 43–71.

Birnbaumer, L., Perez-Reyes, E., Bertrand, P., Gudermann, T., Wei, X.-Y., Kim, H., Castellano, A., and Codina, J. (1991) Molecular diversity and function of G-proteins and calcium channels. *Biol. Reprod.* **44,** 207–224.

Bittman, E. L., Goldman, B. D., and Zucker, I. (1979) Testicular responses to melatonin are altered by lesions of the suprachiasmatic nuclei in golden hamsters. *Biol. Reprod.* **21,** 647–656.

Bourne, H. R., Sanders, D. A., and McCormick, F. (1990) The GTPase superfamily: a conserved switch for diverse cell functions. *Nature* **348,** 125–132.

Bouvier, M., Collins, S., O'Dowd, B. F., Campbell, P. T., de Blasi, A., Kobilka, B. K., MacGregor, G. P., Caron, M. G., and Lefkowitz, R. J. (1989) Two distinct pathways for cAMP-mediated down-regulation of the β_2-adrenergic receptor. *J. Biol. Chem.* **264,** 16,786–17,792.

Boyer, J. L., Hepler, J. R., and Harden, T. K. (1989) Hormone and growth factor receptor-mediated regulation of phospholipase C activity. *Trends Pharmacol. Sci.* 360–364.

Brown, A. M. and Birnbaumer, L. (1988) Direct G protein gating of ion channels. *Am. J. Physiol.* **23,** H401–H410.

Brown, G. M. and Niles, L. P. (1982) Studies on melatonin and other pineal factors, in *Clinical Neuroendocrinology,* (Martini, L. and Besser, G. M., eds.), vol. 2, Academic Press, New York, pp. 205–264.

Brown, R., Kocsis, J., Caroff, S., Amsterdam, J., Winokur, A., Stokes, P., and Frazer, A. (1985) Differences in nocturnal melatonin secretion between melancholic depressed patients and control subjects. *Am. J. Psychiat.* **142,** 811–816.

Brzezinski, A., Seibel, M. M., Lynch, H. J., Deng, M.-H., and Wurtman, R. J. (1987) Melatonin in preovulatory follicular fluid. *J. Clin. Endocrinol. Metab.* **64,** 865–867.

Camps, M., Carozzi, A., Schnabel, P., Scheer, A., Parker, P. J. and Gierschik, P. (1992) Isozyme-selective stimulation of phospholipase C-β2 by G protein βγ-subunits. *Nature* **360,** 684–686.

Capsoni, S., Viswanathan, M., De Oliveira, A. M., and Saavedra, J. M. (1993) Melatonin binding sites in rat circle of Willis are linked to a G protein. *Soc. Neurosci. Abstract* **19,** 1388.

Carlson, L. L., Weaver, D. R., and Reppert, S. M. (1989) Melatonin signal transduction in hamster brain: Inhibition of adenylyl cyclase by a pertussis toxin-sensitive G protein. *Endocrinology* **125,** 2670–2676.

Cassone, V. M. (1990) Effects of melatonin on vertebrate circadian systems. *Trends Neurosci.* **13,** 457–464.

Castillo-Romero, J. L., Vives-Montero, F., Reiter, R. J., and Acuña-Castroviejo, D. (1993) Pineal modulation of the rat caudate-putamen spontaneous neuronal activity: roles of melatonin and vasotocin. *J. Pineal Res.* **15,** 147–152.

Catt, K. J. and Stojilkovic, S. S. (1989) Calcium signalling and gonadotropin secretion. *Trends Endocrinol. Metab.* **1,** 15–20.

Chang, J. P., Graeter, J. and Catt, K. J. (1987) Dynamic actions of arachidonic acid and protein kinase C in pituitary stimulation by gonadotropin releasing hormone. *Endocrinology* **120,** 1837–1845.

Choi, W. C., Gerfen, C. R., Suh, P. G., and Rhee, S. G. (1989) Immunohistochemical localization of a brain isozyme of phospholipase C (PLC III) in astroglia in rat brain. *Brain Res.* **499,** 193–197.

Claustre, Y., Benavides, J., and Scatton, B. (1988) 5-HT$_{1A}$ receptor agonists inhibit carbachol-induced stimulation of phosphoinositide turnover in the rat hippocampus. *Eur. J. Pharmacol.* **149,** 149–153.

Cockcroft, S. (1987) Polyphosphoinositide phosphodiesterase: regulation by a novel guanine nucleotide binding protein, Gp. *Trends Biochem. Sci.* **12,** 75–78.

Coloma, F. M. and Niles, L. P. (1988) Melatonin enhancement of [^{3}H]GABA and [^{3}H]muscimol binding in rat brain. *Biochem. Pharmacol.* **37,** 1271–1274.

Dash, P. K., Karl, K. A., Colicos, M. A. Prywes, R., and Kandel, E. R. (1991) cAMP response element-binding protein is activated by Ca^{2+}/calmodulin—as well as cAMP-dependent protein kinase. *Proc. Natl. Acad. Sci. USA* **88,** 5061–5065.

Davis, T. N. (1992) What's new with calcium? *Cell* **71,** 557–564.

Dogliotti, L., Berruti, A., Buniva, T. Torta, M., Bottini, A., Tapelltini, M., Terzolo, M., Faggiuolo, R., and Angeli, A. (1990) Melatonin and human cancer. *J. Steroid Biochem. Mol. Biol.* **37,** 983–987.

Dollins, A. B., Lynch, H. J., Wurtman, R. J., Deng, M. H. Kischka, K. U. Gleason, R. E., and Lieberman, H. R. (1993) Effect of pharmacological daytime doses of melatonin on human mood and performance. *Psychopharmacology* **112,** 490–496.

Dollins, A. B., Zhdanova, I. V., Wurtman, R. J., Lynch, H. J. and Deng, M. H. (1994) Effect of inducing noctural serum melatonin concentrations in daytime on sleep, mood, body temperature, and performance. *Proc. Natl. Acad. Sci. USA* **91,** 1824–1828.

Dubocovich, M. L. (1983) Melatonin is a potent modulator of dopamine release in the retina. *Nature* **306,** 782–784.

Dubocovich, M. L. (1988) Luzindole (N-0774): a novel melatonin receptor antagonist. *J. Pharmacol. Exp. Ther.* **246,** 902–910.

Dubocovich, M. L. and Takahashi, J. S. (1987) Use of 2-[^{125}I]iodomelatonin to characterize melatonin binding sites in chicken retina. *Proc. Natl. Acad. Sci. USA* **84,** 3916–3920.

Dubrovsky, B. (1993) Effects of adrenal cortex hormones on limbic structures: Some experimental and clinical correlations related to depression. *J. Psychiatr. Neurosci.* **18,** 4–16.

Duncan, M. J., Takahashi, J. S., and Dubocovich, M. L. (1988) 2-[^{125}I]iodomelatonin binding sites in hamster brain membranes: pharmacological characteristics and regional distribution. *Endocrinology* **122,** 1825–1833.

Eason, M. G., Kurose, H., Holt, B. D., Raymond, J. R., and Liggett, S. B. (1992) Simultaneous coupling of α_2-adrenergic receptors to two G proteins with opposing effects: subtype-selective coupling of α_2C10, α_2C4 and α_2C2 adrenergic receptors to G_i and G_s. *J. Biol. Chem.* **267,** 15,795–15,801.

Ebisawa, T., Karne, S., Lerner, M. R., and Reppert, S. M. (1994) Expression cloning of a high-affinity melatonin receptor from *Xenopus* dermal melanophores. *Proc. Natl. Acad. Sci. USA* **91,** 6133–6137.

Eison, A. S. and Mullins, U. L. (1993) Melatonin binding sites are functionally coupled to phosphoinositide hydrolysis in Syrian hamster RPMI 1846 melanoma cells. *Life Sci.* **53,** 393–398.

Eskin, A. and Takahashi, J. S. (1983) Adenylate cyclase activation shifts the phase of a circadian pacemaker. *Science* **220,** 82–84.

Eskin, A., Corrent, G., Lin, C.-Y. and McAdoo, D. J. (1982) Mechanism of shifting the phase of a circadian oscillator by serotonin: involvement of cAMP. *Proc. Natl. Acad. Sci. USA* **79,** 660–664.

Exton, J. H. (1994) Phosphoinositide phospholipases and G proteins in hormone action. *Annu. Rev. Physiol.* **56,** 349–369.

Faillace, M. P., Keller Sarmiento, M. I., Siri, L. N., and Rosenstein, R. E. (1994) Diurnal variations in cyclic AMP and melatonin content of golden hamster retina. *J. Neurochem.* **62,** 1995–2000.

Fain, J. N., Wallace, M. A., and Wojcikiewicz, R. J. H. (1988) Evidence for involvement of guanine nucleotide-binding regulatory proteins in the activation of phospholipases by hormones. *FASEB J.* **2,** 2569–2574.

Federman, A. D., Conklin, B. R., Schrader, K. A., Reed, R. R., and Bourne, H. R. (1992) Hormonal stimulation of adenylyl cyclase through G_i-protein $\beta\gamma$ subunits. *Nature* **356,** 159–161.

Fisch, T. M., Prywes, R., Simon, M. C., and Roeder, R. G. (1989) Multiple sequence elements on the c-*fos* promoter mediate induction by cAMP. *Genes Dev.* **3,** 198–211.

Freissmuth, M., Casey, P. J., and Gilman, A. G. (1989) G Proteins control diverse pathways of transmembrane signalling. *FASEB J.* **3,** 2125–2131.

Gao, B. and Gilman, A. G. (1991) Cloning and expression of a widely distributed (type IV) adenylyl cyclase. *Proc. Natl. Acad. Sci. USA* **88,** 10,179–10,182.

Garfinkel, D., Laudon, M., Nof, D., and Zisapel, N. (1995) Improvement of sleep quality in elderly people by controlled-release melatonin. *Lancet* **346,** 541–544.

Garratt, P. J., Jones, R. And Tocher, D. A. (1995) Mapping the melatonin receptor. 3. Design and synthesis of melatonin agonists and antagonists derived from 2-phenyltryptamines. *J. Med. Chem.* **38,** 1132–1139.

Gauer, F., Masson-Pévet, M., and Pévet, P. (1992) Pinealectomy and constant illumination increase the density of melatonin binding sites in the pars tuberalis of rodents. *Brain Res.* **575,** 32–38.

Gauer, F., Masson-Pévet, M., and Pévet, P. (1993) Melatonin receptor density is regulated in rat pars tuberalis and suprachiasmatic nuclei by melatonin itself. *Brain Res.* **602,** 153–156.

Gerfen, C. R., Choi, W. C., Suh, P. G., and Rhee, S. G. (1988) Phospholipase CI and II brain isozymes: immunohistochemical localization in neuronal systems in rat brain. *Proc. Natl. Acad. Sci. USA* **85,** 3208–3212.

Gillette, M. U. and Prosser, R. A. (1988) Circadian rhythm of the rat suprachiasmatic brain slice is rapidly reset by daytime application of cAMP analogs. *Brain Res.* **474,** 348–352.

Girard, P. R., Mazzei, G. J., and Kuo, J. F. (1986) Immunological quantitation of phospholipid/Ca^{2+}-dependent protein kinase and its fragments. *J. Biol. Chem.* **261,** 370–375.

Gnegy, M. E. (1993) Calmodulin in neurotransmitter and hormone action. *Annu. Rev. Pharmacol. Toxicol.* **32,** 45–70.

Goetzl, E. J., Shames, R. S., Yang, J., Birke, F. W., Lui, Y. F., Albert, P. R., and An, S. (1994) Inhibition of human HL-60 cell responses to chemotactic factors by antisense messenger RNA depletion of G proteins. *J. Biol. Chem.* **269,** 809–812.

Golombek, D. A., Martini, M., and Cardinali, D. P. (1993) Melatonin as an anxiolytic in rats: time dependence and interaction with the central GABAergic system. *Eur. J. Pharmacol.* **237,** 231–236.

Gonzales, R. A. and Crews, F. T. (1985) Guanine nucleotides stimulate production of inositol triphosphate in rat cortical membranes. *Biochem. J.* **232,** 799–804.

Gonzalez-Haba, M. G., Garcia-Maurino, S., Calvo, J. R., Goberna, R., and Guerrero, J. M. (1995) High-affinity binding of melatonin by human circulating T lymphocytes (CD4+). *FASEB J.* **9,** 1331–1335.

Goy, M. F. (1991) cGMP: the wayward child of the cyclic nucleotide family. *Trends Neurosci.* **14,** 293–299.

Green, A., Johnson, J. L., and Milligan, G. (1990) Down-regulation of G_i subtypes by prolonged incubation of adipocytes with an A_1 adenosine receptor agonist. *J. Biol. Chem.* **265,** 5206–5210.

Gudermann, T., Kalkbrenner, F., and Schultz, G. (1996) Diversity and selectivity of receptor-G-protein interaction. *Annu. Rev. Pharmacol. Toxicol.* **36,** 429–459.

Guest, S. J., Hadcock, J. R., Watkins, D. C., and Malbon, C. C. (1990) β_1- and β_2-adrenergic receptor expression in differentiating 3T3-L1 cells: independent regulation at the level of mRNA. *J. Biol. Chem.* **265,** 5370–5375.

Gutowski, S., Smrcka, A., Nowak, L., Wu, D., Simon, M., and Sternweis, P. C. (1991) Antibodies to the α_q subfamily of guanine nucleotide-binding regulatory protein α-subunits attenuate activation of phosphatidylinositol 4,5-biphosphate hydrolysis by hormones. *J. Biol. Chem.* **266,** 20,519–20,524.

Hadcock, J. R. and Malbon, C. C. (1988) Down-regulation of β-adrenergic receptors: agonist-induced reduction in receptor mRNA levels. *Proc. Natl. Acad. Sci. USA* **85,** 5021–5025.

Hadcock, J. R. and Malbon, C. C. (1993) Agonist regulation of gene expression of adrenergic receptors and G proteins. *J. Neurochem.* **60,** 1–9.

Hadcock, J. R., Wang, H.-Y. and Malbon, C. C. (1989) Agonist-induced destabilization of β-adrenergic receptor mRNA. Attenuation of glucocorticoid-induced up-regulation of β-adrenergic receptors. *J. Biol. Chem.* **264,** 19,928–19,933.

Hadcock, J. R., Ros, M., Watkins, D. C., and Malbon, C. C. (1990) Cross-regulation between G protein-mediated pathways: stimulation of adenylyl cyclase increases expression of the inhibitory G protein, $G\alpha_{i2}$. *J. Biol. Chem.* **265,** 14,784–14,790.

Hadcock, J. R., Port, J. D., and Malbon, C. C. (1991) Cross-regulation between G protein-mediated pathways: activation of the inhibitory pathway of adenylyl cyclase increases the expression of β_2-adrenergic receptors. *J. Biol. Chem.* **266,** 11,915–11,922.

Hajak, G., Rodenbeck, A., Staedt, J., Bandelow, B., Huether, G., and Rüther, E. (1995) Nocturnal plasma melatonin levels in patients suffering from chronic primary insomnia. *J. Pineal Res.* **19,** 116–122.

Hallonquist, J. D., Goldbert, M. A., and Brandes, J. S. (1986) Affective disorders and circadian rhythms. *Can. J. Psychiat.* **31,** 259–272.

Hardeland, R., Reiter, R. J., Poeggeler, B., and Tan, D.-X. (1993) The significance of the metabolism of the neurohormone melatonin: antioxidative protection and formation of bioactive substances. *Neurosci. Biobehav. Rev.* **17,** 347–357.

Hausdorff, W. P., Caron, M. G., and Lefkowitz, R. J. (1990) Turning off the signal: desensitization of β-adrenergic receptor function. *FASEB J.* **4,** 2881–2889.

Hazlerigg, D. G., Gonzalez-Brito, A., Lawson, W., Hastings, M. H., and Morgan, P. J. (1993) Prolonged exposure to melatonin leads to time-dependent sensitization of adenylate cyclase and down-regulates melatonin receptors in pars tuberalis cells from ovine pituitary. *Endocrinology* **132**, 285–292.

Hepler, J. R. and Gilman, A. G. (1992) G-proteins. *Trends Biochem. Sci.* 383–387.

Hescheler, J., Rosenthal, W., Trautwein, W., and Schultz, G. (1987) The GTP-binding protein, G_0, regulates neuronal calcium channels. *Nature* **325**, 445–447.

Hill, S. M. and Blask, D. E. (1988) Effects of the pineal hormone melatonin on the proliferation and morphological characteristics of human breast cancer cells (MCF-7) in culture. *Cancer Res.* **48**, 6121–6126.

Holsboer, F. (1995) Neuroendocrinology of mood disorders, in *Psychopharmacology: The Fourth Generation of Progress* (Bloom, F. E. and Kupfer, D. J., eds.), Raven, New York, pp. 957–969.

Houslay, M. D. (1991) 'Crosstalk': a pivotal role for protein kinase C in modulating relationships between signal transduction pathways. *Eur. J. Biochem.* **195**, 9–27.

Houslay, M. D. (1992) G protein linked receptors: a family probed by molecular cloning and mutagenesis procedures. *Clin. Endocrinol.* **36**, 525–534.

Howell, H. E. and Morgan, P. J. (1991) Luzindole (2-benzyl-N-acetyltryptamine), 5-methoxytryptamine, N-acetyltryptamine and 6-methoxy-2-benzoxazolinone activity in ovine pars tuberalis cells, in *Advances in Pineal Research*, vol. 5 (Arendt, J. and Pévet, P., eds.), John Libbey, London, pp. 205–207.

Hu, Z.-H., Yuan, H., Lu, Y., and Pang, S. F. (1991) [^{125}I]iodomelatonin binding sites in spleens of birds and mammals. *Neurosci. Lett.* **125**, 175–178.

Huckle, W. R. and Conn, P. M. (1988) Molecular mechanism of gonadotropin releasing hormone action. II. The effector system. *Endocrine Rev.* **9**, 387–395.

Huganir, R. L. (1987) Biochemical mechanisms that regulate the properties of ion channels, in *Neuromodulation: The Biochemical Control of Neuronal Excitability* (Kaczmarek, L. K. and Levitan, I. B., eds.), Oxford University Press, New York, pp. 64–85.

Iniguez-Lluhi, J., Kleuss, C., and Gilman, A. G. (1993) The importance of G-protein βγ subunits. *Trends Cell Biol.* **3**, 230–236.

Iuvone, P. M. and Gan, J. (1995) Functional interaction of melatonin receptors and D1 dopamine receptors in cultured chick retinal neurons. *J. Neurosci.* **15**, 2179–2185.

Jakobs, K. H., Bauer, S., and Watanabe, Y. (1985) Modulation of adenylate cyclase of human platelets by phorbol ester. Impairment of the hormone-sensitive inhibitory pathway. *Eur. J. Biochem.* **151**, 425–430.

Jouneaux, C., Goldsmith, P., Hanoune, J. and Lotersztajn, S. (1993) Endothelin inhibits the calcium pump and stimulates phosphoinositide phospholipase C in liver plasma membranes via two different G proteins, G_s and G_q. *J. Cardiovasc. Pharmacol.* **22**, S158–S160.

Katada, T., Gilman, A. G., Watanabe, Y., Bauer, S., and Jakobs, K. H. (1985) Protein kinase C phosphorylates the inhibitory guanine-nucleotide-binding regulatory component and apparently suppresses its function in hormonal inhibition of adenylate cyclase. *Eur. J. Biochem.* **151,** 431–437.

Kaziro, Y., Itoh, H., Kozasa, T., Nakafuku, M., and Satoh, T. (1991) Structure and function of signal-transducing GTP-binding proteins. *Annu. Rev. Biochem.* **60,** 349–400.

Kikkawa, U., Takai, Y., Minakuchi, R., Inohara, S., and Nishizuka, Y. (1982) Calcium-activated, phospholipid-dependent protein kinase from rat brain. *J. Biol. Chem.* **257,** 13,341–13,348.

Kilduff, T. S., Landel, H. B., Nagy, G. S., Sutin, E. L., Dement, W. C., and Heller, H. C. (1992) Melatonin influences Fos expression in the rat suprachiasmatic nucleus. *Mol. Brain Res.* **16,** 47–56.

Kobilka, B. (1992) Adrenergic receptors as models for G protein-coupled receptors. *Annu. Rev. Neurosci.* **15,** 87–114.

Kornhauser, J. M., Nelson, D. E., Mayo, K. E., and Takahashi, J. S. (1992) Regulation of *jun*-B messenger RNA and AP-1 activity by light and a circadian clock. *Science* **255,** 1581–1584.

Kosugi, S., Okajima, F., Ban, T., Hidaka, A., Shenker, A., and Kohn, L. D. (1992) Mutation of alanine 623 in the third cytoplasmic loop of the rat thyrotropin (TSH) receptor results in a loss in the phosphoinositide but not cAMP signal induced by TSH and receptor autoantibodies. *J. Biol. Chem.* **267,** 24,153–24,156.

Krause, D. N. and Dubocovich, M. L. (1991) Melatonin receptors. *Annu. Rev. Pharmacol. Toxicol.* **31,** 549–568.

Krupinski, J., Coussen, F., Bakalyar, H. A., Tang, W. J., Feinstein, P. G., Orth, K., Slaughter, C., Reed, R. R., and Gilman, A. G. (1989) Adenylyl cyclase amino acid sequence: possible channel- or transporter-like structure. *Science* **244,** 1558–1564.

Laitinen, J. T. and Saavedra, J. M. (1990a) Characterization of melaton in receptors in the rat suprachiasmatic nuclei: Modulation of affinity with cations and guanine nucleotides. *Endocrinology* **126,** 2110–2115.

Laitinen, J. T. and Saavedra, J. M. (1990b) The chick retinal melatonin receptor revisited: localization and modulation of agonist binding with guanine nucleotides. *Brain Res.* **528,** 349–352.

Laitinen, J. T., Castren, E., Vakkuri, O., and Saavedra, J. M. (1989) Diurnal rhythm of melatonin binding in the rat suprachiasmatic nucleus. *Endocrinology* **124,** 1585–1587.

Laitinen, J. T., Viswanathan, M., Vakkuri, O., and Saavedra, J. M. (1992) Differential regulation of the rat melatonin receptors: Selective age-associated decline and lack of melatonin-induced changes. *Endocrinology* **130,** 2139–2144.

Lee, K.-Y., Ryu, S. H., Suh, P.-G., Choi, W. C., and Rhee, S. G. (1987) Phospholipase C associated with particulate fractions of bovine brain. *Proc. Natl. Acad. Sci. USA* **84,** 5540–5544.

Lefkowitz, R. J., Hausdorff, W. P., and Caron, M. G. (1990) Role of phosphorylation in desensitization of the β-adrenoceptor. *Trends Pharmacol. Sci.* **11**, 190–194.

Levine, M. A., Feldman, A. M., Robishaw, J. D., Ladenson, P. W., Ahn, T. G., Moroney, J. F., and Smallwood, P. M. (1990) Influence of thyroid hormone status on expression of genes encoding G-protein submits in the rat heart. *J. Biol. Chem.* **265**, 3553–3560.

Lewy, A. J. (1983) Biochemistry and regulation of mammalian melatonin production, in *The Pineal Gland* (Relkin, R., ed.), Elsevier Biomedical, New York, pp. 77–128.

Linden J. and Delahunty, T. M. (1989) Receptors that inhibit phosphoinositide breakdown. *Trends Pharmacol. Sci.* **10**, 114–120.

Lissoni, P., Viviani, S., and Bajetta, E. (1986) A clinical study of the pineal gland activity in oncologic patients. *Cancer* **57**, 837–842.

Litosch, I. and Fain, J. N. (1986) Regulation of phosphoinositide breakdown by guanine nucleotides. *Life Sci.* **39**, 187–194.

Liu, F., Yuan, H., Sugamori, K. S., Hamadanizadeh, A., Lee, F. J. S., Pang, S. F., Brown, G. M., Pristupa, Z. B., and Niznik, H. B. (1995) Molecular and functional characterization of a partial cDNA encoding a novel chicken brain melatonin receptor. *FEBS Lett.* **374**, 273–278.

Lopez-Gonzalez, M. A., Calvo, J. R., Osuna, C., Rubio, A., and Guerrero, J. M. (1992) Melatonin potentiates cyclic AMP production stimulated by vasoactive intestinal peptide in human lymphocytes. *Neurosci. Lett.* **136**, 150–152.

MacFarlane, J. G., Cleghorn, J. M., Brown, G. M., and Streiner, D. L. (1991) The effects of exogenous melatonin on the total sleep time and daytime alertness of chronic insomniacs: a preliminary study. *Biol. Psychiatry* **30**, 371–376.

Maes, M. and Meltzer, H. Y. (1995) The serotonin hypothesis of major depression, in *Psychopharmacology: The Fourth Generation of Progress* (Bloom, F. E. and Kupfer, D. J., eds.), Raven, New York, pp. 933–944.

Maestroni, G. J. M. (1993) The immunoneuroendocrine role of melatonin. *J. Pineal Res.* **42**, 221–234.

Maestroni, G. J. M., Conti, A., and Pierpaoli, W. (1986) Role of the pineal gland in immunity. Circadian synthesis and release of melatonin modulates the antibody response and antagonizes the immunosuppressive effect of corticosterone. *J. Neuroimmunol.* **13**, 19–30.

Mahle, C. D., Nolan, R. T., Geissler, M. A., Iben, L. G., Parker, E. M., Baetge, E. E., Hammang, J. P., and Yocca, F. D. (1993) RT2-2 Retinal neuronal cells express high affinity functional melatonin receptors. *Soc. Neurosci. Abstr.* **19, Part 1,** 86.

Malbon, C. C., Rapiejko, P. J., and Watkins, D. C. (1988) Permissive hormone regulation of hormone-sensitive effector systems. *Trends Pharmacol. Sci.* **9,** 33–36.

Manji, H. K. (1992) G proteins: implications for psychiatry. *Am. J. Psychiat.* **149,** 746–760.

Marangos, P., Patel, J., Hirata, F., Sonhein, D., Paul, S. M., Skolnick, P., and Goodwin, F. K. (1981) Inhibition of diazepam binding by tryptophan derivatives including melatonin and its brain metabolite N-acetyl-5-methoxykynurenamine. *Life Sci.* **29,** 259–267.

Martin, J. E., Engel, J. N., and Klein, D. C. (1977) Inhibition of the in vitro pituitary response to luteinizing hormone-releasing hormone by melatonin, serotonin and 5-methoxytryptamine. *Endocrinology* **100,** 675–680.

Mattera, R., Graziano, M. P., Yatani, A., Graf, R., Codina, J., Gilman, A. G., Birnbaumer, L., and Brown, A. M. (1989) Bacterially synthesized splice variants of the α-subunit of the G protein G_s activate both adenylyl cyclase and dihydropyridine-sensitive calcium channels. *Science* **243,** 804–807.

Maus, M., Homburger, V., Bockaert, J., Glowinski, J., and Premont, J. (1990) Pretreatment of mouse striatal neurons in primary culture with 17 beta-estradiol enhances the pertussis toxin-catalyzed ADP-ribosylation of G alpha o, i protein subunits. *J. Neurochem.* **55,** 1244–1251.

Maywood, E. S., Buttery, R. C., Vance, G. H. S., Herbert, J., and Hastings, M. H. (1990) Gonadal responses of the male Syrian hamster to programmed infusions of melatonin are sensitive to signal duration and frequency but not to signal phase nor to lesions of the suprachiasmatic nuclei. *Biol. Reprod.* **43,** 174–182.

McArthur, A. J., Gillette, M. U., and Prosser, R. A. (1991) Melatonin directly resets the rat suprachiasmatic circadian clock in vitro. *Brain Res.* **565,** 158–161.

McClue, S. J., Selzer, E., Freissmuth, M., and Milligan, G. (1992) G_i3 does not contribute to the inhibition of adenylate cyclase when stimulation of an α_2-adrenergic receptor causes activation of both G_i2 and G_i3. *Biochem. J.* **284,** 565–568.

McKenzie, F. R. and Milligan, G. (1990) δ-opioid-receptor-mediated inhibition of adenylate cyclase is transduced specifically by the guanine-nucleotide-binding protein G_{i2}. *Biochem. J.* **267,** 391–398.

Meijer, J. H. and Rietveld, W. J. (1989) Neurophysiology of the suprachiasmatic circadian pacemaker in rodents. *Physiol. Rev.* **69,** 671–707.

Meldrum, E., Parker, P. J., and Carozzi, A. (1991) The Ptd Ins-PLC superfamily and signal transduction. *Biochem. Biophys. Acta.* **1092,** 49–71.

Menendez-Pelaez, A. and Reiter, R. J. (1993) Distribution of melatonin in mammalian tissues: the relative importance of nuclear versus cytosolic localization. *J. Pineal Res.* **15,** 59–69.

Milligan, G. (1993) Mechanisms of multifunctional signalling by G protein linked receptors. *Trends Pharmacol. Sci.* **14,** 239–244.

Milligan, G. and Green, A. (1991) Agonist control of G protein levels. *Trends Pharmacol. Sci.* **12,** 207–209.

Molis, T. M., Spriggs, L. L., and Hill, S. M. (1994) Modulation of estrogen receptor mRNA expression by melatonin in MCF-7 human breast cancer cells. *Mol. Endocrinol.* **8,** 1681–1690.

Moore, R. Y. (1983) Organization and function of a central nervous system circadian oscillator: the suprachiasmatic hypothalamic nucleus. *Fed. Proc.* **42,** 2783–2789.

Morgan, J. I. and Curran, T. (1991) Stimulus-transcription coupling in the nervous system: involvement of the inducible proto-oncogenes *fos* and *jun. Annu. Rev. Neurosci.* **14,** 421–451.

Morgan, P. J., Lawson, W., Davidson, G., and Howell, E. H. (1989a) Guanine nucleotides regulate the affinity of melatonin receptors on the ovine pars tuberalis. *Neuroendocrinology* **50,** 359–362.

Morgan, P. J., Lawson, W., Davidson, G., and Howell, H. E. (1989b) Melatonin inhibits cyclic AMP production in cultured ovine pars tuberalis cells. *J. Mol. Endocrinol.* **3,** R5–R8.

Morgan, P. J., Davidson, G., Lawson, W., and Barrett, P. (1991) Both pertussis toxin-sensitive and insensitive G proteins link melatonin receptor to inhibition of adenylate cyclase in the ovine pars tuberalis. *J. Neuroendocrinol.* **2,** 1–4.

Morgan, P. J., Barrett, P., Howell, H. E., and Helliwell, R. (1994) Melatonin receptors: Localization, molecular pharmacology and physiological significance. *Neurochem. Int.* **24,** 101–146.

Murayama, T. and Ui, M. (1983) Loss of the inhibitory function of the guanine nucleotide regulatory component of adenylate cyclase due to its ADP ribosylation by islet-activating protein, pertussis toxin, in adipocyte membranes. *J. Biol. Chem.* **258,** 3319–3326.

Naranjo-Rodríguez, E. B., Prieto-Gómez, B., and Reyes-Vazquez, C. (1991) Melatonin modifies the spontaneous multiunit activity recorded in several brain nuclei of freely behaving rats. *Brain Res. Bull.* **27,** 595–600.

Nestler, E. J. and Duman, R. S. (1994) G proteins and cyclic nucleotides in the nervous system, in *Basic Neurochemistry: Molecular, Cellular, and Medical Aspects* (Siegel, G. J., Agranoff, B. W., Albers, R. W., and Molinoff, P. B., eds.), Raven, New York, pp. 429–448.

Niles, L. P. (1985) Effects of melatonin on adenylate cyclase activity in rat brain, pineal and retina, in *The Pineal Gland, Endocrine Aspects* (Brown, G. M. and Wainwright, S. D., eds.), Pergamon, Oxford, pp. 283–288.

Niles, L. P. (1989) High-affinity binding sites for melatonin in hamster spleen. *Med. Sci. Res.* **17,** 179–180.

Niles, L. P. (1990): GTP modulates [^{125}I]iodomelatonin binding to a picomolar-affinity site in the Syrian hamster hypothalamus. *Eur. J. Pharmacol. (Mol. Pharmacol. Section),* **189,** 95–98.

Niles, L. P. (1991) Melatonin interaction with the benzodiazepine-GABA receptor complex in the CNS, in *Kynurenine and Serotonin Pathways:*

Progress in Tryptophan Research (Schwarcz, R., Young, S. N., and Brown, R. R., eds.), Plenum, New York, pp. 267–277.

Niles, L. P. and Hashemi, F. (1990a) Pharmacological inhibition of forskolin-stimulated adenylate cyclase activity in rat brain by melatonin, its analogs, and diazepam. *Biochem. Pharmacol.* **40,** 2701–2705.

Niles, L. P. and Hashemi, F. (1990b) Picomolar-affinity binding and inhibition of adenylate cyclase activity by melatonin in Syrian hamster hypothalamus. *Cell. Mol. Neurobiol.* **10,** 553–558.

Niles, L. P. and Peace, C. H. (1990) Allosteric modulation of t[^{35}S] butylbicyclophosphorothionate binding in rat brain by melatonin. *Brain Res. Bull.* **24,** 635–638.

Niles, L. P., Pickering, D. S., and Arciszewski, M. A. (1987a) Effects of chronic melatonin administration on GABA and diazepam binding in rat brain. *J. Neural Transm.* **70,** 117–124.

Niles, L. P., Pickering, D. S., and Sayer, B. G. (1987b) HPLC-purified 2-[^{125}I]iodomelatonin labels multiple binding sites in hamster brain. *Biochem. Biophys. Res. Commun.* **147,** 946–949.

Niles, L. P., Ye, M., Pickering, D. S., and Ying, S.-W. (1991) Pertussis toxin blocks melatonin-induced inhibition of forskolin-stimulated adenylate cyclase activity in the chick brain. *Biochem. Biophys. Res. Commun.* **178,** 786–792.

Niles, L. P., Harjivan, C., and Nanji, A. (1994) Heparin inhibits melatonin binding and signal transduction in chick brain. *Biol. Signals* **3,** 26–33.

Nishizuka, Y. (1986) Studies and perspectives of protein kinase C. *Science* **233,** 305–312.

Olianas, M. C. and Onali, P. (1990) Alteration of the GTP-dependent inhibitory pathway of rat striatal adenylate cyclase by phorbol esters. *Neurochem. Res.* **15,** 1109–1114.

Oriowo, M. A. and Ruffolo, R. R. Jr. (1992) Activation of a single alpha-1-adrenoceptor subtype in rat aorta mobilizes intracellular and extracellular pools of calcium. *Pharmacology* **44,** 139–149.

Oriowo, M. A., Nichols, A. J., and Ruffolo, Jr. R. R. (1992) Receptor protection studies with phenoxybenzamine indicate that a single alpha 1-adrenoceptor may be coupled to two signal transduction processes in vascular smooth muscle. *Pharmacology* **45,** 17–26.

Pang, S. F., Yip, M. K., Liu, H. W., Brown, G. M., and Tsui, H. W. (1980) Diurnal rhythms of immunoreactive N-acetylserotonin and melatonin in the serum of male rats. *Acta Endocrinol.* **95,** 571–576.

Pang, C. S., Brown, G. M., Tang, P. L., Cheng, K. M., and Pang, S. F. (1993) 2-[^{125}I]Iodomelatonin binding sites in the lung and heart: a link between the photoperiodic signal, melatonin, and the cardiopulmonary system. *Biol. Signals* **2,** 228–236.

Persengiev, S. P. (1992) 2-[^{125}I]Iodomelatonin binding sites in rat adrenals: pharmacological characteristics and subcellular distribution. *Life Sci.* **51,** 647–651.

Pickering, D. S. and Niles, L. P. (1989) 2-[^{125}I]Iodomelatonin binding sites in hamster and chick exhibit differential sensitivity to prazosin. *J. Pharm. Pharmacol.* **41,** 356, 357.

Pickering, D. S. and Niles, L. P. (1990) Pharmacological characterization of melatonin binding sites in Syrian hamster hypothalamus. *Eur. J. Pharmacol.* **175,** 71–77.

Pickering, D. S. and Niles, L. P. (1992) Expression of nanomolar-affinity binding sites for melatonin in Syrian hamster RPMI 1846 melanoma cells. *Cell. Signalling* **4,** 201–207.

Pickering, D. S., Niles, L. P., and Jung, C. Y. (1990) Molecular mass of the melatonin receptor in hamster hypothalamus and chicken retina. *Neurosci. Res. Commun.* **6,** 11–18.

Popova, J. S. and Dubocovich, M. L. (1995) Melatonin receptor-mediated stimulation of phosphoinositide breakdown in chick brain slices. *J. Neurochem.* **64,** 130–138.

Prosser, R. A. and Gillette, M. U. (1989) The mammalian circadian clock in the suprachiasmatic nuclei is reset in vitro by cAMP. *J. Neurosci.* **9,** 1073–1081.

Prosser, R. A., McArthur, A. J., and Gillette, M. U. (1989) cGMP induces phase shifts of a mammalian circadian pacemaker at night, in antiphase to cAMP effects. *Proc. Natl. Acad. Sci. USA* **86,** 6812–6815.

Rall, T. W., Sutherland, E. W., and Berthet, J. (1957) The relationship of epinephrine and glucagon to liver phosphorylase. IV. Effect of epinephrine and glucagon on the reactivation of phosphorylase in liver homogenates. *J. Biol. Chem.* **224,** 463–475.

Rasmussen, H. and Barrett, P. Q. (1984) Calcium messenger system: an integrated view. *Physiol. Rev.* **64,** 938–984.

Ray, K., Kunsch, C., Bonner, L. M., and Robishaw, J. D. (1995) Isolation of cDNA clones encoding eight different human G protein γ subunits, including three novel forms designated the γ4, γ10, and γ11 subunits. *J. Biol. Chem.* **270,** 21,765–21,771.

Reiter, R. J. (1991) Pineal melatonin: cell biology of its synthesis and of its physiological interactions. *Endocrine Rev.* **12,** 151–180.

Reiter, R. J. (1996) Functional aspects of the pineal hormone melatonin in combating cell and tissue damage induced by free radicals. *Eur. J. Endocrinol.* **134,** 412–420.

Reithman, C., Gierschik, P., Werdan, K., and Jakobs, K. H. (1991) Role of inhibitory G-protein α-subunits in adenylyl cyclase desensitization. *Mol. Cell. Endocrinol.* **82,** C215–C221.

Reppert, S. M., Weaver, D. R., and Ebisawa, T. (1994) Cloning and characterization of a mammalian melatonin receptor that mediates reproductive and circadian responses. *Neuron* **13,** 1177–1185.

Reppert, S. M., Godson, C., Mahle, C. D., Weaver, D. R., Slaugenhaupt, S. A., and Gusella, J. F. (1995a) Molecular characterization of a second melatonin receptor expressed in human retina and brain: The MEL$_{1b}$ melatonin receptor. *Proc. Natl. Acad. Sci. USA* **92,** 8734–8738.

Reppert, S. M., Weaver, D. R., Cassone, V. M., Godson, C., and Kolakowski, L. F. (1995b) Melatonin receptors are for the birds: Molecular analysis of two receptor subtypes differentially expressed in chick brain. *Neuron* **15,** 1003–1015.

Rhee, S. G. and Choi, K. D. (1992) Regulation of inositol phospholipid-specific phospholipase C isozymes. *J. Biol. Chem.* **267,** 12,393–12,396.

Rivkees, S. A., Carlson, L. L., and Reppert, S. M. (1989a) Guanine nucleotide-binding protein regulation of melatonin receptors in lizard brain. *Proc. Natl. Acad. Sci. USA* **86,** 3882–3886.

Rivkees, S. A., Cassone, V. M., Weaver, D. R., and Reppert, S. M. (1989b) Melatonin receptors in chick brain: characterization and localization. *Endocrinology* **125,** 363–368.

Roerig, S. C., Loh, H. H., and Law, P. Y. (1992) Identification of three separate guanine nucleotide-binding proteins that interact with the δ-opioid receptor in NG108-15 neuroblastoma x glioma hybrid cells. *Mol. Pharmacol.* **41,** 822–831.

Ronnberg, L., Kauppila, A., Leppaluoto, J., Martikainen, H., and Vakkuri, O. (1990) Circadian and seasonal variation in human preovulatory follicular fluid melatonin concentration. *J. Clin. Endocrinol. Metab.* **71,** 493–496.

Rosenstein, R. E., Golombek, D. A., Kanterewicz, B. I., and Cardinali, D. P. (1991) Time-dependency for the bimodal effect of melatonin on calcium uptake in rat hypothalamus. *J. Neural Transm. (Gen. Sect.)* **85,** 243–247.

Rudeen, P. K., Philo, R. C., and Symmes, S. K. (1980) Antiepileptic effects of melatonin in the pinealectomized Mongolian gerbil. *Epilepsia* **21,** 149–154.

Rusak, B. and Morin, L. P. (1976) Testicular responses to photoperiod are blocked by lesions of the suprachiasmatic nuclei in golden hamsters. *Biol. Reprod.* **15,** 366–374.

Ryan, W. A., Franza, B. R., and Gilman, M. Z. (1989) Two distinct cellular phosphoproteins bind to the *c-fos* serum response element. *EMBO J.* **8,** 1785–1792.

Saito, N., Guitart, X., Hayward, M., Tallman, J. F., Duman, R. S., and Nestler, E. J. (1989) Corticosterone differentially regulates the expression of Gs alpha and Gi alpha messenger RNA and protein in rat cerebral cortex. *Proc. Natl. Acad. Sci. USA* **86,** 3906–3910.

Sakaue, M. and Hoffman, B. B. (1991) cAMP regulates transcription of the α2A adrenergic receptor gene in HT-29 cells. *J. Biol. Chem.* **266,** 5743–5749.

Seeman, P. and Van Tol, H. H. M. (1994) Dopamine receptor pharmacology. *Trends Pharmacol. Sci.* **15,** 264–270.

Seltzer, A., Viswanathan, M., and Saavedra, J. M. (1992). Melatonin-binding sites in brain and caudal arteries of the female rat during the estrous cycle and after estrogen administration. *Endocrinology* **130**, 1896–1902.

Shibata, S., Cassone, V. M., and Moore, R. Y. (1989) Effects of melatonin on neuronal activity in the rat suprachiasmatic nucleus *in vitro*. *Neurosci. Lett.* **97**, 140–144.

Sibley, D. R. and Lefkowitz, R. J. (1985) Molecular mechanisms of receptor desensitization using the β-adrenergic receptor-coupled adenylate cyclase system as a model. *Nature* **317**, 124–129.

Sibley, D. R., Benovic, J. L., Caron, M. G., and Lefkowitz, R. J. (1988) Phosphorylation of cell surface receptors: a mechanism for regulating signal transduction pathways. *Endocrine Rev.* **9**, 38–56.

Simon, M. I., Strathmann, M. P., and Gautam, N. (1991) Diversity of G proteins in signal transduction. *Science* **252**, 802–808.

Simonds, W. F., Goldsmith, P. K., Codina, J., Unson, C. G., and Speigel, A. M. (1989) G_{i2} mediates α_2-adrenergic inhibition of adenylyl cyclase in platelet membranes: in situ identification with Gα c-terminal antibodies. *Proc. Natl. Acad. Sci. USA* **86**, 7809–7813.

Smrcka, A. V., Hepler, J. R., Brown, K. O., and Sternweis, P. C. (1991) Regulation of polyphosphoinositide—specific phospholipase C activity by purified G_q. *Science* **251**, 804–807.

Song, Y., Ayre, E. A., and Pang, S. F. (1993) [^{125}I]Iodomelatonin binding sites in mammalian and avian kidneys. *Biol. Signals* **2**, 207–220.

Spadoni, G., Stankov, B., Duranti, A., Biella, G., Lucini, V., Salvatori, A., and Fraschini, F. (1993) 2-substituted 5-methoxy-N-acyltryptamines: synthesis, binding affinity for the melatonin receptor, and evaluation of the biological activity. *J. Med. Chem.* **36**, 4069–4074.

Spiegel, A. M. (1992) Heterotrimeric GTP-binding proteins: an expanding family of signal transducers. *Med. Res. Rev.* **12**, 55–71.

Stankov, B., Biella, G., Panara, C., Lucini, V., Capsoni, S., Fauteck, J., Cozzi, B., and Fraschini, F. (1992) Melatonin signal transduction and mechanism of action in the central nervous system: using the rabbit cortex as a model. *Endocrinology* **130**, 2152–2159.

Stehle, J., Vanecek, J., and Vollrath, L. (1989) Effects of melatonin on spontaneous electrical activity of neurons in rat suprachiasmatic nuclei: an *in vitro* iontophoretic study. *J. Neural Transm. (Gen. Sect.)* **78**, 173–177.

Steinhilber, D., Brungs, M., Werz, O., Wiesenberg, I. S., Danielsson, C., Kahlen, J.-P., Nayeri, S., Schrader, M., and Carlberg, C. (1995) The nuclear receptor for melatonin represses 5-lipoxygenase gene expression in human B lymphocytes. *J. Biol. Chem.* **270**, 7037–7040.

Strathmann, M. and Simon, M. I. (1990) G protein diversity: a distinct class of α-subunits is present in vertebrates and invertebrates. *Proc. Natl. Acad. Sci. USA* **87**, 9113–9117.

Strosberg, A. D. (1991) Structure/function relationship of proteins belonging to the family of receptors coupled to GTP-binding proteins. *Eur. J. Biochem.* **196,** 1–10.

Sugden, D. (1983) Psychopharmacological effects of melatonin in mouse and rat. *J. Pharmacol. Exp. Ther.* **227,** 587–591.

Summers, S. T., Walker, J. M., Sando, J. J., and Cronin, M. J. (1988) Phorbol esters increase adenylate cyclase activity and stability in pituitary membranes. *Biochem. Biophys. Res. Commun.* **151,** 16–24.

Sutherland, E. W. and Rall, T. W. (1958) Fractionation and characterization of a cyclic adenosine ribonucleotide formed by tissue particles. *J. Biol. Chem.* **232,** 1077–1091.

Takahashi, J. S. and Kornhauser, J. M. (1993) Molecular approaches to understanding circadian oscillations. *Annu. Rev. Physiol.* **55,** 729–753.

Tamarkin, L., Baird, C. J. and Almeida, O. F. X. (1985) Melatonin: a coordinating signal for mammalian reproduction? *Science* **227,** 714–720.

Tang, W.-J. and Gilman, A. G. (1991) Type-specific regulation of adenylyl cyclase by G protein βγ subunits. *Science* **254,** 1500–1503.

Tang, W.-J., Krupinski, J., and Gilman, A. G. (1991) Expression and characterization of calmodulin-activated (Type 1) adenylyl cyclase. *J. Biol. Chem.* **266,** 8595–8603.

Taussig, R., Quarmby, L. M., and Gilman, A. G. (1993) Regulation of purified type I and type II adenylylcyclases by G protein βγ subunits. *J. Biol. Chem.* **268,** 9–12.

Taylor, S. S. (1989) cAMP-dependent protein kinase. *J. Biol. Chem.* **264,** 8443–8446.

Taylor, S. J., Chae, H. Z., Rhee, S. G., and Exton, J. H. (1991) Activation of the β1 isozyme of phospholipase C by α-subunits of the G_q class of G protein. *Nature* **350,** 516–518.

Tenn, C. and Niles, L. P. (1993) Physiological regulation of melatonin receptors in rat suprachiasmatic nuclei: diurnal rhythmicity and effects of stress. *Mol. Cell. Endocrinol.* **98,** 43–48.

Tenn, C. C. and Niles, L. P. (1995) Central-type benzodiazepine receptors mediate the antidopaminergic effect of clonazepam and melatonin in 6-hydroxydopamine lesioned rats: Involvement of a GABAergic mechanism. *J. Pharmacol. Exp. Ther.* **274,** 84–89.

Tenn, C. C., Neu, J. M., and Niles, L. P. (1996) PK11195 blockade of benzodiazepine-induced inhibition of forskolin-stimulated adenylate cyclase activity in the striatum. *Br. J. Pharmacol.* **119,** 223–228.

Touitou, Y., Febre-Montange, M., Proust, J., Klinger, E., and Kakache, J. P. (1985) Age and sex associated modification of plasma melatonin concentrations in man-relationship to pathology, malignant or not and autopsy findings. *Acta Endocrinol.* **108,** 135–144.

Turek, F. W. (1996) Melatonin hype hard to swallow. *Nature* **379,** 295–296.

Vacas, M. I., Keller Sarmiento, M. I., and Cardinali, D. P. (1981) Melatonin increases cGMP and decreases cAMP levels in rat medial basal hypothalamus in vitro. *Brain Res.* **225,** 207–211.

Vacas, M. I., Berria, M. I., Cardinali, D. P., and Lascano, E. F. (1984a) Melatonin inhibits β-adrenoceptor-stimulated cyclic AMP accumulation in rat astroglial cell cultures. *Neuroendocrinology* **38,** 176–181.

Vacas, M. I., Keller Sarmiento, M. I., and Cardinali, D. P. (1984b) Pineal methoxyindoles depress calcium uptake by rat brain synaptosomes. *Brain Res.* **294,** 166–168.

Van Dongen, A. M. J., Codina, J., Olate, J., Mattera, R., Joho, R., Birnbaumer, L. and Brown, A. M. (1988) Newly identified brain potassium channels gated by the guanine nucleotide binding protein G_o. *Science* **242,** 1433–1437.

Van Sande, J., Raspé, E., Perret, J., Lejeune, C., Maenhaut, C., Vassart, G., and Dumont, J. E. (1990) Thyrotropin activates both the cyclic AMP and the PIP_2 cascades in CHO cells expressing the human cDNA of TSH receptor. *Mol. Cell Endocrinol.* **74,** R1–R6.

Vanecek, J. and Klein, D. C. (1992) Melatonin inhibits gonadotropin-releasing hormone-induced elevation of intracellular Ca^{2+} in neonatal rat pituitary cells. *Endocrinology* **130,** 701–707.

Vanecek, J. and Vollrath, L. (1989) Melatonin inhibits cyclic AMP and cyclic GMP accumulation in the rat pituitary. *Brain Res.* **505,** 157–159.

Vanecek, J. and Vollrath, L. (1990a) Developmental changes and daily rhythm in melatonin-induced inhibition of 3′,5′-cyclic AMP accumulation in the rat pituitary. *Endocrinology* **126,** 1509–1513.

Vanecek, J. and Vollrath, L. (1990b) Melatonin modulates diacylglycerol and arachidonic acid metabolism in the anterior pituitary of immature rats. *Neurosci. Lett.* **110,** 199–203.

Vanecek, J. Kosar, E., and Vorlicek, J. (1990) Daily changes in melatonin binding sites and the effects of castration. *Mol. Cell. Endocrinol.* **73,** 165–170.

Vesely, D. L. (1981) Melatonin enhances guanylate cyclase activity in a variety of tissues. *Mol. Cell. Biochem.* **35,** 55–58.

Viswanathan, M., Laitinen, J. T., and Saavedra, J. M. (1990). Expression of melatonin receptors in arteries involved in thermoregulation. *Proc. Natl. Acad. Sci. USA* **87,** 6200–6203.

Wahlestedt, C. (1994) Antisense oligonucleotide strategies in neuropharmacology. *Trends Pharmacol. Sci.* **15,** 42–46.

Waldhauser, F., Weiszenbacher, G., Tatzer, E., Gisinger, B., Waldhauser, M., Schemper, M., and Frisch, H. (1988) Alterations in nocturnal serum melatonin levels in humans with growth and aging. *J. Clin. Endocrinol. Metab.* **66,** 648–652.

Walsh, D. A., Perkins, J. P. and Krebs, E. G. (1968) An adenosine 3′,5′-monophosphate-dependent protein kinase from rabbit skeletal muscle. *J. Biol. Chem.* **243,** 3763–3765.

Watson, A. J., Katz, A. and Simon, M. I. (1994) A fifth member of the mammalian G protein β-subunit family. *J. Biol. Chem.* **269,** 22,150–22,156.

Webley, G. E. and Luck, M. R. (1986) Melatonin directly stimulates the secretion of progesterone by human and bovine granulosa cells in vitro. *J. Reprod. Fert.* **78,** 711–717.

Weekley, L. B. (1991) Melatonin induced relaxation of rat aorta: interaction with adrenergic agonists *J. Pineal Res.* **11,** 28–43.

Weekley, L. B. (1993) Effects of melatonin on isolated pulmonary artery and vein: role of the vascular endothelium. *Pulmonary Pharmacol.* **6,** 149–154.

White, B. H., Sekura, R. D. and Rollag, M. D. (1987) Pertussis toxin blocks melatonin-induced pigment aggregation in *Xenopus* dermal melanophores. *J. Comp. Physiol. B* **157,** 153–159.

Williams, L. M., Morgan, P. J., Hastings, M. H., Lawson, W., Davidson, G., and Howell, H. E. (1989) Melatonin receptor sites in the Syrian hamster brain and pituitary. Localization and characterization using [^{125}I]iodomelatonin. *J. Neuroendocrinol.* **1,** 315–320.

Wirz-Justice, A. (1995) Biological rhythms in mood disorders, in *Psychopharmacology: The Fourth Generation of Progress* (Bloom, F. E. and Kupfer, D. J., eds.), Raven, New York, pp. 999–1017.

Withyakumnarnkul, B., Nonaka, K., Santana, C., Attia, A. M. and Reiter, R. J. (1990) Interferon-gamma modulates melatonin production in rat pineal gland in organ culture. *J. Interferon Res.* **10,** 403–411.

Wittkowski, E., Bergmann, M., Hoffmann, K., and Pera, F. (1988) Photoperiod-dependent changes in TSH-like immunoreactivity of cells in the hypophysial pars tuberalis of the Djungarian hamster, *Phodopus sungorus. Cell Tissue Res.* **251,** 183–187.

Wu, D., Lee, C. H., Rhee, S. G., and Simon, M. I. (1992) Activation of phospholipase C by the α-subunits of the G_q and G_{11} proteins in transfected COS-7 cells. *J. Biol. Chem.* **267,** 1811–1817.

Yatani, A., Mattera, R., Codina, J. Graf, R., Okabe, K., Padrell, E., Iyengar, R., Brown, A. M., and Birnbaumer, L. (1988) The G protein-gated atrial K^+ channel is stimulated by three distinct G_i α-subunits. *Nature* **336,** 680–682.

Yie, S.-M., Brown, G. M., Liu, G.-Y., Collins, J. A., Daya, S., Hughes, E. G., Foster, W. G., and Younglai, E. V. (1995a) Melatonin and steroids in human pre-ovulatory follicular fluid: seasonal variations and granulosa cell steroid production. *Human Reprod.* **10,** 50–55.

Yie, S.-M., Niles, L. P., and Younglai, E. V. (1995b) Melatonin receptors on human granulosa cell membranes. *J. Clin. Endocrinol. Metab.* **80,** 1747–1749.

Ying, S. W., Niles, L. P., Pickering, D., and Ye M. (1992) Involvement of multiple sulfhydryl groups in melatonin signal transduction in chick brain. *Mol. Cell. Endocrinol.* **85,** 53–63.

Ying, S.-W., Niles, L. P., and Crocker, C. (1993) Human malignant melanoma cells express high-affinity receptors for melatonin: antiproliferative effects of melatonin and 6-chloromelatonin. *Eur. J. Pharmacol. Mol. Pharmacol. Sect.* **246,** 89–96.

Yous, S., Andrieux, J., Howell, H. E., Morgan, P. J., Renard, P., Pfeiffer, B., Lesieur, D., Guardiola-Lemaitre, B. (1992) Novel naphthalenic ligands with high affinity for the melatonin receptor. *J. Med. Chem.* **35,** 1484–1486.

Yuan, H., Lu, Y., and Pang, S. F. (1991) Binding characteristics and regional distribution of [^{125}I]iodomelatonin binding sites in the brain of the human fetus. *Neurosci. Lett.* **130,** 229–232.

Yung, L. Y., Tsim, S.-T., and Wong, Y. H. (1995) Stimulation of cAMP accumulation by the cloned *Xenopus* melatonin receptor through G_i and G_z proteins. *FEBS Lett.* **372,** 99–102.

Zisapel, N. and Laudon, M. (1983) Inhibition by melatonin of dopamine release in rat hypothalamus: regulation of calcium entry. *Brain Res.* **272,** 378–381.

G Proteins in the Pathophysiology and Treatment of Mood and Anxiety Disorders

Husseini K. Manji

Introduction

Mood and anxiety disorders are common, severe, chronic, and often life-threatening illnesses. Despite well-established genetic diatheses and extensive research, the biochemical abnormalities underlying the predisposition to, and the pathophysiology of, these disorders remain to be clearly established. Early biologic theories regarding the etiology of anxiety disorders and, in particular, mood disorders, have focused on various neurotransmitters, in particular the biogenic amines. In recent years, however, advances in our understanding of the molecular mechanisms underlying neuronal communication have focused research into the role of receptor and postreceptor sites, and indeed, numerous findings from both clinical and preclinical studies suggest that abnormalities in receptor–effector responsiveness may be involved in the pathophysiology of mood and anxiety disorders. Data from a complementary line of research on the cellular mechanism(s) of action of antidepressants, mood-stabilizing agents, and antianxiety agents have also been used to provide clues about the biochemical processes involved in the pathogenesis of these disorders. There has been a growing appreciation that neurotransmitter function might be altered indirectly through alterations in intracellular signaling, and

From: *Neuromethods, Vol. 31: G-Protein Methods and Protocols*
Ed: R. K. Mishra, G. B. Baker, and A. A. Boulton Humana Press Inc.

antidepressants, mood-stabilizing agents, and antianxiety agents might be effective not because they are "catecholaminergic" or "serotonergic" agents per se, but because they alter the postsynaptic signal generated in response to multiple, endogenous neurotransmitters. In this context, signal transduction pathways are in a pivotal position in the central nervous system (CNS), able to affect the functional balance between neurotransmitter systems and have been postulated to explain multiple aspects of the pathophysiology of mood disorders. It is thus not surprising that, over the past several years, research aimed at the elucidation of the molecular mechanisms underlying the therapeutic effects of antidepressants and mood-stabilizing agents have focused on signal transduction pathways, and several independent laboratories have now shown that these drugs, on chronic administration at therapeutically relevant doses, unequivocally affect second-messenger generating systems (reviewed in Manji, 1992; Hudson et al., 1993; Manji et al., 1995a).

Given their widespread and crucial role in the integration, regulation, and amplification of signal transduction pathways, it is not surprising that alterations in the function and/or expression of various G proteins have been recently implicated in a variety of pathophysiological states (*see* Table 1). It should be noted that G protein dysfunction appears to represent the primary pathology in Albright's Hereditary Osteodystrophy, McCune-Albright Syndrome, and certain endocrine tumors; in several of the other conditions, G protein abnormalities are likely secondary, but nonetheless are implicated in the pathophysiology of the condition. We will briefly highlight here some of the clinical conditions with well-characterized G protein dysfunction here.

The first disease in which an intrinsic G protein abnormality was found was pseudohypoparathyroidism (called "pseudohypoparathyroidism" because the individuals have normal [or even increased] levels of parathyroid hormone, but have the manifestations of the disease because of cellular resistance to the hormone). In one form of the illness (termed

Table 1
G Protein Abnormalities in Medical Illnesses

Medical conditions known to arise from a primary G protein
   abnormality
      Albright's Hereditary Osteodystrophy
       (pseudohypoparathyroidism)
      McCune-Albright syndrome
      Certain endocrine tumors
Medical conditions in which the G protein abnormality is likely
   secondary but contributes to the pathophysiology of the
   disorder
      Infection with vibrio cholera or bordetella pertussis
      Hypothyroidism and hyperthyroidism
      Hypocortisolemic and hypercortisolemic states
      Heart failure
      Diabetes
      Chronic cocaine administration
      Bipolar affective disorder

Albright's Hereditary Osteodystrophy), the patients demonstrate a resistance to parathyroid hormone and to a number of hormones utilizing cyclic AMP (cAMP) as the second messenger. In these patients, reduced expression or function (approx 50%) of $G_s\alpha$ has been demonstrated in cultured cells as well as from freshly obtained tissue (Patten et al., 1990; Weinstein et al., 1990). Additionally, distinct mutations have been identified in the gene encoding $G_s\alpha$ in different kindreds with the disease (Patten et al., 1990; Weinstein et al., 1990), thereby providing the first demonstrations of an inherited mutation in a human G protein gene with important implications. The existence of more than one type of mutation resulting in altered function of $G_s\alpha$ demonstrates the genetic heterogeneity present in this autosomal-dominant disease. Moreover, the presence of an identical mutation in the $G_s\alpha$ gene, both in individuals with multiple hormone resistance and in those without hormone resistance, clearly suggests that G_s deficiency is necessary but not sufficient for the full phenotypic expression of the disease (Patten et al., 1990; Weinstein et al., 1990). Thus (as has been suggested for a variety of psychiatric disorders), additional factors, including "modifying genes" and perhaps environmental factors, appear to be involved. In contrast to this "hypofunctional disease," a mutation in the $G_s\alpha$ gene resulting in constitutive activation of G_s in affected tissues has been demonstrated to underlie the pathophysiology of McCune-Albright syndrome, a sporadic disorder in which the subjects manifest such heterogenous symptoms as polyostotic fibrous dysplasia, cafe-au-lait skin pigmentation, and autonomous hyperfunctioning of one or more endocrine glands (gonad, adrenal cortex, thyroid, pituitary somatotrophs) (Weinstein et al., 1991).

Recent studies have also demonstrated significant alterations in G proteins in failing human and animal heart. Thus, a 50% decrease in the apparent concentration and function of Gs in sarcolemma from a canine model of left ventricular failure has recently been reported. In contrast, end-stage idiopathic human congestive failure is associated with increased activity of G_i, suggesting that various manipulations of G_s/G_i stoichiometry may result in similar pathophysiology (Horn et al., 1988; Horn and Bilezkian, 1990). Perhaps more akin to psychiatric research with respect to the use of peripheral blood cells to represent less accessible tissue (CNS, heart), one study (Horn et al., 1988; Horn and Bilezkian, 1990) reported an 80% decrease in lymphocyte G_s levels in subjects with congestive heart failure. More intriguingly, successful treatment with captopril was associated with a significant twofold increase in lymphocyte G_s levels. These examples highlight the diverse range of clinical manifestations arising from alterations in the levels and/or functioning of these key signal transducers. Although G proteins are present in all eukaryotic cells and control various metabolic, humoral, and developmental functions, they may be especially important in the CNS, where they serve the critical roles of first amplifying and "weighting" extracellularly generated neuronal signals and then transmitting these integrated signals to effectors, thereby forming the basis for a complex information-processing network (Ross,

1989; Manji, 1992; Bourne and Nicoll, 1993). Since a single receptor subtype can be coupled to multiple G proteins, and multiple G proteins can converge to activate or inactivate a single effector (Ross, 1989; Taylor, 1990; Manji, 1992; Bourne and Nicoll, 1993), the G protein-coupled interactions form complex networks. The high degree of complexity generated by the interactions of G protein-coupled receptors may be one mechanism by which neurons acquire the flexibility for generating the wide range of responses observed in the nervous system. This has led to the proposal that G proteins may be involved in pathways regulating such diverse vegetative functions as mood, appetite, and wakefulness and by extrapolation, in the molecular mechanisms of action of antidepressants and mood-stabilizing agents. We now turn to the evidence for alterations of signal transduction pathways in the pathophysiology and treatment of mood and anxiety disorders.

Abnormalities in Second Messenger-Generating Systems in Mood Disorders

The future development of selective receptor and second messenger ligands for positron emission tomography (PET) studies may soon permit the direct assessment of CNS receptor and postreceptor sensitivity in humans. To date, studies of receptor and postreceptor function in mood disorders have been limited to indirect research strategies. The most commonly utilized strategy has been to characterize receptor function in readily accessible blood elements. Much clinical research has focused on the activity of the cAMP generating system in mood disorders. Overall, the preponderance of the evidence suggests altered receptor and/or postreceptor sensitivity of the cAMP-generating system in the absence of consistent alterations in the number of receptors themselves, (reviewed in Potter and Manji, 1994; Table 2). Thus, using platelets, several (but not all) studies have demonstrated reduced PGE1 receptor stimulation and $\alpha 2$ inhibition of adenylyl cyclase (AC) in unipolar depression (Siever et al.

Table 2
Adrenergic Receptors and Responsivity in Depression[a]

Experimental parameter	Depressed vs control
Platelet α_2 receptors	
B_{max}—antagonist (yohimbine) binding	No change
B_{max}—agonist (clonidine) binding	Increased
Function	?Subsensitive
Lymphocyte β_2 receptors	
B_{max}—antagonist binding	Inconsistent Findings
Function—cAMP response to isoproteronol	Subsensitive
Neuroendocrine responses	
α_2-agonist (clonidine): MHPG response	No change
ACTH and cortisol	Inconsistent
Growth hormone	Decreased
α_2-antagonist (yohimbine): MHPG response	No change

[a]B_{max}, maximum number of binding sites; α2-, β2-adrenergic receptor subtypes MHPG, 3-methoxy, 4-hydroxyphenylethyleneglycol; ACTH, adrenocorticotropin.

1984; Kafka and Paul, 1986; Kafka et al., 1986; Siever, 1987; Grossman et al., 1993). Several groups have investigated the density of β adrenergic receptors (βAR) in untransformed lymphocytes or leukocytes of untreated depressed patients, yielding fairly inconsistent results. Many groups report a decrease in βAR number (Extein et al., 1979; Pandey et al., 1985, 1986, 1990; Carstens et al., 1987; Magliozzi et al., 1989), whereas others describe an increase or no change when compared to healthy volunteers (Sarai et al., 1982; Zohar et al., 1983; Healy et al., 1985; Mann et al., 1985; Cooper et al., 1985). These studies and possible methodological sources of the differences in results (e.g., methods of tissue preparation, type of ligand used, subtypes of patient populations, length of drug-free interval) have been discussed recently in a very thorough critical appraisal (Werstuik et al., 1990). In contrast to the inconsistent results from binding studies described above, most studies measuring mononuclear leukocytes (MNL) βAR-stimulated AC activity report decreased responsiveness in depressed patients compared to healthy volunteers (Extein et al., 1979; Pandey et al., 1979; Healy et al., 1983; Mann et al.,

1985; Klysner et al., 1987; Ebstein et al., 1988; Halper et al., 1988). One study suggested that this finding correlated with treatment response and that poor responders characteristically had lower pretreatment isoproterenol-stimulated AC activity (Ebstein et al., 1988). The consistently observed decrease in leukocyte βAR function in depression could reflect an inherited abnormality of the βAR/G$_s$/AC complex, as suggested by the findings of Wright et al. (1984), utilizing Epstein-Barr Virus (EBV)-transformed lymphocytes from manic-depressives and controls. However, these findings need to be replicated, and a number of additional confounding factors need to be considered. In this context, twin studies do not show a significant variability of isoproterenol (ISO)-stimulated cAMP production, suggesting that variations in ISO-stimulated cAMP production are most likely caused by "environmental" effects on the number of sensitivity of β-adrenergic receptors (Ebstein et al., 1986). Thus, additional studies are clearly needed to ascertain if the consistent differences in MNL βAR sensitivity in depression truly reflect comparable differences in the brain. Interestingly, bipolar depressives have not consistently demonstrated a similar attenuation of βAR-stimulated AC activity (Berrettini, 1987a,b). These findings are of considerable interest and are compatible with the recent demonstrations of increased levels of Gαs bipolar affective disorder (Young et al., 1994; Manji et al., 1995b; Mitchell et al., 1997).

G Proteins in Mood Disorders

In view of the indirect evidence for abnormalities at postreceptor sites described above, it is not surprising that several independent laboratories have examined G proteins in patients with bipolar affective disorder (Table 3). Schreiber and colleagues (1991) reported "hyperfunctional" G protein function in leukocytes of untreated manic patients by demonstrating that agonist-stimulated binding of [^{3}H]Gpp(NH)p (a stable, nonhydrolysable analog of guanosine triphosphate [GTP]) was enhanced in leukocyte membranes of untreated manic patients compared to controls. These findings suggest the

Table 3
G Proteins in Bipolar Affective Disorder

Three studies examining G proteins in peripheral cells from bipolar
   affective disorder patients have found elevated levels of $G\alpha_s$ (45 kDa)
A postmortem study has also found elevated levels of $G\alpha_s$ in selected
   areas of brain in bipolar affective disorder
It is presently unclear if the elevated $G\alpha_s$ is "primary" or the
   downstream consequence of another abnormality
Lithium treatment does not modify the levels of $G\alpha$ subtypes

presence of increased levels of G proteins and/or enhanced receptor-mediated activation of G proteins in leukocytes from untreated manic subjects. Complementary to these findings, Young and associates (1993, 1994) have reported increased levels of $G\alpha_s$ in bipolar affective disorder patients in two separate studies. They observed increased levels of $G\alpha_s$ in cerebral cortex, but not cerebellum, in postmortem brain tissue from patients with bipolar affective disorder (Young et al., 1993). More recently, the same group of investigators observed significantly higher levels of $G\alpha_s$ in mononuclear leukocytes from depressed bipolar, but not unipolar patients (Young et al., 1994); the latter study did not examine euthymic bipolar affective disorder patients, leaving open the question of the role of the illness state on the levels of $G\alpha_s$ in peripheral cells. Manji and associates (1995b) directly quantitate the levels of the major G protein α-subunits in leukocytes and platelets from both untreated (predominantly manic) and lithium-treated, euthymic bipolar affective disorder patients; in both platelet and leukocyte membranes, this study observed higher levels of the 45-kDa form of $G\alpha_s$ in the overall group of bipolar affective disorder patients (treated or untreated) compared to controls. Thus, in both peripheral cells and postmortem brain tissue, elevations were observed in the predominant subspecies of $G\alpha_s$ present in the tissues examined. Moreover, since both forms arise from the a single gene (Bray et al., 1986), abnormalities in this key component of signal transduction represent a potential molecular basis underlying the predisposition to bipolar affective disorder, and are worthy of further study. Since $G\alpha_s$ is a ubiquitously expressed protein, it may appear counterintuitive that an abnormality in this protein may underlie that pathophysiology of bipolar affective disorder. However, there is already precedence for clinical disorders arising from abnormalities in the levels of $G\alpha_s$, which present with limited clinical manifestations, despite the ubiquitous expression of $G\alpha_s$ (discussed in Spiegel et al., 1992; Manji et al., 1995b). These heterogeneous effects have been postulated to arise from differences in receptor, G protein, and effector stoichiometries in different tissues and to differences in the ability of different cells to compensate for the abnormality. It should be emphasized, however, that there is at present no evidence to suggest that the alterations in the levels of $G\alpha_s$ result from a mutation in the $G\alpha_s$ gene itself (Ram et al., 1997). Indeed, numerous transcriptional and posttranscriptional mechanisms regulate the levels of G protein α-subunits (Milligan and Green, 1991; Hadcock and Malbon, 1993) and the elevated levels of $G\alpha_s$ could potentially represent the sequellae of alterations in any one of these other biochemical pathways. Thus, the elucidation of the mechanism(s) underlying the elevations in the levels of $G\alpha_s$ both in the brain and in peripheral cells may provide important clues regarding the complex biochemical changes in this disorder. In addition to $G\alpha_s$, other G proteins show relatively selective CNS distribution, thereby making them more attractive putative candidates underlying the pathophysiology of bipolar affective disorder. One intriguing G protein is Golf (originally named because it was identified in olfactory neurons) because of its very limited expression primarily in dopamine-rich areas of the CNS, and the location of its gene on chromosome 18 near a region recently showing modest linkage in bipolar affective disorder (Berrettini et al., 1994).

Abnormalities in Second-Messenger-Generating Systems in Anxiety Disorders

Panic Disorder

Patients with panic disorder (PD) experience such symptoms as racing heart, palpitations, and tremor, that are strongly reminiscent of βAR stimulation. Two studies (Nesse et al., 1984; Pohl et al., 1988) have demonstrated that patients with PD have exaggerated behavioral responses to parenteral administration of the β-adrenergic agonist, isoproterenol; however, direct examination of βAR on peripheral blood elements from patients with PD has yielded conflicting results. Three studies have found reduced leukocyte β-adrenoceptor density (Brown et al., 1988; Aronson et al., 1989; Maddock et al., 1993), one study found an increase (Albus et al., 1986), and one using the highly β-adrenoselective ligand ^{125}I-pindolol found no difference (Huzel et al., 1993).

In contrast to the conflicting data with regard to β-adrenoceptors per se, results from several studies suggest a possible alteration distal to the receptor in patients with panic disorder. Thus, attenuated β- or α_2-adrenoceptor regulation of peripheral cell AC have been observed (Charney et al., 1989; Maddock et al., 1993); these findings raise the possibility of dysfunction at the level of G proteins, which have recently been examined more directly (Stein et al., 1996). This study did not observe significant differences in the levels of $G\alpha_s$, $G\alpha_{i1/2}$, or $G\alpha_{q/11}$, or in the pertussis-toxin-catalyzed [^{32}P]ADP-ribosylation in tissues from subjects in the PD group compared to either an anxious comparison group (untreated social phobia patients) or controls. It remains possible, of course, that despite the normal levels of G protein α-subunits, panic disorder is associated with alterations in G protein function. In this context, it is interesting to note that previous reports have suggested the presence of postreceptor abnormalities in two conditions with some clinical overlap (Taylor et al., 1987) with PD. Thus, autonomic nervous system dysfunction has recently been identified in a subset of patients with mitral valve prolapse. These autonomic nervous system abnormalities have been postulated to result from abnormalities in G_s, which result in a "supercoupling" of βAR. The role of G_s in this dysautonomia has been studied by cholate extraction of G_s from erythrocytes and reconstitution into cyc⁻ S49 lymphoma membranes, which have normal receptor and adenylyl cyclase but lack G_s (Davies et al., 1987, 1991). These investigators have found that G_s isolated from the patients produce an abnormal "supercoupling" of β adrenoceptors when reconstituted into cyc$^-$S49 lymphoma, suggesting a fundamental alteration in the $G\alpha_s$ protein itself. However, the levels of $G\alpha_s$ in these patients is normal, and direct cloning and sequencing the α_s cDNA from neutrophils of four symptomatic patients and one control revealed no difference in the α_s cDNA sequence (Balasubramanyam et al., 1991), and suggest that the molecular lesion could be an abnormal posttranslational modification of $G\alpha_s$.

Posttraumatic Stress Disorder (PTSD)

Investigators have also utilized peripheral blood cells to investigate signal transduction pathways in posttraumatic stress disorder (PTSD), and have observed postreceptor alterations. In one study of 19 male subjects with combat-related PTSD and 35 age- and gender-matched healthy controls, Lerer and associates (1990) found significantly lower basal and forskolin-stimulated platelet adenylyl cyclase activity. Interestingly, receptor-stimulated responses were normal, once again suggesting a postreceptor abnormality.

Signal Transduction Pathways as Targets for the Actions of Antidepressants and Mood Stabilizing Agents

Antidepressants and G Proteins

Paralleling the recognition of the typically weeks-long latency to clinical response, studies of antidepressant treatment actions have moved from a focus on disparate acute effects to an interest in possible common chronic changes. Thus, in recent years, research has focused on the effects of chronic administration of antidepressants (ADs) on various aspects of neuronal function, with a wealth of studies demonstrating alterations in the density and/or sensitivity of several neurotransmitter receptor systems. Chronic administration of a variety of antidepressants (including tricyclics [TCAs], monoamine oxidase inhibitors [MAOIs], "atypical ADs," and electroconvulsive shock [ECS]) appear to downregulate/ desensitize the β-adrenergic receptors in rat forebrain (Sulser, 1984; Duman et al., 1985) and enhance the synaptic efficacy of serotonergic neurotransmission via the serotonin$_{1A}$ (5HT$_{1A}$) receptor in rat hippocampus (Blier et al., 1990; Chaput et al., 1991). These effects, however, do not fully explain the clinical efficacy of all antidepressants, and the dissociation between receptor number and their functional responsiveness has led to the investigation of possible postreceptor sites of actions for these drugs, in particular the G proteins and adenylyl cyclase (*see* Table 4). Following chronic treatment, desipramine (DMI) is reported to cause a functional uncoupling of the β receptor from G$_s$ in rat cortex (Okada et al., 1986; Tiong and Richardson, 1990) and to interfere with the breakdown of the βAR high-affinity ternary complex (Tsuchiya et al., 1988), as well as subsequent activation of AC, perhaps by decreasing the affinity of G$_s$ for guanine nucleotides (Yamaoka et al., 1988). Okada and associates (1988) have recently reported that pertussis toxin treatment of rats overcomes DMI-induced βAR desensitization (as assessed by measurements of isoproterenol-stimulated AC activity) without attenuating DMI-induced β receptor downregulation. These studies suggest that antidepressants modify the interactions between the βAR and G$_i$/G$_o$, and are intriguing in view of the recently defined role of G protein βγ subunits in regulating receptor desensitization (Simonds et al., 1993). Using high (50–300 μ*M*) concentrations, it has also recently been demonstrated that a variety of antidepressants interfere with G$_s$ activation of AC in vitro (Yamaoka et al., 1988). Interestingly, greater inhibition was observed when the antidepressants were added prior to (rather than following) the addition of GTP, once again compatible with an antidepressant-induced inhibition of G protein dissociation (Yamaoka et al., 1988). However, given the high doses of antidepressants used, the physiological rele-

Table 4
Evidence for Effects of Antidepressants on G Proteins

Attenuation of receptor and postreceptor stimulated adenylyl cyclase
  activity
Attenuation of agonist-induced [^{3}H] GTP binding
Blockade of β-adrenergic receptor desensitization by pretreatment with
  pertussis toxin
Increase in rat brain and cultured cell β-adrenergic receptor K_L/K_H ratio
Alteration in the immunolabeling of $G\alpha_s$ and $G\alpha_o$ in rat hippocampus
Alteration of $G\alpha_s$, $G\alpha_o$, and $G\alpha_{12}$ mRNA in rat brain

---

[a]Alteration in coupling of $G\alpha_s$ with adenylyl cyclase in rat cortex and in cultured cells.

vance of these findings remains unclear, and requires further study.

Our laboratory has recently investigated the effects of chronic administration of antidepressants on various aspects of G proteins both ex vivo and in vitro. Using immunoblotting we have observed a complex, regionally specific pattern of effects on various G protein α subunits and their mRNA levels following chronic treatment with various antidepressants (Lesch and Manji, 1992). This complex pattern of effects is not altogether surprising in view of the differing pharmacologic profiles of the drugs (and their metabolites) with respect to receptor binding, monoamine reuptake blockade, and enzymatic (MAO) inhibition (Lesch and Manji, 1992). Nevertheless, several interesting, novel common effects have emerged from our studies. Thus, all antidepressants tended to decrease the immunolabeling of $G\alpha_s$ in the hippocampus, consistent with the recent report of postreceptor desensitization of hippocampal adenylyl cyclase activity following chronic desipramine (Tiong and Richardson, 1990), and suggesting a coordinate downregulation of the β-adrenoceptor-G_s-adenylyl cyclase complex by antidepressants (Sugrue, 1983; Sulser, 1984). Paralleling our findings, Duman and associates (1989) have shown that chronic administration of imipramine decreases the levels of $G_s\alpha$ mRNA in rat brain.

We also found that chronic antidepressant treatment tended to decrease the immunolabeling of $G\alpha_{i1/2}$, in hippocampus, whereas all the tricyclics (but not the MAOI) tended to increase the levels of $G\alpha_o$ in the hippocampus. These findings are complementary to the studies of Okada and associates (1988) described above, and suggest the possible involvement of pertussis toxin substrates ($G\alpha_i$, $G\alpha_o$) in mediating DMI's effects.

Moreover, the opposite effects of the ADs on the levels of $G\alpha_i$ and $G\alpha_o$ that we have demonstrated also offer an explanation for the seemingly disparate findings of hippocampal $5HT_{1A}$ "sensitivity" following chronic ADs using functional biochemical or electrophysiological measures (Chaput et al., 1991; Lesch, 1991; Newman et al., 1992). Thus, chronic administration of ADS produces a relative shift in the hippocampal $5HT_{1A}$ receptor response from those mediated by G_i toward those mediated by G_o. In addition to these studies, it has recently been demonstrated that both antidepressants, on chronic administration, produce an enhanced coupling between $G\alpha_s$ and the catalytic unit of adenylyl cyclase (Menkes et al., 1983; Newman and Lerer, 1989; Ozawa and Rasenick, 1989). Taken together these results suggest that antidepressants via their complex effects on G proteins may attenuate β-adrenergic-mediated activation of G_s, while enhancing the effects of agents operating by pathways independent of the β-adrenergic receptor.

The mechanisms by which ADs affect the levels of G proteins remains unknown, but may involve alterations in expression or altered posttranslational modification, resulting in increased or decreased turnover of the proteins. In this context, G proteins are the targets of a number of posttranslational modifications (including ADP-ribosylation, phosphorylation, isoprenylation, and myristolation) that may modify their functional status and may also alter the turnover of the protein. The studies from our laboratory (Lesch and Manji, 1992), as well as those of Duman and associates (1989) suggest that at least some of the changes in G protein α-subunit steady-state concentrations during chronic antidepressant drug treatment may occur at the transcriptional level. We observed a distinct pattern of changes in Gα mRNA expression across brain regions following chronic antidepressant treatment, suggesting that in some cases these transcriptional effects were primary, whereas in other cases they likely represent the compensatory adaptations to primary changes at the level of the protein. Nevertheless, these transcriptional and posttranscriptional effects would be in keeping with the well known latency (days to weeks) of AD response.

To further examine the mechanism(s) by which antidepressants affect β-adrenoceptors and G proteins, we have investigated the effects of antidepressants on these parameters in a system lacking presynaptic input, namely cultured C6 glioma cells. Incubation of rat C6 glioma cells for 5–7 d with desipramine at levels comparable to those achieved in the human CNS produces a significant downregulation of β-adrenoceptors, accompanied by a reduction in the isoproterenol-stimulated cAMP accumulation (Honneger et al., 1986; Fishman and Finberg, 1987; Manji et al., 1991a). Interestingly, the magnitude of the β-receptor downregulation observed is similar to that induced by a variety of ADs in rat brain (Sulser, 1984). Moreover, similar to what has been reported in rat brain (Turkka et al., 1989) the β-adrenoceptor desensitization is accompanied by a significant increase in the K_L/K_H ratio for the receptor. The K_L/K_H ratio reflects the extent to which agonists stabilize the high-affinity ternary complex (composed of agonist, receptor, and G protein, whereby a lower dissociation constant $[K_H]$ favors the ternary complex). Recent studies have shown that the molecular determinants of the βAR involved in the formation of the ternary complex are distinct from those involved in the functional activation of the stimulatory G protein (Hausdorff et al., 1990).

As discussed above, antidepressants have been reported to interfere with the binding of guanine nucleotides to G_s and the subsequent breakdown of the high-affinity ternary complex (Tsuchiya et al., 1988); this would explain the dissociation between the two parameters thought to reflect the same coupling phenomenon, namely, the formation of the high-affinity ternary complex and stimulation of adenylyl cyclase. Unlike the situation observed in rat brain, we have been unable to detect an alteration in the immunolabeling of $G\alpha_s$ or $G\alpha_i$ following antidepressants in vitro (Lesch and Manji, 1992), despite the functional evidence for alteration of G_s function. It is possible that the decrease in the levels of $G\alpha_s$ and $G\alpha_i$ that we have demonstrated in vivo require an intact integrated system (or at least presynaptic input); however, more recent in vivo studies have also demonstrated that antidepressants result in an early "uncoupling" of the β-receptor from G_s, followed by changes in G_s function after chronic (> 3 wk) treatment (Okada et al., 1986; Tiong and Richardson, 1990); thus, the failure to detect any alterations in the immunolabeling of $G\alpha_s$ or $G\alpha_i$ following in vitro may be due simply to the relatively short (5-d) incubation period. Nevertheless, the data suggests that G proteins (in particular G_s and G_o) may be the targets for antidepressant drugs and worthy of further study in the development of novel therapeutic agents.

Similar to most antidepressants, electroconvulsive shock (ECS) desensitizes the β-adrenergic-mediated adenylyl cyclase activity in rat cortex and hippocampus (Sulser, 1984). Fewer studies have investigated the effects of repeated ECS on postreceptor sites. However, repeated ECS is reported to decrease both the agonist-induced [3H] GTP (Avissar et al., 1990) and [3H] forskolin (Gleiter et al., 1989) binding in rat cortex. More recently, using [35S] GTPγS binding, repeated ECS was found to decrease binding in prefrontal cortex and hippocampus, but did not have any effect in the striatum and increasing [35S] GTPγS binding in the amygdala (Nishida et al., 1990).

Lithium and G Proteins

Lithium remains the most widely used treatment for bipolar disorder and this monovalent cation represents one of psychiatry's most important treatments. Although it is far from the perfect drug, it has clearly revolutionized the treatment of this disorder (Goodwin and Jamison, 1990). Lithium not only treats the acute episode of mania, but also reduces the frequency and severity of recurrent episodes of mania and depression in bipolar patients and depression in unipolar patients (Goodwin and Jamison, 1990). Numerous placebo-controlled studies have unequivocally documented the efficacy of lithium in the long-term prophylactic treatment of bipolar disorder, and the beneficial effects appear to involve both a reduction in the number of episodes, as well as in their intensity, with approx 60–80% of all patients with bipolar affective disorder showing at least a partial response. Abundant experimental evidence has shown that lithium attenuates receptor-mediated adenylyl cyclase activity and PI turnover in rodents and in humans, in the absence of consistent changes in the density of the receptors themselves (Bunney and Garland-Bunney, 1987; Mork et al., 1992; Jope and Williams, 1994; Lenox and Manji, 1995). The first direct evidence that G proteins may be the targets of lithium's actions was provided by Avissar and colleagues (1988), who reported that lithium dramatically eliminated isoproterenol- and carbachol-induced increases in [^{3}H]GTP binding to various G proteins in rat cerebral cortical membranes. These effects were reported to occur both in vitro in the presence of 0.6 mM LiCl and in washed cortical membranes from rats treated with lithium carbonate for 2–3 wk, suggesting that the function of several G proteins (e.g., G_s, G_i, G_o) might be modified by this mood-stabilizing drug. In animals withdrawn from lithium for 2 d, the agonist-stimulated response ([^{3}H]GTP binding) returned. Although these studies have been of considerable heuristic interest and the preponderence of data do suggest an action of chronic lithium at the level of G proteins (*see* Table 5), such a direct action of lithium on G protein function has been difficult to replicate in light of a number of methodological problems (discussed in Lenox and Manji, 1995; Manji et al., 1995a). Moreover, the fact that routine assays of agonist-stim-

Table 5
Evidence for Lithium's Effects on G Proteins

Attenuation of receptor-stimulated adenylyl cyclase activity
Attenuation of receptor-mediated and GTPγS-mediated PI
  turnover
Attenuation of agonist-induced [^{3}H]GTP binding
Reversal of the effects of lithium by increasing GTP
Increase in lymphocyte and rat brain β adrenergic receptor
  K_L/K_H ratio
Increase in pertussis-toxin-catalyzed [^{32}P]ADP ribosylation in
  platelets and in rat brain
Reduction in $G\alpha_s$, $G\alpha_{i1}$, and $G\alpha_{i2}$ mRNA in rat cortex

ulated PI hydrolysis in brain slices are conducted in the presence of 10 mM LiCl suggests that the lithium ion does not directly exert any major effects on G protein function. There is, however, considerable evidence that chronic lithium administration affects G protein function, and numerous investigations have addressed the role of G proteins in the attenuated receptor-mediated AC activity and PI responses observed after chronic lithium administration in rodents and in humans (reviewed in Bunney and Garland-Bunney, 1987; Jope and Williams, 1994; Lenox and Manji, 1995).

Our laboratory has utilized in vivo microdialysis measurements of cAMP to assess lithium's effects on G proteins in the intact animal, since with previous in vitro and ex vivo studies the particular tissue preparation affected the results obtained. We found that chronic lithium treatment produced a significant increase in basal and postreceptor-stimulated (cholera toxin or forskolin) AC activity, but attenuated the β-adrenergic-mediated effect in rat frontal cortex or hippocampus (Manji et al., 1991b; Masana et al., 1992). Interestingly, chronic lithium treatment resulted in an almost absent cAMP response to pertussis toxin; taken together, these results suggest a lithium-induced attenuation of G_i function and of the β-adrenergic receptor/G_s interaction. To examine this more directly, we have measured the levels of various G protein α subunits after chronic lithium. We have been unable to demonstrate any alterations in the amounts of α_s (52 kDa), α_s (45 kDa), α_{i1-3}, or α_o (Masana et al., 1992). Moreover, lithium treatment resulted in a significant increase in pertussis-toxin-catalyzed [^{32}P]ADP-ribosylation in both frontal cortex and hippocampus. Since pertussis toxin selectively ADP-ribosylates the undissociated, inactive aβ heterotrimeric form of G_i (Ui, 1990), these results suggest that lithium activates G_i via a stabilization of the inactive conformation (discussed in detail below).

In order to examine lithium's effects on the $\beta AR/G_s$ interactions in more detail, we have measured the ratio of low and high affinity β-adrenergic receptor binding dissociation constants K_L/K_H (which corresponds to the degree of stabilization of the high affinity state). The attenuated functional responsiveness of the βAR after chronic lithium treatment was accompanied by an increase in the K_L/K_H ratio for the βAR (Turkka et al., 1992). Taken together, these results suggest that one of lithium's effects may be to stabilize the high affinity ternary complex, perhaps by interfering with GTP binding. If dissociation of G protein subunits were inhibited by lithium, this could both decrease β-adrenergic stimulated AC activity while simultaneously producing a relative stabilization of the receptor in a high affinity state.

Most of the studies of lithium's actions on G proteins (and indeed most of the current knowledge about it neurochemical effects) derives from animal and in vitro models. However, since lithium is clinically relevant only in humans, it is imperative to ascertain which of its actions in animals and in vitro generalize to humans. To overcome the potentially confounding, significant effects of alterations in mood–state-dependent biochemical and neuroendocrine parameters, our recent studies have utilized normal volunteers. Any changes would presumably be a generalizable effect of lithium in humans (although the magnitude of the effect may depend on the pre-existing "set-point" of the substrate). In a series of clinical investigations we examined the effects of 2 wk of lithium administration to healthy volunteers at "therapeutic levels" on dynamic binding parameters, AC activity, and G protein measures (Manji et al., 1991c; Risby et al., 1991; Hsiao et al., 1992). A particularly striking, novel finding of the study was lithium's differential effects on AC activity in platelets and lymphocytes. Basal and postreceptor-stimulated platelet AC activity was increased in platelets, but in the same subjects no significant effects were found in lymphocyte AC activity (indeed, basal AC activity in lymphocytes tended to decrease) (Risby et al., 1991). Since the inhibitory G protein, G_i, appears to exert significant inhibitory effects on platelets but not lymphocytes, these tissue-specific effects of lithium are compatible with an attenuation of G_i function. Similar to our findings in rat brain, we found no alterations in the levels of platelet G proteins (G_s or G_i), but observed a significant 40% increase in the pertussis-toxin-catalyzed [^{32}P]ADP-ribosylation in platelet membranes following chronic lithium (Hsiao et al., 1992), once again suggesting a stabilization of the inactive undissociated $\alpha\beta\gamma$ heterotrimeric form of G_i. Once again as observed in rat cortex and hippocampus, chronic lithium treatment produced a significant increase in the lymphocyte βAR K_L/K_H ratio (Risby et al., 1991), in the absence of any alterations in the density for the receptor.

Modulation of G Proteins by Lithium: A Synthesis

The data reviewed strongly argue for an effect of chronic lithium on the signal-transducing G proteins in both humans and rodents (*see* Table 5). Interestingly, for both G_s and G_i, lithium's major effects appear to be a stabilization of the heterotrimeric, undissociated ($\alpha\beta\gamma$) conformation of the G protein. Thus, one might postulate that there is a built-in temporal and spatial selectivity to lithium's actions, since it would exert its major effects on those neurotransmitter systems/ neuronal pathways that are most active and thereby are undergoing the highest rate of guanine–nucleotide exchange.

The allosteric modulation of G proteins might serve to explain lithium's long-term prophylactic efficacy in protecting susceptible individuals from spontaneous-, stress-, and drug (e.g., antidepressant, stimulant)-induced cyclic affective episodes. Additionally, although speculative, it might be postulated that these effects on G proteins represent the mechanistic basis by which chronic lithium administration blocks the development of supersensitive dopaminergic, adrenergic, and cholinergic receptors. Thus, many (but not all) previous studies have demonstrated that chronic lithium administration is able to block the development of lesion- or drug-induced supersensitive receptor responses (discussed in Bunney and Garland-Bunney, 1987; Lenox and Manji, 1995). These supersensitive responses have been assayed by biochemical, electrophysiological, and behavioral studies; interestingly, most receptor-binding studies have not generally demonstrated lithium-induced alterations in the density of the receptors themselves, suggesting that lithium exerts its effects at postreceptor sites, in particular at the level of the signal-transducing G proteins. Such a contention also receives support from the observation that lithium's effects appear to involve different classes of receptors coupled both to adenylyl cyclase and phospholipase C isozymes; however, it is clear that lithium exerts effects on additional intracellular targets, and it is likely that these effects also contribute to the prevention of supersensitive receptor responses.

At present, the molecular mechanism(s) underlying lithium's effects on G proteins remains to be fully elucidated. There is some evidence to suggest that lithium has acute, in vitro effects on G proteins (Avissar et al., 1988; Greenwood and Jope, 1994), but many of chronic lithium's effects persist after washing of the membranes, and are therefore likely be attributable to an indirect posttranslational modification of the G proteins (Jope and Williams, 1994; Lenox and Manji, 1995). Long-term effects of chronic lithium (and arguably the therapeutically relevant effects) may more likely be attributable to an indirect posttranslational modification of the G proteins.

Since G protein function is known to be regulated by phosphorylation (Sagi-Eisenberg, 1989; Houslay 1991) one potential mechanistic explanation for lithium's effects on G protein subunit dissociation is phosphorylation by PKC, an enzyme responsible for considerable convergence (*see* Table 6) and "crosstalk" between multiple second messenger systems (Houslay, 1991). At present, the possible effects of chronic lithium on the absolute levels of $G\alpha_s$ and $G\alpha_i$ remain unclear—two independent laboratories have not observed any alterations Li et al., 1991; Hsiao et al., 1992) whereas another one has reported small but significant decreases in the levels of $G\alpha_{i1}$, $G\alpha_{i2}$ in rat frontal cortex (Colin et al., 1991). However, chronic lithium administration has been reported to reduce the mRNA levels of several G protein α-subunits in rat brain, including $G\alpha_{i1}$, $G\alpha_{i2}$ (Colin et al., 1991; Li et al., 1991), and $G\alpha_s$ (Li et al., 1991), suggesting that lithium produces complex transcriptional and posttranscriptional effects after chronic administration, many of which may be mediated via its effects on PKC (Manji and Lenox, 1994).

Valproic Acid and G Proteins

Valproic acid (VPA) is a broad spectrum anticonvulsant drug with demonstrated efficacy in the treatment of manic-depressive illness (Bourgeois, 1989; McElroy et al., 1992; Bowder, 1996). Whereas the anticonvulsant effects of VPA are usually observed quite rapidly, its therapeutic effects in the treatment of manic-depressive illness requires chronic administration, often with a lag time of onset of action of several days. The anticonvulsant effects of VPA have been postulated to involve its biochemical actions on voltage-dependent Na^+ channels and/or inhibitory and excitatory amino acid systems (Rogawski and Porter, 1990; Löscher, 1993). The mechanism(s) underlying the antimanic effects of VPA have not yet been elucidated, but have been postulated to involve biochemical effects observed after chronic but acute administration (Post et al., 1992).

Table 6
Evidence for Lithium's Effects on Protein Kinase C

Biphasic effects on phorbol-ester-mediated neurotransmitter release in
rat hippocampus
Cross-desensitization with the effects of phorbol esters
Reduced in vitro PKC-mediated phosphorylation of major PKC
substrates (80 kDa, 43 kDa)
Reduced [3H]PDBu binding in hippocampal structures, notably CA1 and
subiculum
Reduced phorbol-ester-mediated Na^+/H^+ exchange in cultured cells
Reduced immunolabeling of PKC α and PKC ϵ in rat brain and cultured
cells

As discussed already, lithium (the prototypic drug used
in the treatment of manic-depressive illness) has been demonstrated to significantly attenuate βAR-mediated cAMP accumulation both in vivo and in vitro. These effects of lithium
have generally been attributed to an effect at the level of the
interaction of the receptor with the stimulatory G protein (G_s)
(Newman and Belmaker 1987; Mork and Geisler, 1989a,b;
Manji, 1992). However, postreceptor stimulation of AC (e.g.,
with forskolin, fluoride, or nonhydrolyzable analogs of GTP)
has also been shown to be inhibited by lithium in slices and
membranes from rat cerebral cortices (Andersen and Geisler,
1984; Geisler et al., 1985; Mork and Geisler, 1987; Newman
and Belmaker, 1987). Interestingly, the effects of chronic
lithium on inhibition of βAR-stimulated cAMP production
persists after washing of the membranes and are reversed by
increasing concentrations of GTP (Mork and Geisler, 1989b);
these results suggest that the physiologically relevant effects
of lithium (that is, those seen on chronic drug administration,
and not reversed immediately on drug discontinuation) may
be exerted at the level of signal transducing G proteins at a
GTP responsive step.

Together, these effects of chronic lithium at the level of
signal-transducing G proteins, and the fact that elevated levels of $G\alpha_s$ have been observed both in postmortem brain tissue (Young et al., 1993) and in peripheral cells (Young et al.,

1994; Manji et al., 1995b; Mitchell et al., 1997) from subjects with bipolar affective disorder, suggest that alterations in the levels and/or function of certain G proteins may underlie the pathophysiology of bipolar affective disorder, and as such may represent targets for drugs used in the treatment of the disorder.

Recent studies have examined the effects of VPA on components of the PKC and βAR-coupled cAMP-generating system (Chen et al., 1994, 1996). Cultured rat C6 glioma cells were chosen because the components of the βAR-coupled cAMP-generating system have been well characterized in these cells (Fishman and Finberg, 1987; Manji et al., 1991a), and more importantly, because chronic incubation of these cells with VPA (up to 1 mM) does not have any cytotoxic effects (Martin and Regan, 1988), thereby making them a suitable model for the study of chronic drug effects. Chronic VPA produced a significant alteration of the βAR-coupled cAMP generating system. These effects were observed at concentrations of VPA similar to those attained in the plasma in the clinical treatment of neuropsychiatric disorders, and were not accompanied by any toxic effects on C6 glioma cells. Since the clinical antimanic effects of VPA require chronic treatment (McElroy et al., 1992; Bowden et al., 1994), the effects of VPA on βAR-binding parameters are evident after several days of treatment. In contrast to what has been observed with chronic lithium treatment (discussed above), it was found that chronic VPA produced a significant reduction in the density of βARs. Interestingly, the decrease in number of βARs (approx 30%) was accompanied by an even greater decrease in ISO-stimulated cAMP production (52%). Chronic exposure of C6 cells to VPA also resulted in significantly lower levels of cAMP in response to both $MnCl_2$ (43% lower) and forskolin (61% lower), suggesting that chronic VPA also exerts effects at the βAR/G_s interaction, or at postreceptor sites (e.g., G_s, AC).

To elucidate the mechanism(s) underlying VPA's effects on postreceptor-stimulated cAMP production, immunolabeling with selective antibodies and toxin-catalyzed [^{32}P]ADP-ribosylation was used to characterize G proteins (Chen et al., 1996). It was found that chronic, but not acute, VPA incubation induced a marked decrease in the levels of $G\alpha_s$ 45 but not any other G protein α subunits examined ($G\alpha_s$ 52, $G\alpha_{i1-2}$, $G\alpha_o$, or $G\alpha_{q/11}$). It is now well established that the different forms of $G\alpha_s$ arise as a result of alternative splicing of a single gene, resulting in a differential relative expression of the subtypes in various tissues. At present, the mechanism(s) underlying VPA's selective effects on the levels of $G\alpha_s$ 45 is unknown; however, it is noteworthy that differential regulation of $G\alpha_s$ 45 has previously been observed in other systems (Chen et al., 1996). In contrast to the immunolabeling of $G\alpha_s$, chronic VPA incubation resulted in a decrease in the cholera toxin (CT) catalyzed [^{32}P]ADP-ribosylation of both G_s 52 and 45. These results suggest that chronic VPA results in a modification of $G\alpha_s$ 52, and renders it less susceptible to CT-catalyzed [^{32}P]ADP-ribosylation, without affecting the absolute levels of the protein. Consistent with the lack of effect on the immunolabeling of $G\alpha_{i1-2}$ or $G\alpha_o$, chronic VPA did not produce any alterations in pertussis-toxin-catalyzed [^{32}P]ADP-ribosylation. At present, the mechanism(s) by which VPA produces these effects is unknown. Whatever the underlying mechanism(s), given the evidence for the G proteins in the pathophysiology of manic-depressive illness as well as the effects of lithium on the $\beta AR/G_s/$adenylyl cyclase system, these effects may play a role in VPA's therapeutic effects, and are worthy of further study.

Regulation of Signal Transduction Pathways by Hormones: Implications for Mood and Anxiety Disorders

The CNS is a major target for the actions of hormones (including thyroid hormones, glucocorticoids and gonadal steroids) (Nemeroff, 1987), but the biochemical alterations ultimately responsible for producing the effects remain unclear. Malbon et al. (1988) coined the term "permissive hormones" to describe agents, such as glucocorticoids and thyroid hormones, that modulate the actions of a variety of agents acting through cAMP. In recent years it has become increasingly clear that many hormones modulate the activity of multiple interacting and overlapping neurotransmitter systems; signal transduction pathways are in a pivotal position in the CNS, able to affect the functional balance between neurotransmitter systems and thus represent attractive targets to explain the mechanisms by which hormones modulate multiple neurotransmitter systems. It is thus not surprising that, in recent years, investigators have examined the effects of hormonal manipulations on signal transduction pathways, and a growing body of evidence suggests that the modulatory effects of various hormones are exerted through regulation of second messenger pathways. In this context, recent evidence from rodent and cell culture studies point to signal transduction pathways as targets for gonadal steroids, and it has been demonstrated that Estradiol produces an uncoupling of various receptors (including dopamine D2, α_2 adrenergic, and μ opioid receptors) from their respective G proteins; these effects appear to be mediated, at least in part, by a stabilization of the G protein (G_i and/or G_o) in the inactive, undissociated $\alpha\beta\gamma$ heterotrimeric conformation. (Maus et al., 1989, 1990; Lagrange et al., 1994).

In addition to their effects on the cAMP generating system, considerable evidence also suggests a potent effect of gonadal steroids on the phosphoinositide (PI) second messenger generating system. Thus, estradiol pretreatment significantly modulates receptor-mediated PI hydrolysis in osteoblasts and osteoblast-like cells (Tokuda et al., 1992; Lieberherr et al., 1993), modulates receptor-mediated activation of phospholipase C in myometrial cells (Phaneuf et al., 1995), and regulates the activity of phospholipase C in the human breast cancer cell line MCF-7 (Graber et al., 1993). Since many of these effects occur in the absence of alterations in the receptors themselves, it is likely that estradiol exerts its effects at a postreceptor site. A growing body of evidence also suggests that estradiol exerts significant effects on PKC. Thus, estradiol has been demonstrated to modulate the activity of PKC in several systems, including brain (Sidorkina et al., 1988; Maizels et al., 1992; Maeda and Lloyd, 1993; Rebas et al., 1995); moreover, at least some studies suggests that estradiol exerts isozyme-specific effects on PKC, and on PKC–mRNA levels (Maizels et al., 1992; Thompson et al., 1993).

Perhaps the hormonal system postimplicated in mood and anxiety disorders is the thyroid hormone system, and these is considerable evidence that many systemic effects of thyroid hormone (e.g., heart rate) are secondary to increased sensitivity of adrenergic signal transduction pathways. The expression of α and β adrenergic receptors is altered in many tissues by hypo- and hyperthyroidism. Thyroid dysfunction also leads to postreceptor abnormality, which appears to be mediated, at least in part, by alterations in the expression of G protein subunits (Levine et al., 1990). Thus, in vivo alterations of hormone levels have been demonstrated to alter the steady state levels of several G protein subunits and thereby regulate the overall sensitivity of transmembrane signaling. In this context, perinatal hypothyroidism has been found to significantly reduce the levels of $G\alpha_s$ (Wong et al., 1994), whereas treatment of adult rats for 3 d with T3 reduces the abundance of the α subunits of G_{i1} and G_{i2} in synaptosomal membranes isolated from the cerebral cortex by 50% (Orford et al., 1992). Overall, although a variety of tissue-specific effects of thyroid hormone alterations on G proteins have been reported, the most consistent finding is that of an increase in the levels of $G\alpha_i$ in many tissues, including brain in the hypothyroid state; these effects may play a major role in the well-established cognitive and behavioral manifestations of hypothyroidism.

Many of the effects of glucocorticoids may also depend on altered signal transduction (Ros et al., 1989; Paulssen et al., 1992). The regulation of G protein expression by glucocorticoids has been studied in several tissues, most of which show an increase in the levels of $G\alpha_s$ following exposure to excess levels of glucocorticoids. It has been demonstrated that 7-d administration of corticosterone increases the steady-state levels of both $G\alpha_s$ protein and mRNA levels, while decreasing the levels of $G\alpha_i$ and its mRNA levels in cerebral cortex (Saito et al., 1989). Moreover, adrenalectomy without corticosterone replacement results in a significant 20% decrease in $G\alpha_s$ mRNA, suggesting in vivo physiological regulation.

Abundant evidence clearly supports an interaction between thyroid, gonadal, and adrenal hormones and psychiatric disorders, in particular mood and anxiety disorders. Thus, primary disorders of both thyroid and hypothalamic-pituitary-adrenal (HPA) axis have been linked with mood and anxiety symptoms. Additionally, there is general agreement that more than 50% of patients with major depression exhibit hyperactivity of the HPA axis, as evidenced by hypersortisolemia, ACTH hypersecretion, and nonsuppression of plasma cortisol after dexamethasone administration. Similarly, at least some studies have shown the administration of thyroid hormones to be beneficial in the treatment of refractory depression and rapid cycling bipolar affective disorder. Although a variety of biochemical effects may contribute to the CNS manifestations of abnormalities in thyroid and corticosteroid status, alterations in second messenger generating systems (with the inherent amplification of receptor responses) by these permissive hormones are an attractive mechanism. Similarly, the effects of gonadal hormones on G proteins and PKC isozymes, resulting in subsequent modification of signal transduction, may be one mechanism by which biological maturation (puberty or menopause) triggers the expression of certain psychotic illnesses (e.g., bipolar affective disorder in late adolescence, melancholic depression in late adulthood).

Alterations in Signal Transduction Pathways by Kindling and Stimulant-Induced Behavioral Sensitization

There are few well accepted animal models of psychiatric disorders, but two that have been of considerable heuristic value in the study of mood and anxiety disorders have been kindling and behavioral sensitization. In both of these paradigms, there is increased responsivity (either behavioral or electrophysiological) to repeated "low-dose" stimulation (pharmacological or electrical) over time. Indeed, the sensitized behavioral response has been reported to persist for months and years after discontinuation of drug administration. Considerable evidence implicates long-term alterations in midbrain dopaminergic transmission in the development of behavioral sensitization, but the cellular mechanism(s) underlying the long-term changes in excitability observed in kindled or stimulant-sensitized animals have not been fully elucidated.

In recent years, an appreciation of the major role of signal transduction pathways in the regulation of neuronal excitability has led to extensive research into their involvement in kindling and behavioral sensitization, and a growing body of evidence implicates alterations in both PKC and certain G proteins (especially G_i and G_o) (Steketee and Kalivas, 1991; Steketee et al., 1991). Based on the evidence that behavioral sensitization may involve a disinhibition of mesocorticolimbic dopamine neurons, Kalivas and associates (Steketee and Kalivas, 1991; Steketee et al., 1991) have undertaken an elegant series of studies to investigate the role of the G proteins, that couple somatodendritic dopamine D2 autoreceptors, and γ-aminobutyric acid B ($GABA_B$) receptors, which increase K^+ efflux. They have demonstrated that injection of pertussis toxin (which inactivates G_i and G_o) produced a significant augmentation of psychostimulant-induced motor activity and dopamine release in the nucleus accumbens, thereby implicating A10 dopamine neurons and G proteins in the development of behavioral sensitization. In keeping with these studies, specific decreases in $G\alpha_i$ and $G\alpha_o$ levels have been observed in response to chronic cocaine in the ventral tegmental area and nucleus accumbens, once again consistent with the hypothesis that regulation of G proteins represents part of the biochemical changes that underlie the effects of chronic psychostimulants (Nestler et al., 1990).

Interestingly, the long-term electrophysiological changes observed with electrical kindling also appear to involve pertussis-sensitive G proteins. Thus, injection of pertussis toxin into the amygdala of kindled rats markedly modifies kindled seizures, effects that appear to be mediated via an alteration in the afterdischarge threshold (Watanabe et al., 1991). In addition to these robust effects on G proteins, several studies have also implicated alterations in PKC activity as mediators of long-term alterations in neuronal excitability in the CNS following kindling. In particular, dramatic increases in membrane-associated PKC have been observed in the bilateral hippocampus (HIPP) up to 4 wk and in the amygdala/pyriform cortex (AM/PC) at 4 wk after the last kindled seizure (Daigen et al., 1991).

It is striking that both behavioral sensitization and kindling produce robust alterations in pertussis-toxin-sensitive G proteins and membrane-associated PKC activity in limbic structures, since lithium also targets the very same biochemical targets, and these models have considerable validity as models of bipolar affective disorder and mania. Thus, although considerable caution obviously needs to be employed when extrapolating from animal models, the available data suggests that using these models to screen for pharmacological agents modulating pertussis-toxin-sensitive G proteins and PKC activity offers much promise for the development of new treatments for bipolar affective disorder.

Concluding Remarks

Signal transduction through specific receptors plays a central role in neuronal activation. There has been a dramatic increase in our understanding the mechanisms by which the binding of a neurotransmitter to a receptor at the cell surface is translated into a biochemical (and ultimately physiologic) effect, and in our appreciation of the critical role of signal transduction in the pathophysiology and treatment of a variety of neuropsychiatric disorders. Regulation of signal transduction within critical regions of the brain affects the intracellular signal generated by multiple neurotransmitter systems; these pathways thus represent attractive putative mediators of the pathophysiology of mood and anxiety disorders, since the behavioral and physiological manifestations of the illnesses are complex (and include disruption of behavior, circadian rhythms, neurophysiological features of sleep, and neuroendocrine and biochemical regulation within the brain) and likely are mediated by a network of interconnected neurotransmitter pathways. There is now abundant direct and indirect evidence implicating signal transduction pathways in bipolar affective disorder; the evidence for abnormalities in second-messenger-generating systems in unipolar depression and anxiety disorders is currently indirect, but nonetheless, compelling.

It is also becoming increasingly clear that for many refractory patients with these disorders, new drugs simply mimicking the "traditional" drugs that directly or indirectly alter neurotransmitter levels and those that bind to cell-surface receptors may be of limited benefit, because such strategies implicitly assume that the target receptor(s) are functionally intact, and that altered synaptic activity will thus be transduced to modify the postsynaptic "throughput" of the system. However, the existence of abnormalities in signal transduction pathways suggests that for patients refractory to conventional medications, improved therapeutics may only be obtained by the direct targeting of postreceptor sites. Recent discoveries concerning a variety of mechanisms involved in the formation and inactivation of second messengers offers the promise for the development of novel pharmacological agents designed to "site specifically" target signal transduction pathways. The challenge for the next era in neuropsychopharmacology is to transform the knowledge gained from advances in neurobiology, cellular physiology, and molecular pharmacology into clinical use. In this context, it is also noteworthy that antidepressants and mood-stabilizing agents require chronic (days to weeks) administration for therapeutic efficacy; biochemical changes requiring such prolonged administration of a drug suggest the possibility of alterations at the genomic level that may be mediated in large part by the activation and inactivation of subsets of genes with temporal specificity. These changes in gene expression are likely mediated by the downstream effects of converging signal transduction pathways, which are modulated by lithium and antidepressants. In sum, the rapid technological advances in both biochemistry and molecular biology has greatly enhanced our ability to understand the complexities of the regulation of neuronal function; these advances hold much promise for the development of novel improved therapeutics for mood and anxiety disorders.

Acknowledgment

The author would like to acknowledge the invaluable contributions of William Z. Potter and Guang Chen. The author would also like to thank Celia Knobelsdorf for outstanding editorial assistance.

References

Albus, M., Bondy, B., and Ackenheil, M. (1986) Adrenergic receptors on blood cells: relation to the pathophysiology of anxiety. *Clin. Neuropharmacol.* **9 (Suppl.),** 359–361.

Andersen, P. H. and Geisler, A. (1984) Lithium inhibition of forskolin-stimulated adenylate cyclase. *Neuropsychobiology* **12,** 1–3.

Aronson, T. A., Carasiti, I., McBane, D., Whitaker-Azmitia, P. (1989) Biological correlates of lactate sensitivity in panic disorder. *Biol. Psychiatry* **26,** 463–477.

Avissar, S., Schreiber, G., Aulakh, C. S., et al. (1990) Carbamazepine and electroconvulsive shock attenuate beta-adrenoceptor and muscarinic cholinoceptor coupling to G proteins in rat cortex. *Eur. J. Pharmacol.* **189,** 99–103.

Avissar, S., Schreiber, G., Danon, A., et al. (1988) Lithium inhibits adrenergic and cholinergic increases in GTP binding in rat cortex. *Nature* **331,** 440–442.

Balasubramanyam, A., Davies, A. O., Codina, J., Birnbaumer, L. (1991) Abnormal Gs function in mitral valve prolapse dysautonomia is not associated with abnormal alpha S cDNA sequence. *Life Sci.* **48,** 789–793.

Berrettini, W. H., Bardakjian, J., Cappellari, C. B., Barnett, A. L. Jr., Albright, A., Nurnberger, J. I., Jr. Gershon, E. S. (1987a) Skin fibroblast beta-adrenergic receptor function in manic-depressive illness. *Biol. Psychiatry* **22(12),** 1439–1443.

Berrettini, W. H., Cappellari, C. B., Nurnberger, J. I. Jr., Gershon, E. S. (1987b) Beta-adrenergic receptors on lymphoblasts. A study of manic-depressive illness. *Neuropsychobiology* **17(1–2),** 15–18.

Berrettini, W. H., Ferraro, T. N., Goldin, L. R. et al. (1994) Chromosome 18 DNA markers and manic-depressive illness: evidence for a susceptibility gene. *Proc. Natl. Acad. Sci. USA* **91(13),** 5918–5921.

Blier, P., de Montigny, C., Chaput, Y. (1990) A role for the serotonin system in the mechanisms of action of antidepressant treatments: preclinical evidence. *J. Clin. Psychiatry* **51S,** 14–20.

Bourgeois, B. F. D. (1989) Valproate: clinical use, in *Anitepleptic,* third edition (Levy, R., Mattson, R., Meldrum, B., Penry, J. K., and Dreifuss, F. E., eds.), Raven, New York, pp. 663–641.

Bourne, H. R., Nicoll, R. (1993) Molecular machines integrate coincident synaptic signals. *Cell* **(Suppl. 72)** 65–75.

Bowden, C. L. (1996) Role of newer medications for bipolar disorder. *J. Clin. Psychopharmacol.* **16,** 48S–56S.

Bowden, C. L., Brugger, A. M., Swann, A. C., et al. (1994) Efficacy of divalproex vs. lithium and placebo in the treatment of mania. *JAMA* **271,** 918–924.

Bray, P., Simons, C., Guo, B., Puckett, C., Kamholz, J., Spiegel, A., Nirenberg, M. (1986) Human cDNA clones for four species of $G_{\alpha s}$ signal transduction protein. *Proc. Natl. Acad. Sci. USA* **83,** 8893–8897.

Brown, S. L., Charney, D. S., Woods, S. W., Heninger, G. R., and Tallman, J. (1988) Lymphocyte beta-adrenergic receptor binding in panic disorder. *Psychopharmacology* **94,** 24–28.

Bunney, W. E. and Garland-Bunney, B. L. (1987) Mechanism of action of lithium in affective illness: basic and clinical implications, in *Psychopharmacology: The Third Generation of Progress* (Meltzer, H. Y., ed.), Raven, New York, pp. 553–565.

Carstens, M. E., Engelbrecht, A. H., Russell, V. A., Aalberg, C., Gagiano, C. A., Chalton, D. O., et al. (1987) β-adrenoreceptors on lymphocytes of patients with major depressive disorders. *Psychiatry Res.* **20,** 239–248.

Chaput, Y., de Montigny, C., Blier, P. (1991) Presynaptic and postsynaptic modifications of the serotonin system by long-term administration of antidepressant treatments: an in vivo electrophysiologic study in the rat. *Neuropsychopharmacology* **5,** 219–224.

Charney, D. S., Innis, R. B., Duman, R. S., et al. (1989) Platelet alpha-2-receptor binding and adenylate cyclase activity in panic disorder. *Psychopharmacology* **98,** 102–107.

Chen, G., Manji, H. K., Wright, C. B., Hawver, D., and Potter, W. Z. (1996) Effects of valproic acid on β-adrenergic receptors, G proteins and adenylyl-cyclase in rat C6 glioma cells. *Neuropsychopharmacology,* **15,** 271–280.

Chen, G., Manji, H. K., Hawver, D. B., et al. (1994) Chronic sodium valproate selectively decreases protein kinase Cα and ε in vitro. *J. Neurochem.* **63,** 2361–2364.

Colin, S. F., Chang, H. C., Mollner, S., et al. (1991) Chronic lithium regulates the expression of adenylate cyclase and Gi-protein alpha subunit in rat cerebral cortex. *Proc. Natl. Acad. Sci. USA* **88,** 10,634–10,637.

Cooper S. J., Kelly J. G., King D. J. (1985) Adrenergic receptors in depression: effects of electro-convulsive therapy. *Br. J. Psychiatry* **147,** 23–29.

Daigen, A., Akiyama, K., Otsuki, S. (1991) Long-lasting change in the membrane-associated protein kinase C activity in the hippocampal kindled rat. *Brain Res.* **545(1–2),** 131–136.

Davies, A. O., Su, C. J., Balasubramanyam, A., et al. (1991) Abnormal guanine-nucleotide regulatory protein in "MVP Dysautonomia": evidence from reconstitution of Gs. *J. Clin. Endocrinol. Metab.* **72,** 867–875.

Davis, J. M. and Bersnahan, D. B. (1987) Psychopharmacology in clinical psychiatry, in *American Psychiatric Association Annual Review* (Hales, R. E. and Frances, A. J. eds.), vol. 6 American Psychiatric Press, Washington, DC, 159–187.

Duman, R. S., Terwilliger, R. Z., Nestler, E. J. (1989) Chronic antidepressanty regulation of Gsα and cyclic AMP-dependent protein kinase. *Pharmacologist* **31,** 182.

Duman, R. S., Strada, S. J., Enna, S. J. (1985) Effect of imipramine and adrenocorticotropin administration on the rat brain norepinephrine-coupled cyclic nucleotide generating system. Alteration in alpha and beta components. *J. Pharmacol. Exp. Ther.* **234,** 409–414.

Ebstein, R. P., Lerer, B., Shapira, B., et al. (1988) Cyclic AMP second-messenger signal amplification in depression. *Br. J. Psychiatry* **152,** 665–669.

Ebstein, R., Oppenheim, G., Ebstein, B. S., et al. (1986) The cyclic AMP second messenger system in man: the effects of heredity, hormones, drugs, aluminum, age and disease on signal amplification. *Prog. Neuropsychopharmacol. Biol. Psychiatry* **10,** 323–353.

Extein, I., Tallman, J., Smith, C. C., Goodwin, F. K. (1979) Changes in lymphocyte β-adrenergic receptors in depression and mania. *Psychiatry Res.* **1,** 191–197.

Fishman, P. H., Finberg, J. P. (1987) Effect of the tricyclic antidepressant desipramine on beta-adrenergic receptors in cultured rat glioma C6 cells. *J. Neurochem.* **49,** 282–289.

Geisler, A., Klysner, R., Andersen, P. H. (1985) Influence of lithium in vitro and in vitro and in vivo on the catecholamine-sensitive cerebral adenylate cyclase systems. *Acta Pharmacol. Toxicol.* **56,** 80–97.

Gleiter, C. H., Deckert, J., Nutt, D. J., et al. (1989) Electroconvulsive shock (ECS) and the adenosine neuromodulatory system: effect of single and repeated ECS on the adenosine A1 and A2 receptors, adenylate cyclase, and the adenosine uptake site. *J. Neurochem.* **52**, 641–646.

Goodwin, F. K. and Jamison, K. R. (1990) *Manic-Depressive Illness.* Oxford University Press, New York.

Graber, R., Sumida C., Vallette, G., and Nunez, E. A. (1993) Rapid and long-term effects of 17 beta-estradiol on PIP_2-phospholipase C-specific activity of MCF-7 cells. *Cell Signal* **5(2)**, 181–186.

Greenwood, A. F. and Jope, R. S. (1994): Brain G protein proteolysis by calpain: Enhancement by lithium. Brain Res. **636**, 320–326.

Grossman, F., Manji, H. K., Potter, W. Z. (1993) Platelet α_2-adrenoreceptors in depression: a critical examination. *J. Psychopharmacol.* **7**, 4–18.

Hadcock, J. R., Malbon, C. C. (1993) Agonist regulation of gene expression of adrenergic receptors and G proteins. *J. Neurochem.* **60**, 1–9.

Halper, J. P., Brown, R. P., Sweeney, J. A., et al. (1988) Blunted beta-adrenergic responsivity of peripheral blood mononuclear cells in endogenous depression: isoproterenol dose response studies. *Arch. Gen. Psychiatry* **45**, 241–244.

Hausdorff, W. P., Hnatowich, M., O'Dowd, F. O., Caron, M. G., Lefkowitz, R. J. (1990) A mutation of the β_2-adrenergic receptor impairs agonist activation of adenylyl cyclase without affecting high affinity agonist binding. Distinct molecular determinants of the receptor are involved in physical coupling to and functional activation of G_2. *J. Biol. Chem.* **265**, 1388–1394.

Healy, D., Carney, P. A., Leonard, B. E. (1983) Monoamine-related markers of depression: changes following treatment. *J. Psychiatr. Res.* **17**, 251–260.

Healy, D., Halloran, A. O., Carney, P. A., et al. (1985) Peripheral adrenoreceptors and serotonin receptors in depression: changes associated with response to treatment with trazodone on amitriptyline. *J. Affective Disord.* **9**, 285–296.

Honegger, U. E., Disler, B., Wiesmann, U. N. (1986) Chronic exposure of human cells in culture to the tricyclic antidepressant desipramine reduces the number of beta-adrenoceptors. *Biochem. Pharmacol.* **35**, 1899–1902.

Horn, E. M., Bilezkian, J. P. (1990) Mechanism of abnormal transmembrane signaling of the beta-adrenergic receptor in congestive heart failure. *Circulation* **80**, 271–283.

Horn, E. M., Corwin, S. J., Steinberg, S. F., et al. (1988) Reduced lymphocyte stimulatory guanine nucleotide regulatory protein and beta-adrenergic receptors in congestive heart failure and reversal with angiotensin converting enzyme inhibitor therapy. *Circulation* **78**, 1373–1379.

Houslay, M. D. (1991) "Crosstalk": a pivotal role for protein kinase C in modulating relationships between signal transduction pathways. *Eur. J. Biochem.* **195,** 9–27.

Hsiao, J. K., Manji, H. K., Chen, G. A., et al. (1992) Lithium administration modulates platelet Gi in humans. *Life Sci.* **50,** 227–233.

Hudson, C. J., Young, L. T., Li, P. P., Warsh, J. J. (1993) CNS signal transduction in the patho-physiology and pharmotherapy of affective disorders and schizophrenia. *Synapse* **13,** 278–293.

Huzel, L. L., Delaney, S. M. and Stein, M. B. (1993) Lymphocyte β-adrenergic receptors in panic disorder: findings with a highly selective ligand and relationship to clinical parameters. *J. Anxiety Disorders* **7,** 349–357.

Jope, R. S., Williams, M. B. (1994) Lithium and brain signal transduction systems. *Biochem. Pharmacol.* **77,** 429–441.

Kafka, M. S., Paul, S. M. (1986) Platelet α_2-adrenergic receptors in depression. *Arch. Gen. Psychiatry* **43,** 91–95.

Kafka, M. S., Numberger, J. I., Siever, L. J., et al. (1986) Alpha$_2$-adrenergic receptor function in patients with unipolar and bipolar affective disorder. *J. Affect. Disord.* **10,** 163–169.

Klysner, R., Geisler, A., Rosenberg, R. (1987) Enhanced histamine and β-adrenoreceptor mediated cyclic AMP formation in leukocytes from patients with endogenous depression. *Affect. Disord.* **13,** 227–232.

Lagrange, A. H., Ronnekleiv, O. K., Kelly, M. J. (1994) The potency of μ-opioid hyperpolarization of hypothalamic arcuate neurons is rapidly attenuated by 17 β-estradiol. *J. Neurosci.* **14(10),** 6196–6204.

Lenox, R. H. and Manji, H. K. (1995) Lithium, in *American Psychiatric Press Textbook of Psychopharmacology* (Nemeroff, C. B. and Schatzberg, A. F., eds.), American Psychiatric Press, Washington, DC, pp. 303–349.

Lerer, B., Bleich, A., Bennett, E. R., et al. (1990) Platelet adenylate cyclase and phospholipase C activity in posttraumatic stress disorder. *Biol. Psychiatry* **27** 735–740.

Lesch, K. P., Manji, H. K. (1992) Signal-transducing G proteins and antidepressant drugs: evidence for modulation of alpha subunit gene expression in rat brain. *Biol. Psychiatry* **32,** 549–579.

Lesch, K. P. (1991) 5-HT$_{1A}$ receptor responsivity in anxiety disorders and depression. *Prog. Neuro-Psychopharmacol. Biol. Psychiatry* **15,** 723–733.

Levine, M. A., Feldman, A. M., Robishaw, J. D., et al. (1990) Influence of thyroid hormone status on expression of genes encoding G protein subunits in the rat heart. *J. Biol. Chem.* **265,** 3553–3560.

Li, P. P., Tam, Y. K., Young, L. T. et al. (1991) Lithium decreases Gs, Gi-1 and Gi-2 alpha-subunit mRNA levels in rat cortex. *Eur. J. Pharmacol.* **206,** 165, 166.

Lieberherr, M., Grosse, B., Kachkache, M., Balsan, S. (1993) Cell signaling and estrogens in female rat osteoblasts: a possible involvement of unconventional nonnuclear receptors. *J. Bone Miner. Res.* **8(11),** 1365–1376.

Löscher, W. (1993) Effects of the antiepileptic drug valproate on metabolism and function of inhibitory and excitatory amino acids in the brain. *Neurochem. Res.* **18,** 485–502.

Maddock, R. J., Carter, C. S., Magliozzi, J. R., and Gietzen, D. W. (1993) Evidence that decreased function of lymphocyte beta adrenoreceptors reflects regulatory and adaptive processes in panic disorder with agoraphobia. *Am. J. Psychiatry* **150,** 1219–1225.

Maeda, T., Lloyd, R. V. (1993) Protein kinase C activity and messenger RNA modulation by estrogen in normal and neoplastic rat pituitary tissue. *Lab. Invest.* **68(4),** 472–480.

Magliozzi, J. R., Gietzen, D., Maddock, R. J., Haack, D., Doran, A. R., Goodman, T., et al. (1989) Lymphocyte β-adrenoreceptor density in patients with unipolar depression and normal controls. *Biol. Psychiatry* **26,** 15–25.

Maizels, E. T., Miller, J. B., Cutler, R. E. Jr., Jackiw, V., Carney, E. M., Mizuno, K., Ohno, S., Hunzicker-Dunn, M. (1992) Estrogen modulates Ca(2+)-independent lipid-stimulated kinase in the rabbit corpus luteum of pseudopregnancy. Identification of luteal estrogen-modulated lipid-stimulated kinase as protein kinase C delta. *J. Biol. Chem.* **267(24),** 17,061–17,068.

Malbon, C. C., Rapiejko, P. J., Watkins, D. C. (1988) Permissive hormone regulation of hormone-sensitive effector systems. *Trends Pharmacol. Sci.* **9,** 33–36.

Manji, H. K., Potter, W. Z., Lenox, R. H. (1995a) Signal transduction pathways. Molecular targets for lithium's actions. *Arch. Gen. Psychiatry* **52(7),** 531–543.

Manji, H. K., Chen, G., Shimon, H., et al. (1995b) Guanine nucleotide-binding proteins in bipolar affective disorder. Effects of long-term lithium treatment. *Arch. Gen. Psychiatry* **52,** 135–144.

Manji, H. K. and Lenox, R. H. (1994) Long-term action of lithium: a role for transcriptional and posttranscriptional factors regulated by protein kinase C. *Synapse* **16,** 11–28.

Manji, H. K. (1992) G proteins: implications for psychiatry. *Am. J. Psychiatry* **149,** 746–760.

Manji, H. K., Chen, G., Bitran, J. A., et al. (1991a) Chronic exposure of C6 glioma cells to desipramine desensitizes β-adrenoceptors, but increases Kl/Kh ratio. *Eur. J. Pharmacol. Mol. Pharmacol. Sec.* **206,** 159–162.

Manji, H. K., Bitran, J. A., Masana, M. I., et al. (1991b) Signal transduction modulation by lithium: cell culture, cerebral microdialysis and human studies. *Psychopharmacol. Bull.* **27,** 199–208.

Manji, H. K., Hsiao, J. K., Risby, E. D., et al. (1991c) The mechanisms of action of lithium. *Arch. Gen. Psychiatry* **48,** 505–512.

Mann, J. J., Brown, R. P., Halper, J. P., et al. (1985) Reduced sensitivity of lymphocyte beta-adrenergic receptors in patients with endogenous depression and psychomotor agitation. *N. Engl. J. Med.* **313,** 715–720.

Martin, M. L., Regan, C. M. (1988) The anticonvulsant sodium valproate specifically induces the expression of a rat glial heat shock protein which is identified as the collagen type IV receptor. *Brain Res.* **459,** 131–137.

Masana, M. I., Bitran, J. A., Hsiao, J. K., et al. (1992) In vivo evidence that lithium inactivates Gi modulation of adenylate cyclase in brain. *J. Neurochem.* **59,** 200–205.

Maus, M., Homburger, V., Bockaert, J., et al. (1990) Pretreatment of mouse striatal neurons in primary culture with 17 beta-estradiol enhances the pertussis toxin-catalyzed ADP-ribosylation of G alpha o,i protein subunits. *J. Neurochem.* **55,** 1244–1251.

Maus, M., Bertrand, P., Drouva, S., et al. (1989) Differential modulation of D1 and D2 dopamine-sensitive adenylate cyclases by 17 beta-estradiol in cultured striatal neurons and anterior pituitary cells. *J. Neurochem.* **52,** 410–418.

McElroy, S. L., Keck, J. P. E., Pope, J. H. G., Hudson, J. I. (1992) Valproate in the treatment of bipolar disorder: literature review and clinical guidelines. *J. Clin. Psychopharmacol.* **12,** 42S–52S.

Menkes, D. B., Rasenick, M. M., Wheeler, M. A., et al. (1983) Guanosine triphosphate activation of brain adenylate cyclase: enhancement by long-term antidepressant treatment. *Science* **219,** 65–67.

Milligan, G., Green, A. (1991) Agonist control of G protein levels. *Trends Pharmacol. Sci.* **12,** 207–209.

Mitchell, P. B., Manji, H. K., Chen, G., Jolkovsky, L. A., Smith-Jackson, E., Denicoff, K., Schmidt M., Potter W. Z. (1997) High levels of $G_s\alpha$ in platelets of euthymic patients with bipolar affective disorder. *Am. J. Psychiatry* **154,** 218–223.

Mork, A., Geisler, A., and Hollund, P. (1992) Effects of lithium on second messenger systems in the brain. *Pharmacol. Toxicol.* **71(S1),** 4–17.

Mork, A., Geisler A. (1989a) Effects of GTP on hormone-stimulated adenylate cyclase activity in cerebral cortex, striatum, and hippocampus from rats treated chronically with lithium. *Biol. Psychiatry* **26,** 279–288.

Mork, A., Geisler, A. (1989b) The effects of lithium in vitro and ex vivo on adenylate cyclase in brain are exerted by distinct mechanisms. *Neuropharmacology,* **28,** 307–311.

Mork, A., Geisler, A. (1987) Mode of action on the catalytic unit of adenylate cyclase from rat brain. *Pharmacol. Toxicol.* **60,** 241–248.

Nemeroff, C. B. (1987) *Handbook of Clinical Psychoneuroendocrinology.* Guilford, New York.

Nesse, R. M., Cameron, O. G., Curtis, G. C., McCann, D. S., Huber-Smith, M. J. (1984) Adrenergic function in patients with panic anxiety. *Arch. Gen. Psychiatry* **41,** 771–776.

Nestler, E. J., Terwilliger, R. Z., Walker, J. R. et al. (1990) Chronic cocaine treatment decreases levels of the G protein subunits Gi alpha and Go alpha in discrete regions of rat brain. *J. Neurochem.* **55(3),** 1079–1082.

Newman, M. E., Lerer, B., Lichtenberg, P., Shapira, B. (1992) Platelet adenylate cyclase activity in depression and after clomipramine and lithium treatment. *Psychopharmacology* **109,** 231–234.

Newman, M. E. and Lerer, B. (1989) Post-receptor-mediated increases in adenylate cyclase activity after chronic antidepressant treatment: relationship to receptor desensitization. *Eur. J. Pharmacol.* **162,** 345–352.

Newman, M. E., Belmaker, R. H. (1987) Effects of lithium in vitro and ex vivo on components of the adenylate cyclase system in membranes from the cerebral cortex of the rat. *Neuropharmacology* **26,** 211–217.

Nishida, A., Kaiya, H., Tohmatsu, T., et al. (1990) Electroconvulsive treatment: effects on phospholipase C activity and GTP binding activity in rat brain. *J. Neural Trans. Gen. Sect.* **81,** 121–130.

Okada, F., Tokumitsu, Y., Ui, M. (1988) Possible involvement of pertussis toxin substrates (Gi, Go) in desipramine-induced refractoriness of adenylate cyclase in cerebral cortices of rats. *J. Neurochem.* **51,** 194–199.

Okada, F., Tokumitsu, Y., Ui, M. (1986) Desensitization of beta-adrenergic receptor-coupled adenylate cyclase in cerebral cortex after in vivo treatment of rats with desipramine. *J. Neurochem.* **47,** 454–459.

Orford, M. R., Leung, F. C., Milligan, G., Saggerson, E. D. (1992) Treatment with triiodothyronine decreases the abundance of the alpha-subunits of Gi1 and Gi2 in the cerebral cortex. *J. Neurol. Sci.* **112(1–2),** 34–37.

Ozawa, H., Rasenick, M. M. (1989) Coupling of the stimulatory GTP-binding protein Gs to rat synaptic membrane adenylate cyclase is enhanced subsequent to chronic antidepressant treatment. *Mol. Pharmacol.* **36,** 803–808.

Pandey, G. N., Pandey, S. C., and Davis, J. M. (1990) Peripheral adrenergic receptors in affective illness and schizophrenia. *Pharmacol. Toxicol.* **3,** 13–36.

Pandey, G. N., Davis, J. M. (1986) Leukocyte β-adrenergic receptors: a marker for central β-adrenergic receptor function in depression. *Clin. Neuropharmacol.* **14, 9 (Suppl.),** 353–355.

Pandey, G. N., Janicak, P., Davis, J. M. (1985) Studies of beta-adrenergic receptors in leukocytes of patients with affective illness and effects of antidepressant drugs. *Psychopharmacol. Bull.* **21,** 603–609.

Pandey, G. N., Dysken, M. W., Garver, D. L., et al. (1979) Beta-adrenergic receptor function in affective illness. *Am. J. Psychiatry* **136,** 675–678.

Patten, J. L., Johns, D. R., Valle, D., Eil, C., Gruppuso, P. A., Steele, G., Smallwood, P. M., Levine, M. A. (1990) Mutation in the gene encoding the stimulatory G protein of adenylate cyclase in Albright's hereditary osteodystrophy. *N. Engl. J. Med.* **322,** 1412–1419.

Paulssen, R. H., Paulssen, E. J., Gautvik, K. M., Gordeladze, J. O. (1992) Modulation of G proteins and second messenger responsiveness by steroid hormones in GH3 rat pituitary tumour cells. *Acta Physiol. Scand.* **146(4)**, 511–518.

Phaneuf, S., Europe-Finner, G. N., MacKenzie, I. Z., Watson, S. P., and Lopez Bernal, A. (1995) Effects of oestradiol and tamoxifen on oxytocin-induced phospholipase C activation in human myometrial cells. *J. Reprod. Fertil.* **103(1)**, 121–126.

Pohl, R., Yeragani, V. K., Balon, R., Rainey, J. M., Lycaki, H., Ortiz, A., Berchou, R., Weinberg, P. (1988) Isoproterenol-induced panic attacks. *Biol. Psychiatry* **24**, 891–902.

Post, R. M., Weiss, S. R., Chuang, D. M. (1992) Mechanisms of action of anticonvulsants in affective disorders: comparisons with lithium. *J. Clin. Psychopharmacol.* **12**, 23S–35S.

Potter, W. Z. and Manji, H. K. (1994) Affective disorders and adrenergic function: an update. *Clin. Biochem.* **40**, 279–287.

Ram, A., Guedj, F., Cravchik, A., Weinstein, L., Cao, Q., Badner, J., Goldin, L. R., Grisaru, N., Manji, H. K., Belmaker, R. H., Gershon E. S., Gejman, P. V. 1997 No abnormality in the gene for the G protein stimulatory α subunit in patients with bipolar disorder. *Arch. Gen. Psychiatry* **54**, 44–48.

Rebas, E., Lachowicz, A., Lachowicz, L. (1995) Estradiol and pregnenolone sulfate could modulate PMA-stimulated and Ca2+/calmodulin-dependent synaptosomal membrane protein phosphorylation from rat brain in vivo. *Biochem. Biophys. Res. Commun.* **207**, 606–612.

Risby, E. D., Hsiao, J. K., Manji, H. K., et al. (1991) The mechanisms of action of lithium. *Arch. Gen. Psychiatry* **48**, 513–524.

Rogawski, M. A. and Porter, R. J. (1990) Antiepileptic drugs: pharmacological mechanisms and clinical efficacy with consideration of promising development stage compounds. *Pharmacol. Rev.* **42**, 223–286.

Ros, M., Northup, J. K., Malbon, C. C. (1989) Adipocyte G proteins and adenylate cyclase. Effects of adrenalectomy. *Biochem. J.* **257**, 737–744.

Ross, E. M. (1989) Signal sorting and amplification through G protein-coupled receptors. *Neuron* **3**, 141–152.

Sagi-Eisenberg, R. (1989) GTP-binding proteins as possible targets for protein kinase C action. *Trends Biochem. Sci.* **14**, 355–357.

Saito, N., Guitart, X., Hayward, M., et al. (1989) Corticosterone differentially regulates the expression of Gs alpha and Gi alpha messenger RNA and protein in rat cerebral cortex. *Proc. Natl. Acad. Sci. USA* **86**, 3906–3910.

Sarai, K., Nakahara, T., Kanagana, K., et al. (1982) Lymphocyte beta-adrenergic receptor function in affective disorders. *Adv. Biosci.* **40**, 161–165.

Schreiber, G., Avissar, S., Danon, A., Belmaker, R. H. (1991) Hyperfunctional G proteins in mononuclear leukocytes of patients with mania. *Biol. Psychiatry.* **29**, 273–280.

Sidorkina, O. M., Morozova, T. M., Rau, V. A. (1988) Translocation of protein kinase C under the action of estradiol from the cytosol into cell membranes and activation of the enzyme in target cells. *Biokhimiya* **53**, 406–412.

Siever, L. J., Uhde, T. W., Jimerson, D. C., Lake, C. R., Siberman, E. R., Post, R. M., et al. (1984) Differential inhibitory noradrenergic responses to clonidine in 25 depressed patients and 25 normal control subjects. *Am. J. Psychiatry* **141**, 733–741.

Siever, L. J. (1987) Role of noradrenergic mechanisms in the etiology of the affective disorders, in *Psychopharmacology: The Third Generation of Progress* (Meltzer, H. Y., ed.), Raven, New York, pp. 493–504.

Simonds, W. F., Manji, H. K., Lupas, A., Garritsen, A. (1993) G proteins and βARK: a new twist for the coiled-coil. *Trends Biochem. Sci.* **18**, 315–317.

Spiegel, A. M. Shenker, A., Weinstein, L. S. (1992) Receptor-effector coupling by G proteins: implications for normal and abnormal signal transduction. *Endocr. Rev.* **13**, 536–565.

Stein, M. B., Chen, G., Potter, W. Z., and Manji, H. K. (1996) G protein level quantification in platelets and leukocytes from patients with panic disorder. *Neuropsychopharmacology* **15**, 180–186.

Steketee, J. D., Kalivas, P. W. (1991) Sensitization to psychostimulants and stress after injection of pertussis toxin into the A10 dopamine region. *J. Pharmacol. Exp. Ther.* **259(2)**, 916–924.

Steketee, J. D., Striplin, C. D., Murray, T. F., et al. (1991) Possible role for G proteins in behavioral sensitization to cocaine. *Brain Res.* **545(1–2)**, 287–291.

Sugrue, M. F. (1983) Chronic antidepressant therapy and associated changes in central monoaminergic function. *Pharmacol. Ther.* **21**, 1–37.

Sulser, F. (1984) Antidepressant treatments and regulation of norepinephrine-receptor-coupled adenylate cyclase systems in brain. *Adv. Biochem. Psychopharmacol.* **39**, 249–261.

Taylor, A. A,, Mares, A., Pool J. L., et al. (1987) Mitral valve prolapse with symptoms of beta-adrenergic hypersensitivity. Beta2-adrenergic supercoupling with desensitization on isoproterenol exposure. *Am. J. Med.* **82**, 193–201.

Taylor, C. W. (1990) The role of G proteins in transmembrane signaling. *Biochem. J.* **272**, 1–13.

Thomson, F. J., Johnson, M. S., MacEwan, D. J., Mitchell, R. (1993) Oestradiol-17 beta modulates the actions of pharmacologically distinct forms of protein kinase C in rat anterior pituitary cells. *J. Endocrinol.* **136(1)**, 105–117.

Tiong, A. H., Richardson, J. S. (1990) Beta-adrenoceptor and post-receptor components show different rates of desensitization to desipramine. *Eur. J. Pharmacol.* **188**, 411–415.

Tokuda, H., Yoneda, M., Oiso, Y., Kozawa, O. (1992) Inhibitory effect of 17 beta-estradiol on prostaglandin E2-induced phosphoinositide hydrolysis in osteoblast-like cells. *Prostaglandins* **43(3)**, 271–280.

Tsuchiya, F., Ikeda, H., Hatta, Y., et al. (1988) Effects of desipramine administration on receptor adenylate cyclase coupling in rat cerebral cortex. *Jpn. J. Psychiatr. Neurol.* **42**, 858–860.

Turkka, J. G., Gurguis, W. Z., Potter, W. Z., and Linnoila, M. (1989) Ten day administration of desipramine produces an increase in K_L/K_H for β-receptors in rat hippocampus. *Eur. J. Pharmacol.* **167**, 87–91.

Turkka, J., Bitran, J. A., Manji, H. K., Linnoila, M., and Potter, W. Z. (1992) Effects of chronic lithium on agonist and antagonist binding to β adrenergic receptors of rat brain. *Lithium* **3**, 43–47.

Ui, M. (1990) Pertussis toxin as a valuable probe for G protein involvement in signal transduction, in *ADP-Ribosylation Toxins and G Proteins* (Moss, J. and Vaughn, M., eds.), American Society for Microbiology, Washington, DC, pp. 36–52.

Watanabe, Y., Mori, N., Uemura, S. (1991) Suppression of kindled seizure following intraamygdaloid injection of pertussis toxin in rats. *Neurosci. Lett.* **130(2)**, 199–202.

Weinstein, L. S., Shenker, A., Gejman, P. V., Merino, M. J., Friedman, E., Spiegel, A. M. (1991) Activating mutations of the stimulatory G protein in the McCune-Albright syndrome. *N. Engl. J. Med.* **325**, 1688–1695.

Weinstein, L. S., Gejman, P. V., Friedman, E., Kadowaki, T., Collins, R. M., Gershon, E. S., Spiegel, A. M. (1990) Mutations of the Gs alpha-subunit gene in Albright hereditary osteodystrophy detected by denaturing gradient gel electrophoresis. *Proc. Natl. Acad. Sci. USA* **87**, 8287–8290.

Werstiuk, E. S., Steiner, M., and Burns, T. (1990) Studies on leukocyte beta-adrenergic receptors in depression: a critical appraisal. *Life Sci.* **47**, 85–105.

Wong, C. C., Warsh, J. J., Sibony, D., Li, P. P. (1994) Differential ontogenetic appearance and regulation of stimulatory G protein isoforms in rat cerebral cortex by thyroid hormone deficiency. *Brain Res. Dev. Brain Res.* **79**, 136–139.

Wright, A. F., Crichton, D. N., Loudon, J. P., Morten, J. E., Steel, C. M. (1984) Beta-adrenoceptor binding defects in cell lines from families with manic-depressive disorder. *Ann. Hum. Genet.* **48(3)**, 201–214.

Yamaoka, K., Nanba, T., Nomura, S. (1988) Direct influence of antidepressants on GTP binding protein of adenylate cyclase in cell membranes of the cerebral cortex of rats. *J. Neural. Trans.* **71**, 165–175.

Young, L. T., Li, P. P., Kamble, A., Siu, K. P., Warsh, J. J. (1994) Mononuclear leukocyte levels of G proteins in depressed patients with bipolar disorder or major depressive disorder. *Am. J. Psychiatry* **151**, 594–596.

Young, L. T., Li, P. P., Kish, S. J., et al. (1993) Cerebral cortex G_s a protein levels and forskolin-stimulated cyclic AMP formation are increased in bipolar affective disorder. *J. Neurochem.* **61**, 890–898.

Zohar, J., Bannet, J., Drummer, D., Fisck, R., Epstein, R. P., and Belmaker, R. H. (1983) The responses of lymphocyte beta-adrenergic receptors to chronic propranolol treatment in depressed patients, schizophrenic patients, and normal controls. *Biol. Psychiatry* **18,** 553–560.

Determination of Adenosine Receptor–G Protein Coupling

Significance for Psychiatric and Neurological Disorders

Anna Lorenzen

Introduction

The purine nucleoside adenosine has been shown to act as a neuromodulator in many areas of the mammalian brain. One of the main actions of adenosine in the central nervous system is the modulation of the release of a variety of neurotransmitters. Adenosine acts as a general depressant in the brain. In contrast, adenosine receptor antagonists show somnolytic and stimulant properties. Although purinergic psychopharmaceuticals are rare, the nonselective adenosine receptor antagonist caffeine may currently be considered the most frequently used nonprescription drug (Fredholm, 1995). Caffeine is of therapeutical benefit in states of migraine headache and headache of nonvascular origin. It is used as an adjuvant in analgesic medications, for the treatment of idiopathic apnea in premature neonates, and for seizure prolongation during courses of electroconvulsive therapy for the treatment of depression (Sawynok, 1995).

The actions of adenosine are mediated through four different subtypes of G protein-coupled receptors, A1, A2a, A2b, and A3, coupled to a wide variety of effector systems. These receptor subtypes show different localizations in the brain and can be differentiated pharmacologically by the use of subtype-specific ligands. Clearly, for future therapeutic applica-

From: *Neuromethods, Vol. 31: G Protein Methods and Protocols*
Ed: R. K. Mishra, G. B. Baker, and A. A. Boulton Humana Press Inc.

tions of adenosine receptor ligands in state of disease, more selective drugs than caffeine are required in order to facilitate a higher specificity of the desired effects and to minimize side effects mediated by the different receptor subtypes. In order to understand better the mechanism of action of adenosinergic drugs, it seems reasonable to monitor all components of the signaling cascade of adenosine receptors during short- and long-term application of these compounds, because quantitative as well as qualitative changes in drug effects depending on the application period have been reported.

Physiological Functions of Adenosine Receptors in Brain

Adenosine-A1 Receptors

Adenosine-A1 receptors have a relatively widespread distribution in the human brain. Highest concentrations of this receptor subtype have been determined in the hippocampus, the cerebral cortex, the striatum, and the medial and anterior nuclei of the thalamus. Low levels were measured in the brainstem, the spinal cord, and in the cerebellar cortex (Fastboom et al., 1987). During cerebral ischemia or hypoxia, adenosine levels are increased (Hagberg et al., 1987), and adenosine acts as an endogenous neuroprotective agent in these states (Rudolphi et al., 1992; Martin et al., 1994). Via A1 receptors, adenosine inhibits the release of acetylcholine (Pedata et al., 1983; Carter et al., 1995), noradrenaline (Jonzon and Fredholm, 1984), dopamine (Harms et al., 1979; Michaelis et al., 1979), 5-hydroxytryptamine (Harms et al., 1979), glutamate (Dolphin and Archer, 1983), glutamate-induced release of aspartate (Drejer et al., 1987), and possibly the release of GABA (Harms et al., 1979; Hollins and Stone, 1980; Dolphin and Archer, 1983). Postsynaptic mechanisms, which alter neuronal excitability, also contribute to the beneficial effects of adenosine A1 receptor stimulation in ischemia (Proctor and Dunwiddie, 1987). In addition, acting via A1 receptors, adenosine is a natural anticonvulsant (Dragunow et al., 1985). The effects of antagonists of A1 receptors are opposite to the effects of adenosine. They promote seizures and prolong seizure duration (Dragunow et al., 1985), and deteriorate ischemic brain damage (Dux et al., 1990). However, when applied chronically, adenosine receptor antagonists exert opposite effects (Rudolphi et al., 1989; Georgiev et al., 1993). Caffeine applied chronically increases the number of adenosine A1 receptors and the quantity of the α-subunit of the inhibitory G protein, G_i. In addition, this treatment leads to an increased coupling of the receptor to G protein(s) and to an increased coupling of the receptor–G protein complex to adenylyl cyclase (Ramkumar et al., 1988). It is not understood at present if these long-term beneficial effects of A1 receptor antagonists are owing to changes in receptor densities or other adaptive changes.

Adenosine-A2 Receptors

Two different subtypes of A2 receptors exist in the brain. They are pharmacologically distinct (Daly et al., 1983) and have different amino acid sequences, as deduced from cDNA cloning (Fink et al., 1992; Stehle et al., 1992). The A2b receptor has been shown to be present throughout the brain (Daly et al., 1983), but its physiological role is essentially unknown. In contrast, the A2a receptor shows a more restricted localization in the caudate, putamen, nucleus accumbens, and the olfactory tubercle (Jarvis and Williams, 1989; Jarvis et al., 1989a; Parkinson and Fredholm, 1990). It is colocalized with dopamine-D2 receptors, but not D1 receptors, on striatal neurons (Schiffmann et al., 1991; Fink et al., 1992). This colocalization provides the anatomical basis for the behavioral effects of adenosine A2a receptor activation. Adenosine A2a receptors interact with dopamine-D2-receptors, which results in psychomotor depressant effects of adenosine agonists and psychomotor stimulant effects of adenosine antagonists, e.g., methylxanthines (for review, *see* Ferré et al., 1992). The effects elicited by A2a receptor stimulation are similar to the effects of neuroleptic drugs (Bridges et al., 1987; Heffner et al., 1987), which are

D2 receptor antagonists. On a molecular level, agonist occupation of A2a receptors leads to a decrease in affinity of D2 receptors both in the high- and in the low-affinity state for agonists, with no change in D2 antagonist affinity (Ferré et al., 1991). The A2a-selective agonist (2-*p*-carboxyethyl)phenyl-amino-5′-*N*-carboxamidoadenosine (CGS 21680) induced an increased coupling of the D2 receptor to G proteins (Ferré et al., 1993). The mechanism of this action of A2a agonists is not known at present; however, it seems distinct from the mechanism of guanine nucleotides (Ferré et al., 1993). Taken together, experimental evidence supports the notion that owing to their selective effects on D2 receptors, adenosine-A2a receptor agonists could be effective drugs in the treatment of schizophrenia.

Adenosine-A3 Receptors

Recently, a third subtype of the adenosine receptor family, the A3 receptor, was cloned using the polymerase chain reaction (PCR) with degenerate primers (Zhou et al., 1992; Salvatore et al., 1993). In the brain, activation of A3 receptors causes behavioral depressant effects (Jacobson et al., 1993). Acutely applied selective agonists of the A3 receptor, e.g., N^6-(3-iodobenzyl)-5′-methylcarboxamidoadenosine (IB-MECA), increase postischemic damage in the brain (Von Lubitz et al., 1994) and protect against chemically induced seizures (Von Lubitz et al., 1995). When applied chronically, A3 agonists showed protective effects against ischemia-induced neuronal damage and mortality, and reduced postepileptic mortality (Von Lubitz et al., 1994, 1995).

Signal Transduction by Adenosine Receptors

On a molecular level, the effects of the adenosine-A1 receptor subtype are mediated by different effector systems. This receptor subtype inhibits adenylyl cyclase (Van Calker et al., 1978; Cooper et al., 1980). The physiological role of inhibition of this enzyme by adenosine is not fully understood, since the inhibition of transmitter release by A1 receptors proceeds independently of cyclic AMP (cAMP) (Dunwiddie and Fredholm, 1985; Fredholm and Lindgren, 1987). Furthermore, activation of A1 receptors leads to an inhibition of calcium currents (Scott and Dolphin, 1987; Scholz and Miller, 1991), to a stimulation of potassium efflux (Trussell and Jackson, 1985, 1987) through direct interaction with G proteins, and to an inhibition of inositol phospholipid hydrolysis (Kendall and Hill, 1988; Nakahata et al., 1991). The relative importance of these mechanisms is not yet clear. Whereas some aspects of the inhibitory actions of adenosine may proceed independently of the inhibition of calcium currents (Scholz and Miller, 1992), the modulation of ATP-sensitive potassium channels by adenosine has recently been reported to be an essential mechanism in adenosine-induced preconditioning against cerebral ischemia (Heurteaux et al., 1995).

The modulation of effector systems by adenosine-A1 receptors proceeds via interaction with G proteins of the G_i and G_o subtype. A1 receptors from bovine brain have been copurified with pertussis toxin-sensitive G proteins (Munshi and Linden, 1989). These G proteins have been identified as G_{i1}, G_{i2}, and G_o (Munshi et al., 1991). A1 receptors from bovine (Freissmuth et al., 1991a) and human brain (Jockers et al., 1994) may interact with recombinant G_{i1}, G_{i2}, G_{i3}, and G_o with typical rank orders of potency. The functional role of a possible interaction of the A1 receptor with G_{i3}, in brain, however, has to be evaluated carefully, since previous studies have demonstrated the absence of the G_{i3} protein from bovine brain (Goldsmith et al., 1988). The relative contribution of the different G protein subtypes activated by the A1 receptor to the physiological effects is not known at present. Preliminary evidence indicates that G_o is the G protein subtype that links L-type calcium channels to the A1 receptor, and activated G_o may also act as a GTPase-activating protein (Sweeney and Dolphin, 1995), as has been described for other G protein effector systems (Bourne and Stryer, 1992).

In contrast to A1 receptors, both A2a and A2b receptors have been shown to interact with a single effector system and a single G protein subtype. Both receptors activate adenylyl cyclase via the stimulatory G protein, G_s (Van Calker et al., 1979; Prémont et al., 1979; Stehle et al., 1992; Marala and Mustafa, 1993).

The cloned A3 receptor has been shown to inhibit adenylyl cyclase through a pertussis toxin-sensitive G protein in stably transfected CHO cells (Zhou et al., 1992). In a cultured mast cell line (RBL-2H3 cells), stimulation of A3 receptors induces an increase in the levels of inositol-1,4,5-triphosphate and a consecutive rise in intracellular Ca^{2+} concentration (Ramkumar et al., 1993). In stably transfected CHO cells, A3 receptors interact with G_{i2}, G_{i3}, as well as G_q and/or G_{i1} G proteins, as shown recently by stimulation of 4-azidoanilido-[α-^{32}P]-guanosine-5′-triphosphate (AA-[α-^{32}P]GTP) incorporation into the respective α-subunits (Palmer et al., 1995). However, prolonged treatment with agonists induced a downregulation of only $G\alpha_{i3}$ and β-subunits (Palmer et al., 1995). No studies of A3 receptor–G protein interactions in brain have been performed so far. The signal transduction mechanisms of adenosine receptors through various G proteins and effectors are summarized in Table 1.

Investigation of Adenosine Receptor–G Protein Coupling

The interaction of adenosine receptors with G proteins has been studied by direct radioligand binding experiments of the receptor and by measurement of the effects of adenosine receptor ligands on the activity of G proteins. G protein activity was monitored by the study of the guanine nucleotide exchange reaction, which reflects the activation of the G protein, or by measurement of GTPase activity, which represents the off-switch step of G protein activation consecutive to the guanine nucleotide exchange reaction. All approaches can give estimates of the affinity and the efficacy of the receptor ligand employed.

Table 1

Signal Transduction by Adenosine Receptors in Brain

	Receptor subtype			
	A_1	A_{2a}	A_{2b}	A_3
G protein	G_{i1}, G_{i2}, G_{i3},[a] G_o	G_s	G_s	G_{i2},[b] G_{i3},[b] G_q/G_{11}[b]
Effector	cAMP$\downarrow$,	cAMP$\uparrow$	cAMP$\uparrow$	cAMP$\downarrow$,[c]
	gK$^+\uparrow$, gCa$^{2+}\downarrow$,			IP$_3$/Ca$^{2+}\uparrow$[d]
	IP$_3$/Ca$^{2+}\downarrow$			

[a]Interaction of adenosine-A1 receptors with G_{i3} has been shown in reconstituted systems, not in native brain membranes (Freissmuth et al., 1991a; Jockers et al., 1994).

[b]Interaction of A3 receptors with these G proteins was demonstrated in a stably transfected CHO cells (Palmer et al., 1995).

[c]Determined for A3 receptors stably transfected in CHO cells (Zhou et al., 1992; Salvatore et al., 1993).

[d]Detected in the RBL-2H3 mast cell line (Ramkumar et al., 1993).

Radioligand Binding Studies of Adenosine Receptors

It is generally accepted that adenosine receptors, like other receptors that activate their effectors through G proteins, may show two different affinities for agonists, depending on their G protein-coupling state. In the G protein-coupled state, the receptor displays a high affinity (K_H) for agonists. When the receptor is uncoupled from G proteins, e.g., in the presence of GTP, it is in a low-affinity state (K_L) for agonists (Gilman, 1987). Antagonists bind to both the G protein-coupled and uncoupled state with high affinity. The affinity of antagonists is regulated in a way distinct from agonists. Although it is generally assumed that binding of antagonists is not affected by the interaction of receptors with G proteins, experimental evidence for adenosine-A1 receptors demonstrates that this may not be the case.

In receptor binding studies, either selective agonists or antagonists are applicable. For example, 2-chloro-N^6-[³H]-cyclopentyladenosine ([³H]CCPA; Klotz et al., 1989) as an agonist or the antagonist 1,3-[³H]dipropyl-8-cyclopentylxanthine ([³H]DPCPX; Bruns et al., 1987; Lohse et al., 1987a) has been used for the A1 receptor subtype. For A2a receptors, [³H]CGS 21680 (Jarvis et al., 1989b) is currently the most selective agonist, and ¹²⁵I-AB-MECA (Gallo-Rodriguez et al., 1994; Olah et al., 1994) is the radioligand agonist preferred for A3 receptors.

Although direct characterization of the high- as well as the low-affinity states of the A1 receptor has been performed using a radiolabeled agonist (Lohse et al., 1984; Stiles, 1988), the use of agonist radioligands is usually impaired by high amounts of nonspecific binding at higher radioligand concentrations or in the presence of GTP, when the majority of receptor sites are in the low-affinity state for agonists. This problem is circumvented by the use of an antagonist radioligand. The lack of selective high-affinity antagonist radioligands for A2 and A3 receptors, however, has greatly hampered the investigation of receptor–G protein coupling in native membranes for these subtypes. Antagonist radioligand binding studies of A2a receptors have been performed with the nonselective antagonist [³H]xanthine amine congener ([³H]XAC) in the presence of 25–50 nM DPCPX, which selectively blocks the binding to A1 receptors (Ji et al., 1991, 1992). Alternatively, high- and low-affinity states of the A2a receptor were detectable with a radioiodinated high-affinity agonist 2-[2-(4-amino-3-[¹²⁵I]iodophenyl)ethylamino]adenosine (¹²⁵I-APE; Luthin et al., 1995a) and its 4-azido-derivative as a photoaffinity agonist ligand (Luthin et al., 1995b). Possibly the introduction of the A2a antagonist 8-(3,4-dimethoxystyryl)-1,3-dipropyl-7-[³H]methylxanthine ([³H]KF 17837S; Nonaka et al., 1994) will improve this situation.

In competition studies of antagonist binding by agonists, the affinities of agonists for the high- (K_H) and low- (K_L) affinity states of the receptor are determined. For antagonists, this experimental setting yields a single affinity state. In addition, this method also gives an estimate of the percentage of the receptors in the G protein-coupled or uncoupled state. This approach has been used in numerous studies of A1 receptor binding (e.g., Lohse et al., 1984; Stiles, 1988; Goodman et al.,

1982; Lorenzen et al., 1993). In addition, the effects of modulators of ligand receptor binding or of receptor–G protein interaction, e.g., mono- and divalent cations (Stiles, 1988; Fastboom and Fredholm, 1990; Johansson et al., 1992; Goodman et al., 1982; Lorenzen et al., 1993), guanine nucleotides (Yeung and Green, 1983; Stiles, 1988; Parkinson and Fredholm, 1990; Fastboom and Fredholm, 1990; Goodman et al., 1982; Lorenzen et al., 1993), pertussis toxin (Van der Ploeg, 1992), and *N*-ethylmaleimide (Yeung and Green, 1983; Fredholm et al., 1985; Lorenzen et al., 1993), have been investigated. Since agonists, but not antagonists, differentiate between two affinity states of receptors, the change in the affinity induced by GTP, the GTP shift, can be used to assess the intrinsic activity of A1 receptor ligands (Van der Wenden et al., 1995a,b), an approach applied previously for various other G protein-coupled receptors (Kent et al., 1980; Waelbroeck et al., 1982; Ehlert, 1985; Lahti et al., 1992). Whereas the affinity of antagonists is barely altered by the addition of GTP, full agonists of the A1 receptor show GTP shifts of ~6, and intermediate values were determined for partial agonists (Van der Wenden et al., 1995a,b).

Some controversy exists concerning the influence of the G protein-coupling state of the A1 receptor on antagonist binding. Endogenous adenosine, which is physiologically present in tissues and in membrane preparations, may modulate antagonist binding properties. In spite of the addition of adenosine deaminase, which degrades and inactivates adenosine, vesicular pools of adenosine, which are not accessible to the enzyme, may persist in membrane and solubilized preparations (Prater et al., 1992). Exogenously added GTP decreases the affinity of adenosine by shifting the receptors into the low-affinity, G protein-uncoupled state, which results in an increased binding of antagonists to the receptor. This increase in antagonist binding is characterized by an increase in affinity, in maximum binding capacity, or both (Yeung and Green, 1983; Stiles, 1988; Klotz et al., 1990; Lorenzen et al., 1993), and may be attributed to changes in the affinity of the receptor protein owing to uncoupling from the G protein, or to the dissociation of endogenous adenosine from the receptor. When the metabolically stable GTP analog guanosine-5′-[γ-thio]triphosphate (GTPγS) was added to affinity-purified A1 receptors reconstituted with G proteins of the G_i and G_o subtype, antagonist binding was increased (Freissmuth et al., 1991b). However, since the receptors were eluted from the affinity column with 20 mM adenosine and it was not shown that the receptor preparation used for antagonist binding was essentially free of this nucleoside, these data do not unanimously prove that the effects of GTP on antagonist binding are independent of contaminating adenosine.

Modulation of G Protein Activity by Adenosine Receptors

Occupation of G protein-coupled receptors by agonists catalyzes the exchange of GDP bound to the α-subunit of the G protein by GTP. GTP ligation of the α-subunit causes dissociation from the βγ complex and subsequent interaction of the activated α-subunit with effectors. The activation is terminated by hydrolysis of GTP by the GTPase activity of the G protein α-subunit (Gilman, 1987). Three steps of this activation–inactivation cycle have been monitored to study G protein–adenosine receptor interactions.

The release of [^{32}P]GDP by A2a and A1 agonists was studied after prelabeling of G proteins in bovine striatal membranes with [α-^{32}P]GTP either in the presence of isoproterenol, a β-adrenergic agonist, or in the presence of epinephrine and the β-adrenergic antagonist propranolol (Marala and Mustafa, 1993). Under optimal conditions, the A1-selective agonist R-PIA (R-N^6-phenylisopropyladenosine) and the A2a-selective agonist CGS 21680 induced a 30% increase in the release of [^{32}P]GDP from G_i and G_s proteins, respectively, above nonstimulated release levels. Agonist-dependent release of [^{32}P]GDP was dependent on the presence of a second guanine nucleotide. This is the first direct evidence of G_s

activation by A2a receptors. Owing to the low signal-to-noise ratio, however, more detailed characterizations of the A2a receptor–G protein interaction or of agonist potencies and efficacies were not possible.

Agonist-induced labeling of the activated α-subunit of G proteins has been performed with labeled GTP analogs, e.g., the metabolically stable [^{35}S]GTPγS (Freissmuth et al., 1991b; Lorenzen et al., 1993) or the photoaffinity probe AA-[α-^{32}P]GTP (Palmer et al., 1995). Whereas stimulation of [^{35}S]GTPγS binding by receptor agonists in membranes measures a composite response of all G protein subtypes activated, the individual G protein subtypes activated by a receptor can be identified by the use of AA-[α-^{32}P]GTP, which is covalently bound after UV radiation, in combination with immunoprecipitation of α-subunits with subtype-specific antisera (Offermanns et al., 1991; Laugwitz et al., 1994).

Stimulation of [^{35}S]GTPγS binding, which has been utilized for many different receptor systems (Hilf et al., 1989; Gierschik et al., 1991; Tian et al., 1994; Lorenzen et al., 1993; Traynor and Nahorski, 1995), is a simple and sensitive method to study the interaction of ligands with receptors, and to determine potencies and efficacies of ligands independently from the various second messenger systems activated. When A1 receptor–G protein coupling is studied in rat and bovine brain membranes, assays of [^{35}S]GTPγS binding utilize only 1–7.5 µg of membrane protein/tube, which facilitates the investigation of very small samples.

Basically, the Tris-HCl- (50 mM) buffered reaction mixture for measurement of A1 receptor-stimulated [^{35}S]GTPγS binding is performed in a 100-µL assay volume containing 0.3–0.5 nM [^{35}S]GTPγS, 1 mM dithiothreitol, 1 mM EDTA, 5 mM MgCl$_2$, 10 µM GDP, 100 mM NaCl, 0.2 U/mL adenosine deaminase, 0.2% bovine serum albumin, membranes (routinely 2–3 µg), and adenosine receptor ligands. Nonspecific binding, which amounts to 1% of the radioligand, is determined in the presence of 10 µM unlabeled GTPγS. Incubations are routinely performed for 2 h at 25°C, and are terminated by filtration of the samples through glass fiber filters and subsequent washing of the filters with Tris buffer (Lorenzen et al., 1993).

Binding of [^{35}S]GTPγS critically depends on the presence of free Mg^{2+} ions (Higashijima et al., 1987). Mg^{2+} ions exerted a biphasic effect on [^{35}S]GTPγS binding and stimulation of G protein activation by A1 receptors in bovine brain membranes. Low concentrations (1–10 μM) of this ion were necessary to detect [^{35}S]GTPγS binding to the membranes, whereas higher concentrations (100 μM–10 mM) facilitated agonist stimulation of G protein activation (Lorenzen et al., 1993). Second, for the detection of agonist-mediated effects, addition of GDP, or another second unlabeled guanine nucleotide (GTP or GDPβS, but not GTPγS or GMP) was necessary (Fig. 1). GDP reduced basal binding more than agonist-stimulated binding, and optimal signal to noise ratios were measured around 10 μM concentration of this nucleotide. This requirement for GDP may relate to the possibility that G protein α-subunits exist in a guanine nucleotide-free form in isolated membranes (Wieland et al., 1992), or to the activation of G proteins by agonist-unliganded receptors (Hilf and Jakobs, 1992). The addition of NaCl further decreases basal binding, but to a smaller extent than agonist-stimulated binding of [^{35}S]GTPγS, giving rise to more clearly measurable agonist effects on G protein activation. In the presence of 100 mM NaCl and 10 μM GDP, full agonists of adenosine A1 receptors lead to a two- to threefold increase in G protein labeling by [^{35}S]GTPγS.

Although exerting seemingly similar effects, GDP and NaCl act at different sites. Their mechanisms of action were investigated in radioligand binding assays of the A1 receptor and in [^{35}S]GTPγS binding experiments under identical conditions (Table 2). Whereas NaCl decreases the affinity of both the high-affinity state (determined as EC_{50} value for stimulation of [^{35}S]GTPγS binding and as K_H value in competition experiments) as well as the low-affinity state (K_L in competition experiments) of the receptor, GDP solely affects the high-affinity state of the A1 receptor.

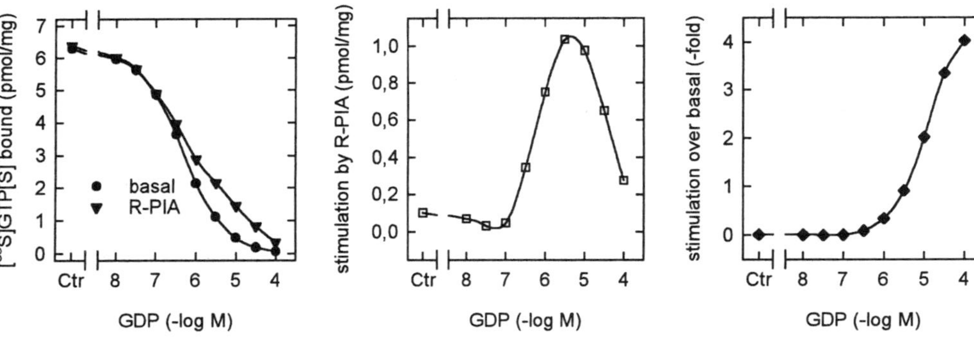

Fig. 1. Regulation by GDP of basal and adenosine A1 receptor-stimulated binding of [^{35}S]GTPγS. Two micrograms of bovine cerebral cortical membranes were incubated at 25°C for 2 h in the absence or presence of 10 μM R-PIA in a reaction mixture containing 50 mM Tris-HCl, 0.3 nM [^{35}S]GTPγS, 1 mM dithiothreitol, 1 mM EDTA, 5 mM MgCl$_2$, 100 mM NaCl, 0.2 U/mL adenosine deaminase, 0.2% bovine serum albumin, and GDP in the indicated concentrations. (Left) Absolute [^{35}S]GTPγS binding values. (Middle) agonist-induced stimulation (absolute values). (Right) Agonist stimulation in relation to basal values.

Table 2
Effect of NaCl and GDP on A1 Receptor-Binding and G Protein
Activation of the Agonist R-PIA[a]

Addition		[35S]GTPγS	[3H]DPCPX binding		
GDP	NaCl	EC_{50}	K_H	K_L	$\%R_H$
10 μM	—	0.35	0.59	9.78	61.3
10 μM	100 mM	1.94	1.15	29.19	41.8
—	100 mM	n.a.[b]	0.48	24.31	80.5
10 μM	100 mM	1.84	0.97	23.25	49.3

[a]Receptor binding (K_H and K_L values for [3H]DPCPX competition, percentage of receptors in the high-affinity state, $\% R_H$) and EC_{50} values for G protein activation ([35S]GTPγS binding) by R-PIA were determined under identical conditions in bovine brain membranes in the presence of 1 mM EDTA, 5 mM MgCl$_2$, with or without 10 μM GDP, and in the absence or presence of 100 mM NaCl. Data are quoted from Lorenzen et al., (1993) and are given in nmol/L.
[b]n.a., not applicable.

In these experimental settings, ligand affinities and relative efficacies are easily determined (Fig. 2). Whereas R-PIA, NECA (5'-*N*-ethylcarboxamidoadenosine), and ClA (2-chloroadenosine) are full A1 agonists of approximately the same efficacy, but different affinities, the ligand 5'-deoxy-5'-methylthioadenosine (MeSA) is a partial agonist with a 10-fold lower efficacy than R-PIA. Removal of the 2'-hydroxyl group from the ribose from the parent compound 2-chloroadenosine leads to a complete loss of agonist efficacy in cladribine (2-chloro-2'-deoxyadenosine). Rather, cladribine shows antagonist characteristics, since it inhibits basal binding of [35S]GTPγS, which may be owing to antagonism of residual endogenous adenosine, or to inhibition of basal activation of G proteins by agonist-unliganded adenosine receptors.

Finally, the GTPase activity of the G protein α-subunit terminates the activation of the G protein. Thus, in contrast to the release of GDP or binding of GTP, measurement of modulation of GTPase activity represents the off-switch step. Whereas [35S]GTPγS is barely released after binding to G proteins owing to its high GTPase stability, measurement of GTPase activity of the α-subunit permits determination of

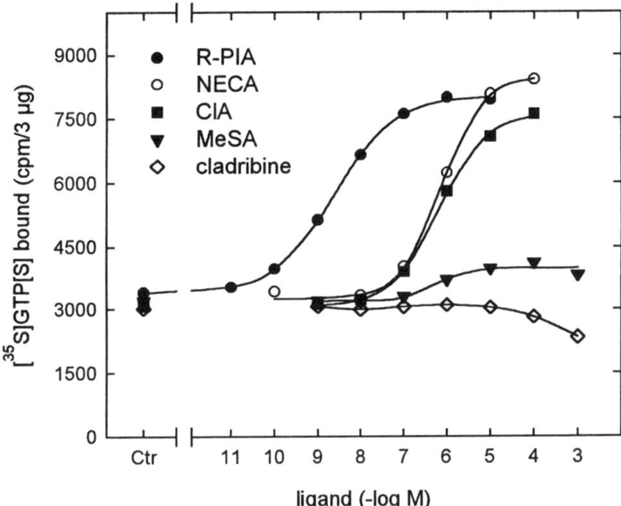

Fig. 2. Stimulation of G protein activation by adenosine-A1 receptors in bovine cerebral cortical membranes in the presence of ligands of different efficacies of the A1 receptor. R-PIA, *R-N*[6]-phenylisopropyladenosine; NECA, 5'-*N*-ethylcarboxamidoadenosine; CRA, 2-chloroadenosine; MeSA, 5'-deoxy-5'-methylthioadenosine; cladribine, 2-chloro-2'-deoxy-adenosine.

repeated activation of identical α-subunit molecules by one agonist stimulus. Although a better signal-to-noise ratio in GTPase measurements in comparison to [35S]GTPγS binding experiments would have to be expected owing to this amplification mechanism, increase in GTPase activity by A1 receptor agonists above unstimulated levels was only approx 20–50% (Haussleithner et al., 1985; Sweeney and Dolphin, 1995). In contrast to [35S]GTPγS binding studies, however, this technique facilitates the study of synergistic activation of G proteins by receptors and GTPase activating proteins, e.g., effector molecules (Bourne and Stryer, 1992; Sweeney and Dolphin 1995). This synergistic activation of agonist-stimulated GTPase induces a more rapid GTP turnover, which leads, on one hand, to a more rapid termination of the activation, but also increases the possibility of re-activation of the same α-subunit, which would considerably enhance the signal amplification of adenosine receptors by G protein activation.

Relevance for Psychiatric and Neurological Disorders

The interactions of adenosine receptors with G proteins are of interest with respect to the therapeutic and side effects of drugs, as well as to the etiology and therapy of diseases affecting the central nervous system. These interactions are dynamic, since they are influenced by the duration of exposure to adenosine receptor ligands, which has been shown by qualitative changes in drug effects depending on the duration of application in the prophylaxis of ischemia and seizures (Dragunow et al., 1985; Rudolphi et al., 1989; Dux et al., 1990; Georgiev et al., 1993).

Several drugs affect the signal transduction through adenosine receptors. The interaction of brain purines with benzodiazepine receptors was observed more than 15 years ago (Skolnick et al., 1980). Benzodiazepines, e.g., diazepam, triazolam, and Ro 15-1788, enhance the inhibitory actions of adenosine A1 receptors by inhibition of adenosine transporters, which increases the extracellular adenosine concentration (Phillis et al., 1981; Morgan and Stone, 1986; Phillis and Stair, 1987; O'Regan and Phillis, 1988). Ro 15-1788 also has direct inhibitory actions on adenosine receptors (Phillis and Stair, 1987). Chronic treatment with diazepam decreases the density of A2, but not of A1 receptors (Hawkins et al., 1988). Carbamazepine, which is used as an anticonvulsant and in affective psychoses, is a competitive antagonist at adenosine A1 receptors (Weir et al., 1990; Van Calker et al., 1991) and possibly at A2b receptors, but not at A2a receptors (Van Calker et al., 1991). Chronic carbamazepine treatment induces an increase in the density of A1 receptors in cerebral cortex, hippocampus, and cerebellum, which even persists for 2 mo after discontinuation of the drug (Marangos et al., 1985, 1987). In addition to their actions at the GABA–receptor complex, the barbiturates are selective competitive antagonists at adenosine A1 receptors (Lohse et al., 1985), and this mechanism of action probably explains their excitatory side effects (Lohse et al., 1987b,c).

Possible alterations of adenosine receptor signaling have been investigated in some diseases. In membranes from striatum from Alzheimer's disease patients, photoaffinity labeling of A2a receptors with an agonist radioligand did not reveal any differences from normal human subjects in receptor size and number (Ji et al., 1992). It is interesting to note that activation of adenylyl cyclase via the G_s protein as well as the direct activation of this enzyme by forskolin was impaired in the hippocampus and cerebellum from patients with Alzheimer's disease (Schnecko et al., 1994). These findings would imply that in spite of apparently "normal" A2a receptors, signal transduction through this receptor subtype is impaired owing to the reduced activity of this effector system. In contrast, inhibition of adenylyl cyclase by adenosine A1 receptors was not affected (Schnecko et al., 1994). Selective antagonists of A1 receptors are currently under development as cognition enhancers. Owing to their proconvulsant and cardiac stimulatory side effects, however, most compounds have proved to be too hazardous, especially in patients with impaired cardiac function. The analog of DPCPX, KFM 19, seems to be devoid of side effects on the heart (Schingnitz et al., 1991). Again, as pointed out above for the use of adenosine-A1 receptor agonists and antagonists for the treatment and prophylaxis of seizures and ischemia, the alterations in the signaling system of this receptor after prolonged treatment for chronic cognitive deficits, e.g., in Alzheimer's disease, rather than the acute effects are of high importance for the prognosis of a patient. Monitoring of all compounds of the signaling system will be of considerable importance to detect changes in sensitivity to drugs and to determine qualitative changes of drug effects over prolonged application periods even in preclinical research.

References

Bourne, H. R. and Stryer, L. (1992) The target sets the tempo. *Nature (Lond.)* **358**, 541–543.

Bridges, A. J., Moos, W. H., Szotek, D. L., Trivedi, B. K., Bristol, J. A., Heffner, T. G., Bruns, R. F., and Downs, D. A. (1987) N^6-(2,2-

Diethyl)adenosine, a novel adenosine receptor agonist with antipsychotic-like activity. *J. Med. Chem.* **30,** 1709–1711.

Bruns, R. F., Fergus, J. H., Badger, E. W., Bristol, J. A., Santay, L. A., Hartman, J. D., Hays, S. J., and Huang, C. C. (1987) Binding of the A1-selective adenosine antagonist 8-cyclopentyl-1,3-dipropylxanthine to rat brain membranes. *Naunyn-Schmiedeberg's Arch. Pharmacol.* **335,** 59–63.

Carter, A. J., O'Connor, W. T., Carter, M. J., and Ungerstedt, U. (1995) Caffeine enhances acetylcholine release in the hippocampus in vivo by a selective interaction with adenosine A1 receptors. *J. Pharmacol. Exp. Ther.* **273,** 637–642.

Cooper, D. M. F., Londos, C., and Rodbell, M. (1980) Adenosine receptor-mediated inhibition of rat cerebral cortical adenylate cyclase by a GTP-dependent process. *Mol. Pharmacol.* **18,** 598–601.

Daly, J. W., Butts-Lamb, P., and Padgett, W. (1983) Subclasses of adenosine receptors in the central nervous system: Interaction with caffeine and related methylxanthines. *Cell. Mol. Neurobiol.* **3,** 69–80.

Dolphin, A. C. and Archer, E. R. (1983) An adenosine agonist inhibits and a cyclic AMP analogue enhances the release of glutamate but not GABA from slices of rat dentate gyrus. *Neurosci. Lett.* **43,** 49–54.

Dragunow, M., Goddard, G. V., and Laverty, R. (1985) Is adenosine an endogenous anticonvulsant? *Epilepsia* **26,** 480–487.

Drejer, J., Frandsen, A., Honoré, T., and Schousboe, A. (1987) Adenosine inhibits glutamate stimulated [^{3}H]D-aspartate release from cerebellar granule cells. *Neurochem. Int.* **11,** 77–81.

Dunwiddie, V. and Fredholm, B. B. (1985) Adenosine modulation of synaptic responses in rat hippocampus: Possible role of inhibition or activation of adenylate cyclase. *Adv. Cyclic Nucleotide Protein Phosphorylation Res.* **19,** 259–272.

Dux, E., Fastboom, J., Ungerstedt, U., Rudolphi, K., and Fredholm, B. B. (1990) Protective effect of adenosine and a novel xanthine derivative propentofylline on the cell damage after bilateral carotid occlusion. *Brain Res.* **516,** 248–256.

Ehlert, F. J. (1985) The relationship between muscarinic receptor occupancy and adenylate cyclase inhibition in the rabbit myocardium. *Mol. Pharmacol.* **28,** 410–421.

Fastboom, J. and Fredholm, B. B. (1990) Regional differences in the effect of guanine nucleotides on agonist and antagonist binding to adenosine A1-receptors in rat brain, as revealed by autoradiography. *Neuroscience* **34,** 759–769.

Fastboom, J., Pazos, A., Probst, A., and Palacios, J. M. (1987) Adenosine A$_1$ receptors in the human brain: A quantitative autoradiographic study. *Neuroscience* **22,** 827–839.

Ferré, S., Snaprud, P., and Fuxe, K. (1993) Opposing actions of an adenosine A$_2$ receptor agonist and a GTP analogue on the regulation of dopamine D_2 receptors in rat neostriatal membranes. *Eur. J. Pharmacol. (Mol. Pharmacol. Section)* **244**, 311–315.

Ferré, S., von Euler, G., Johansson, B., Fredholm, B., and Fuxe, K. (1991) Stimulation of high-affinity adenosine A2 receptors decreases the affinity of dopamine D2 receptors in rat striatal membranes. *Proc. Natl. Acad. Sci. USA* **88**, 7238–7241.

Ferré, S., Fuxe, K., von Euler, G., Johansson, B., and Fredholm, B. B. (1992) Adenosine-dopamine interactions in the brain. *Neuroscience* **51**, 501–512.

Fink, J. S., Weaver, D. R., Rivkees, S. A., Peterfreund, S. A., Pollack, A. E., Adler, E. M., and Reppert, S. M. (1992) Molecular cloning of the rat A_2 adenosine receptor: selective co-expression with D_2 dopamine receptors in rat striatum. *Mol. Brain Res.* **14**, 186–195.

Fredholm, B. B. (1995) Adenosine, adenosine receptors and the actions of caffeine. *Pharmacol. Toxicol.* **76**, 93–101.

Fredholm, B. B. and Lindgren, E. (1987) Effects of N-ethylmaleimide and forskolin on noradrenaline release from rat hippocampal slices. Evidence that prejunctional adenosine and α-receptors are linked to N-proteins but not to adenylate cyclase. *Acta Physiol. Scand.* **130**, 95–105.

Fredholm, B. B., Lindgren, E., and Lindström, K. (1985) Treatment with N-ethylmaleimide selectively reduces adenosine receptor-mediated decreases in cyclic AMP accumulation in rat hippocampal slices. *Br. J. Pharmacol.* **86**, 509–513.

Freissmuth, M., Schütz, W., and Linder, M. E. (1991a) Interactions of the bovine brain A_1-adenosine receptor with recombinant G protein α-subunits. Selectivity for $rG_{i\alpha-3}$. *J. Biol. Chem.* **266**, 17,778–17,783.

Freissmuth, M., Selzer, E., and Schütz, W. (1991b) Interactions of purified bovine brain A_1 adenosine receptors with G proteins. Reciprocal modulation of agonist and antagonist binding. *Biochem. J.* **275**, 651–656.

Gallo-Rodriguez, C., Ji, X.-D., Melman, N., Siegman, B. D., Sanders, L. H., Orlina, J., Fischer, B., Pu, Q., Olah, M. E., van Galen, P. J. M., Stiles, G. L., and Jacobson, K. A. (1994) Structure-activity relationships of N^6-benzyladenosine-5'-uronamides as A_3-selective adenosine agonists. *J. Med. Chem.* **37**, 636–646.

Georgiev, V., Johansson, B., and Fredholm, B. B. (1993) Long-term caffeine treatment leads to a decreased susceptibility to NMDA-induced clonic seizures in mice without changes in adenosine A_1 receptor number. *Brain Res.* **612**, 271–277.

Gierschik, P., Moghtader, R., Straub, C., Dieterich, K., and Jakobs, K. H. (1991) Signal amplification in HL-60 granulocytes. *Eur. J. Biochem.* **197**, 725–732.

Gilman, A. G. (1987) G proteins: transducers of receptor-generated signals. *Annu. Rev. Biochem.* **56**, 615–649.

Goldsmith, P., Rossiter, K., Carter, A., Simonds, W., Unson, C. G., Vinitsky, R., and Spiegel, A. M. (1988) Identification of the GTP-binding protein encoded by G_{i3} complementary DNA. *J. Biol. Chem.* **263,** 6476–6479.

Goodman, R. R., Cooper, M. J., Gavish, M., and Snyder, S. H. (1982) Guanine nucleotide and cation regulation of the binding of [³H]cyclohexyladenosine and [³H]diethylphenylxanthine to adenosine A_1 receptors in brain membranes. *Mol. Pharmacol.* **21,** 329–335.

Hagberg, H., Andersson, P., Lacarewicz, J., Jacobson, I., Butcher, S., and Sandberg, M. (1987) Extracellular adenosine, inosine, hypoxanthine, and xanthine in relation to tissue nucleotides and purines in rat striatum during transient ischemia. *J. Neurochem.* **49,** 227–231.

Harms, H. H., Warden, G., and Mulder, A. H. (1979) Effects of adenosine on depolarization-induced release of various radiolabelled neurotransmitters from slices of rat corpus striatum. *Neuropharmacology* **18,** 577–580.

Haussleithner, V., Freissmuth, M., and Schütz, W. (1985) Adenosine-receptor-mediated stimulation of low-K_m GTPase in guinea pig cerebral cortex. *Biochem. J.* **232,** 501–504.

Hawkins, M., Pan, W., Stefanovich, P., and Radulovacki, M. (1988) Desensitation of adenosine A_2 receptors in the striatum of the rat following chronic treatment with diazepam. *Neuropharmacology* **27,** 1131–1140.

Heffner, T. G., Wiley, J. N., Williams, A. E., Bruns, R. F., Coughenour, L. L., and Downs, D. A. (1987) Comparison of the behavioral effects of adenosine agonists and dopamine antagonists in mice. *Psychopharmacology* **98,** 31–37.

Heurteaux, C., Lauritzen, I., Widmann, C. and Lazdunski, M. (1995) Essential role of adenosine, adenosine A_1 receptors, and ATP-sensitive K^+ channels in cerebral ischemic preconditioning. *Proc. Natl. Acad. Sci. USA* **92,** 4666–4670.

Higashijima, T., Ferguson, K. M., Sternweis, P. C., Smigel, M. D., and Gilman, A. G. (1987) Effects of Mg^{2+} and the $\beta\gamma$-subunit complex on the interactions of guanine nucleotides with G proteins. *J. Biol. Chem.* **262,** 762–766.

Hilf, G. and Jakobs, K. H. (1992) Agonist-independent inhibition of G protein activation by muscarinic acetylcholine receptor antagonists in cardiac membranes. *Eur. J. Pharmacol.* **225,** 245–252.

Hilf, G., Gierschik, P., and Jakobs, K. H. (1989) Muscarinic acetylcholine receptor-stimulated binding of guanosine-5'-O-(3-thiotriphosphate) to guanine-nucleotide-binding proteins in cardiac membranes. *Eur. J. Biochem.* **186,** 725–731.

Hollins, C. and Stone, T. W. (1980) Adenosine inhibits γ-amino-butyric acid release from slices of rat cerebral cortex. *Br. J. Pharmacol.* **69,** 107–112.

Jacobson, K. A., Nikodijevic, O., Shi, D., Gallo-Rodriguez, C., Olah, M. E., Stiles, G. L., and Daly, J. W. (1993) A role for central A_3-adenosine receptors. Mediation of behavioral depressant effects. *FEBS Lett.* **336,** 57–60.

Jarvis, M. F. and Williams, M. (1989) Direct autoradiographic localization of adenosine A_2 receptors in the rat brain using the A_2-selective agonist, [^{3}H]CGS 21680. *Eur. J. Pharmacol.* **168,** 243–246.

Jarvis, M. F., Jackson, R. H., and Williams, M. (1989a) Autoradiographic characterization of the high-affinity adenosine A_2 receptors in the rat brain. *Brain Res.* **484,** 111–118.

Jarvis, M. F., Schulz, R., Hutchison, A. J., Do, U. H., Sills, M. A., and Williams, M. (1989b) [^{3}H]CGS 21680, a selective A_2 adenosine receptor agonist directly labels A_2 receptors in rat brain. *J. Pharmacol. Exp. Ther.* **251,** 888–893.

Ji, X.-D., Stiles, G. L., and Jacobson, K. A. (1991) [^{3}H]XAC (xanthine amine congener) is a radioligand for A_2-adenosine receptors in rabbit striatum. *Neurochem. Int.* **18,** 207–213.

Ji, X.-D., Stiles, G. L., van Galen, P. J. M., and Jacobson, K. A. (1992) Characterization of human striatal A_2-adenosine receptors using radioligand binding and photoaffinity labeling. *J. Recept. Res.* **12,** 149–169.

Jockers, R., Linder, M. E., Hohenegger, M., Nanoff, C., Bertin, B., Strosberg, A. D., Marullo, S., and Freissmuth, M. (1994) Species difference in the G protein selectivity of the human and bovine A_1-adenosine receptor. *J. Biol. Chem.* **269,** 32,077–32,084.

Johansson, B., Parkinson, F. E., and Fredholm, B. B. (1992) Effects of mono- and divalent ions on the binding of the adenosine analogue CGS 21680 to adenosine A_2 receptors in rat striatum. *Biochem. Pharmacol.* **44,** 2365–2370.

Jonzon, B. and Fredholm, B. B. (1984) Adenosine receptor mediated inhibition of noradrenaline release from slices of the rat hippocampus. *Life Sci.* **35,** 1971–1979.

Kendall, D. A. and Hill, S. J. (1988) Adenosine inhibition of histamine-stimulated inositol phospholipid hydrolysis in mouse cerebral cortex. *J. Neurochem.* **50,** 497–502.

Kent, R. S., de Lean, A., and Lefkowitz, R. J. (1980) A quantitative analysis of beta-adrenergic receptor interactions: Resolution of high and low affinity states of the receptor by computer modeling of ligand binding data. *Mol. Pharmacol.* **17,** 14–23.

Klotz, K.-N., Lohse, M. J., Schwabe, U., Cristalli, G., Vittori, S., and Grifantini, M. (1989) 2-Chloro-N^6-[^{3}H]cyclopentyladenosine ([^{3}H]CCPA)— a high affinity agonist radioligand for A_1 adenosine receptors. *Naunyn-Schmiedeberg's Arch. Pharmacol.* **340,** 679–683.

Klotz, K.-N., Keil, R., Zimmer, F. J., and Schwabe, U. (1990) Guanine nucleotide effects on 8-cyclopentyl-1,3-[^{3}H]dipropylxanthine binding to membrane-bound and solubilized A_1 adenosine receptors of rat brain. *J. Neurochem.* **54,** 1988–1994.

Lahti, R. A., Figur, L. M., Piercey, M. F., Ruppel, P. L., and Evans, D. L. (1992) Intrinsic activity determinations at the dopamine D2 guanine nucleotide-binding protein-coupled receptor: utilization of receptor state binding affinities. *Mol. Pharmacol.* **42,** 432–438.

Laugwitz, K.-L., Spicher, K., Schultz, G., and Offermanns, S. (1994) Identification of receptor-activated G proteins: Selective immunoprecipitation of photolabeled G protein α subunits. *Methods Enzymol.* **237,** 283–294.

Lohse, M. J., Lenschow, V., and Schwabe, U. (1984) Two affinity states of R_i adenosine receptors in brain membranes. *Mol. Pharmacol.* **26,** 1–9.

Lohse, M. J., Klotz, K.-N., Jakobs, K. H., and Schwabe, U. (1985) Barbiturates are selective antagonists at A_1 adenosine receptors. *J. Neurochem.* **45,** 1761–1770.

Lohse, M. J., Klotz, K.-N., Lindenborn-Fotinos, J., Reddington, M., Schwabe, U., and Olsson, R. A. (1987a) 8-Cyclopentyl-1,3-dipropyl-xanthine (DPCPX)—a selective high affinity antagonist radioligand for A_1 adenosine receptors. *Naunyn-Schmiedeberg's Arch. Pharmacol.* **336,** 204–210.

Lohse, M. J., Brenner, A. S., and Jackisch, R. (1987b) Pentobarbital antagonizes the A_1 receptor-mediated inhibition of hippocampal neurotransmitter release. *J. Neurochem.* **49,** 189–194.

Lohse, M. J., Böser, S., Klotz, K.-N., and Schwabe, U. (1987c) Affinities of barbiturates for the GABA-receptor complex and A_1 adenosine receptors: a possible explanation of their excitatory effects. *Naunyn-Schmiedeberg's Arch. Pharmacol.* **336,** 211–217.

Lorenzen, A., Fuss, M., Vogt., H., and Schwabe, U. (1993) Measurement of guanine nucleotide binding protein activation by A_1 adenosine receptor agonists in bovine brain membranes: stimulation of guanosine-5′-O-(3-[^{35}S]thio)triphosphate binding. *Mol. Pharmacol.* **44,** 115–123.

Luthin, D. R., Olsson, R. A., Thompson, R. D., Sawmiller, D. R., and Linden, J. (1995a) Characterization of two affinity states of adenosine A_{2a} receptors with a new radioligand, 2-[2-(4-amino-3-[^{125}I]iodophenyl)-ethylamino]adenosine. *Mol. Pharmacol.* **47,** 307–313.

Luthin, D. R., Lee, K. S., Okonkwo, D., Zhang, P. J., and Linden, J. (1995b) Photoaffinity labeling with 2-[2-(4-azido-3-[I-125]iodophenyl)ethylamino]adenosine of A_{2a} adenosine receptors in rat brain. *J. Neurochem.* **65,** 2072–2079.

Marala, R. B. and Mustafa, S. J. (1993) Direct evidence for the coupling of A_2-adenosine receptor to stimulatory guanine nucleotide-binding-protein in bovine brain striatum. *J. Pharmacol. Exp. Ther.* **266,** 294–300.

Marangos, P. J., Weiss, S. R. B., Montgomery P., Patel, J., Narang, P. K., Cappabianca, A. M., and Post, R. M. (1985) Chronic carbamazepine treatment increases brain adenosine receptors. *Epilepsia* **26,** 493–498.

Marangos, P. J., Montgomery, P., Weiss, S. R. B., Patel, J., and Post, R. M. (1987) Persistent upregulation of brain adenosine receptors in response to chronic carbamazepine treatment. *Clin. Neuropharmacol.* **10,** 443–448.

Martin, R. L., Lloyd, H. G. E., and Cowan, A. I. (1994) The early events of oxygen and glucose deprivation: setting the scene for neuronal death? *Trends Neurosci.* **17,** 251–257.

Michaelis, M. L., Michaelis, E. K., and Myers, S. L. (1979) Adenosine modulation of synaptosomal dopamine release. *Life Sci.* **24,** 2083–2092.

Morgan, P. F. and Stone, T. W. (1986) Inhibition by benzodiazepines and beta-carbolines of brief (5 seconds) synaptosomal accumulation of [^{3}H]-adenosine. *Biochem. Pharmacol.* **35,** 1760–1762.

Munshi, R. and Linden, J. (1989) Co-purification of A_1 adenosine receptors and guanine nucleotide-binding proteins from bovine brain. *J. Biol. Chem.* **264,** 14,853–14,589.

Munshi, R., Pang, I.-H., Sternweis, P., and Linden, J. (1991) A_1 Adenosine receptors of bovine brain couple to guanine nucleotide-binding proteins G_{i1}, G_{i2}, and G_o. *J. Biol. Chem.* **266,** 22,285–22,289.

Nakahata, N., Abe, M. T., Matsuoka, I., Ono, T., and Nakanishi, H. (1991) Adenosine inhibits histamine-induced phosphoinositide hydrolysis mediated via pertussis toxin-sensitive G protein in human astrocytoma cells. *J. Neurochem.* **57,** 963–969.

Nonaka, H., Mori, A., Ichimura, M., Shindou, T., Yanagawa, K., Shimada, J., and Kase, H. (1994) Binding of [^{3}H]KF17387S, a selective adenosine A_2 receptor antagonist, to rat brain membranes. *Mol. Pharmacol.* **46,** 817–822.

Offermanns, S., Schultz, G., and Rosenthal, W. (1991) Identification of receptor-activated G proteins with photoreactive GTP analog, [α-^{32}P]GTP azidoanilide. *Methods Enzymol.* **195,** 286–301.

Olah, M. E., Gallo-Rodriguez, C., Jacobson, K. A., and Stiles, G. L. (1994) ^{125}I-4-Aminobenzyl-5'-N-methylcarboxamidoadenosine, a high affinity radioligand for the rat A_3 adenosine receptor. *Mol. Pharmacol.* **45,** 978–982.

O'Regan, M. H. and Phillis, J. W. (1988) Potentiation of adenosine-evoked depression of rat cerebral cortical neurons by triazolam. *Brain Res.* **445,** 376–379.

Palmer, T. M., Gettys, T. W., and Stiles, G. L. (1995) Differential interaction with and regulation of multiple G proteins by the rat A_3 adenosine receptor. *J. Biol. Chem.* **270,** 16,895–16,902.

Parkinson, F. E. and Fredholm, B. B. (1990) Autoradiographic evidence for G protein coupled A_2-receptors in rat neostriatum using [^{3}H]-CGS 21680 as a ligand. *Naunyn-Schmiedeberg's Arch. Pharmacol.* **342,** 85–89.

Pedata, F., Antonelli, T., Lambertini, L., Beani, L., and Pepeu, G. (1983) Effect of adenosine, adenosine triphosphate, adenosine deaminase, dipyridamole, and aminophylline on acetylcholine release from electrically-stimulated brain slices. *Neuropharmacology* **22**, 609–614.

Phillis, J. W. and Stair, R. E. (1987) Ro 15-1788 both antagonizes and potentiates adenosine-evoked depression of cerebral cortical neurons. *Eur. J. Pharmacol.* **136**, 151–156.

Phillis, J. W., Wu, P. H., and Bender, A. S. (1981) Inhibition of adenosine uptake into rat brain synaptosomes by the benzodiazepines. *Gen. Pharmacol.* **12**, 67–70.

Prater, M. R., Taylor, H., Munshi, R., and Linden, J. (1992) Indirect effect of guanine nucleotides on antagonist binding to A_1 adenosine receptors: occupation of cryptic binding sites by endogenous vesicular adenosine. *Mol. Pharmacol.* **42**, 765–772.

Prémont, J., Perez, M., Tassin, J.-P., Thierry, A.-M., Hervé, D., and Bockaert, J. (1979) Adenosine-sensitive adenylate cyclase in rat brain homogenates: kinetic characteristics, specificity, topographical, subcellular and cellular distribution. *Mol. Pharmacol.* **16**, 790–804.

Proctor, W. R. and Dunwiddie, T. V. (1987) Pre- and postsynaptic actions of adenosine in the in vitro rat hippocampus. *Brain Res.* **426**, 187–190.

Ramkumar, V., Bumgarner, J. R., Jacobson, K. A., and Stiles, G. L. (1988) Multiple components of the A_1 adenosine receptor-adenylate cyclase system are regulated in rat cerebral cortex by chronic caffeine ingestion. *J. Clin. Invest.* **82**, 242–247.

Ramkumar, V., Stiles, G. L., Beaven, M. A., and Ali, H. (1993) The A_3 adenosine receptor is the unique adenosine receptor which facilitates release of allergic mediators in mast cells. *J. Biol. Chem.* **268**, 16,887–16,890.

Rudolphi, K. A., Keil, M., Fastboom, J., and Fredholm, B. B. (1989) Ischaemic damage in gerbil hippocampus is reduced following upregulation of adenosine (A_1) receptors by caffeine treatment. *Neurosci. Lett.* **103**, 275–280.

Rudolphi, K. A., Schubert, P., Parkinson, F. A., and Fredholm, B. B. (1992) Neuroprotective role of adenosine in cerebral ischemia. *Trends Pharmacol. Sci.* **13**, 439–445.

Salvatore, C. A., Jacobson, M. A., Taylor, H. E., Linden, J., and Johnson, R. G. (1993) Molecular cloning and characterization of the human A_3 adenosine receptor. *Proc. Natl. Acad. Sci. USA* **90**, 10,365–10,369.

Sawynok, J. (1995) Pharmacological rationale for the clinical use of caffeine. *Drugs* **49**, 37–50.

Schiffmann, S. N., Jacobs, O., and Vanderhaeghen, J.-J. (1991) Striatal restricted adenosine A_2 receptor (RDC8) is expressed by enkephalin but not by substance P neurons: an in situ hybridization study. *J. Neurochem.* **57**, 1062–1067.

Schingnitz, G., Küfner-Mühl, U., Ensinger, H., Lehr, E., and Kuhn, F. J. (1991) Selective A_1-antagonists for treatment of cognitive deficits. *Nucleosides Nucleotides* **10,** 1067–1076.

Schnecko, A., Witte, K., Bohl, J., Ohm, T., and Lemmer, B. (1994) Adenylyl cyclase activity in Alzheimer's disease brain: stimulatory and inhibitory and signal transduction pathways are differentially affected. *Brain Res.* **644,** 291–296.

Scholz, K. P. and Miller, R. J. (1991) Analysis of adenosine actions on Ca^{2+} currents and synaptic transmission in cultured rat hippocampal pyramidal neurones. *J. Physiol. (Lond.)* **435,** 373–393.

Scholz, K. P. and Miller, R. J. (1992) Inhibition of quantal transmitter release in the absence of calcium influx by a G protein-linked adenosine receptor at rat hippocampal synapses. *Neuron* **8,** 1139–1150.

Scott, R. H. and Dolphin, A. C. (1987) Inhibition of calcium currents by an adenosine analogue 2-chloroadenosine, in *Topics and Perspectives in Adenosine Research* (Gerlach, E., and Becker, B. F., eds.), Springer-Verlag, Berlin, pp. 549–558.

Skolnick, P., Paul, S. M., and Marangos, P. J. (1980) Purines as endogenous ligands of the benzodiazepine receptor. *Fed. Proc.* **39,** 3050–3055.

Stehle, J. H., Rivkees, S. A., Lee, J. J., Weaver, D. R., Deeds, J. D., and Reppert, S. M. (1992) Molecular cloning and expression of the cDNA for a novel A_2-adenosine receptor subtype. *Mol. Endocrinol.* **6,** 384–393.

Stiles, G. L. (1988) A_1 Adenosine receptor-G protein coupling in bovine brain membranes: effects of guanine nucleotides, salt, and solubilization. *J. Neurochem.* **51,** 1592–1598.

Sweeney, M. I. and Dolphin, A. C. (1995) Adenosine A_1 agonists and the Ca^{2+} channel agonist Bay K 8644 produce a synergistic stimulation of the GTPase activity of G_o in rat frontal cortical membranes. *J. Neurochem.* **64,** 2034–2042.

Tian, W.-T., Duzic, E., Lanier, S. M., and Deth, R. C. (1994) Determinants of α_2 adrenergic receptor activation of G proteins: evidence for a precoupled receptor/G protein state. *Mol. Pharmacol.* **45,** 524–531.

Traynor, J. R. and Nahorski, S. R. (1995) Modulation by μ-opioid agonists of guanosine-5'-O-(3-[^{35}S]thio)triphosphate binding to membranes from human neuroblastoma SH-SY5Y cells. *Mol. Pharmacol.* **47,** 848–854.

Trussell, L. O. and Jackson, M. B. (1985) Adenosine-activated potassium conductance in cultured striatal neurons. *Proc. Natl. Acad. Sci. USA* **82,** 4857–4861.

Trussell, J. O. and Jackson, M. B. (1987) Dependence of an adenosine-activated potassium current on a GTP-binding protein in mammalian central neurons. *J. Neurosci.* **7,** 3306–3316.

Van Calker, D., Müller, M., and Hamprecht, B. (1978) Adenosine inhibits the accumulation of cyclic AMP in cultured brain cells. *Nature (Lond.)* **276,** 839–841.

Van Calker, D., Müller, M., and Hamprecht, B. (1979) Adenosine regulates via two different types of receptors, the accumulation of cyclic AMP in cultured brain cells. *J. Neurochem.* **33,** 999–1005.

Van Calker, D., Steber, R., Klotz, K.-N., and Greil, W. (1991) Carbamazepine distinguishes between adenosine receptors that mediate different second messenger responses. *Eur. J. Pharmacol. (Mol. Pharm. Section)* **206,** 285–290.

Van der Wenden, E. M., Hartog-Witte, H. R., Roelen, H. C. P. F., von Frijtag Drabbe Künzel, J. K., Pirovano, I. M., Mathôt, R. A. A., Danhof, M., van Aerschot, A., Lidaks, M. J., Ijzerman, A. P., and Soudijn, W. (1995a) 8-Substituted adenosine and theophylline 7-riboside analogues as potential partial agonists for the adenosine A_1 receptor. *Eur. J. Pharmacol. (Mol. Pharm. Section)* **290,** 189–199.

Van der Wenden, E. M., von Frijtag, Drabbe Künzel, J. K., Mathôt, R. A. A., Danhof, M., Ijzerman, A. P., Soudijn, W. (1995b) Ribose-modified adenosine analogues as potential partial agonists for the adenosine receptor. *J. Med. Chem.* **38,** 4000–4006.

Van der Ploeg, I., Parkinson, F. E., and Fredholm, B. B. (1992) Effect of pertussis toxin on radioligand binding to rat brain adenosine A_1 receptors. *J. Neurochem.* **58,** 1221–1229.

Von Lubitz, D. K., Lin, R. C., Popik, P., Carter, M. F., and Jacobson, K. A. (1994) Adenosine A_3 receptor and cerebral ischemia. *Eur. J. Pharmacol.* **263,** 59–67.

Von Lubitz, D. K., Carter, M. F., Deutsch, S. I., Lin, R. C., Mastropaolo, J., Meshulam, Y., and Jacobson, K. A. (1995) The effects of adenosine A_3 receptor stimulation on seizures in mice. *Eur. J. Pharmacol.* **275,** 23–29.

Waelbroeck, M., Robberecht, P., Chatelain, P., and Christophe, J. (1982) Rat cardiac muscarinic receptors. I. Effects of guanine nucleotides on high- and low-affinity binding sites. *Mol. Pharmacol.* **21,** 581–588.

Weir, R. L., Anderson, S. M., and Daly, J. W. (1990) Inhibition of N^6-[^{3}H]cyclohexyladenosine binding by carbamazepine. *Epilepsia* **31,** 503–512.

Wieland, T., Kreiss, J., Gierschik, P., and Jakobs, K. H. (1992) Role of GDP in formyl-peptide-receptor-induced activation of guanine-nucleotide-binding proteins in membranes of HL 60 cells. *Eur. J. Biochem.* **205,** 1201–1206.

Yeung, S.-M. H. and Green, R. D. (1983) Agonist and antagonist affinities for inhibitory adenosine receptors are reciprocally affected by 5′-guanylylimidodiphosphate or N-ethylmaleimide. *J. Biol. Chem.* **258,** 2334–2339.

Zhou, Q.-Y., Li, C., Olah, M. E., Johnson, R. A., Stiles, G. L., and Civelli, O. (1992) Molecular cloning and characterization of an adenosine receptor: the A_3 adenosine receptor. *Proc. Natl. Acad. Sci. USA* **89,** 7432–7436.

G Proteins and Mood Disorders

Jun-Feng Wang and L. Trevor Young

Introduction

A neurobiological basis for mood disorders has long been postulated, but is yet to be conclusively established. Earlier studies on the monoaminergic neurotransmitter systems in mood disorder have been very suggestive, although not conclusive, of alterations in these systems (noradrenergic, dopaminergic, serotonergic, and cholinergic) possibly owing to changes in receptor sensitivity (Post and Ballenger, 1984). These data have resulted in a recent and relatively extensive field of research investigating mechanisms that regulate receptor responsivity, which has focused to a large extent on the G protein-coupled signal transduction pathways.

Following is a review of evidence regarding the molecular pharmacology of mood stabilizing and antidepressant medications, as they relate to regulation of G proteins and their coupling to specific signal transduction systems. A discussion of clinical findings obtained in subjects with mood disorders (bipolar disorder (BD) and major depression) follows. As will be demonstrated, there is substantial evidence of dysfunction in guanine nucleotide binding protein (G protein) coupled signal transduction pathways in patients with mood disorders, which may be, at least in part, compensated for by pharmacologic treatment.

From: *Neuromethods, Vol. 31: G Protein Methods and Protocols*
Ed: R. K. Mishra, G. B. Baker, and A. A. Boulton Humana Press Inc.

Guanine Nucleotide Binding Proteins

G proteins are a family of signal transduction proteins that play a critical role in the transduction of information across the plasma membrane by coupling receptors to effectors (Gilman 1987; Birnbaumer et al., 1990). Neurotransmitters, modulators, hormones, and other extracellular messengers bind to a family of membrane receptors that couple to intracellular events through G proteins. Binding of neurotransmitters to their specific receptors results not only in ion flux through membrane channels, but also in the activation of a variety of enzymes, i.e., adenylyl cyclase (AC) and phospholipase C (PLC), which then regulate cell function through the production of second messengers. These second messengers include cyclic AMP (cAMP), calcium, diacylglycerol, and the inositol polyphosphates among others. These molecules activate a number of related protein kinases that lead to phosphorylation of various membranes and cytosolic proteins, which not only induce intracellular instantaneous reactions like neuronal excitability, depression, and secretion of hormones, but also lead to regulation of gene expression by activating a variety of transcription factors (*see* Fig. 1). These intricate systems result in transduction and amplification of the original extracellular signal.

G proteins are heterotrimers consisting of an α-subunit which binds and hydrolyzes GTP, and a complex of regulatory β- and γ-subunits tightly associated in membranes (Backlund et al., 1990; Federman et al., 1992; Spiegel et al., 1992; Birnbaumer, 1993). A multiplicity of α-subunits provides for the coupling of a wide variety of neuromodulatory receptors to the same or different intracellular second messenger systems and ion channels. Mammals have more than 20 different G protein α-subunits and are highly conserved in amino acid sequence in G_1–G_5 regions (Hepler and Gilman, 1992; Raymond, 1995). The α-subunits can be divided into four subgroups on the basis of amino acid sequence homology: $G\alpha_s$, $G\alpha_i$, $G\alpha_q$, and $G\alpha_{12}$ subfamilies. It is generally assumed that the α-subunits selectively activate appropriate effector

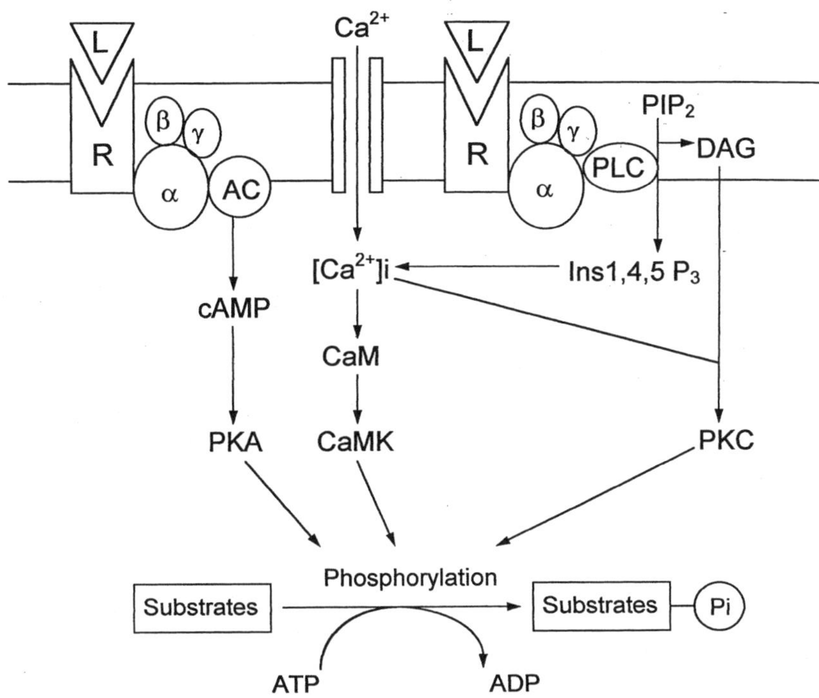

Fig. 1. Signal transduction pathways. Ligands (L) bind to receptors (R) that interact with G proteins composed of α-, β-, and γ-subunits. Then G proteins regulate second messenger signal transduction systems through activation of adenylyl cyclase (AC) and phospholipase C (PLC). Second messengers, including cAMP, 1,4,5-inositol trisphosphate (Ins[1,4,5]P$_3$), diacylglycerol (DAG), and Ca^{2+}, lead to changes in the activity of cAMP-dependent protein kinase (PKA), protein kinase C (PKC), and Ca^{2+}/calmodulin-dependent protein kinase (CaMK), which regulate cell function by phosphorylating various substrates.

proteins. For example, G$_s$ stimulates AC, G$_i$ inhibits AC, and G$_q$ regulates the phosphoinositide metabolism. Some G protein α-subunits are substrates of bacterial toxin-catalyzed ADP ribosylation (Moss and Vaughan, 1988). Cholera and pertussis toxins can catalyze the transfer of the ADP ribose moiety of nicotinamide adenine dinucleotide (NAD) to specific residues within the α-subunits by ADP ribosyltransferase (Tamir and Gill, 1988).

The G proteins have two conformations: an active form bound to GTP and an inactive form bound to GDP. Receptor activation induces a conformational change in receptor-associated GTP protein. First of all, GDP dissociates from its binding site on the α-subunit and GTP is bound. The GTP-bound α-subunit activates intracellular effector enzymes. The GTPase activity of α-subunit quickly hydrolyzes GTP back to GDP, which results in reassociation of G protein subunits and receptors, and termination of second messenger production. The interconversion of GDP/GTP is the key to regulating G protein activity. In this cycle, the exchange of GTP for GDP activates the G protein (Neer, 1995; Raymond, 1995; Rens-Domiano and Hamm, 1995).

β- and γ-subunits act as a dimer. In general, agonist-occupied receptors cannot stimulate GTP/GDP exchange on an α-subunit without the presence of the βγ dimer. The βγ-subunit modulates GTP/GDP exchange directly, and also anchors α-subunits to the plasma membrane (Birnbaumer et al., 1990). Recent evidence has also indicated important roles of β- and γ-subunits in signal transduction pathways. For example, βγ-subunits activate muscarinic K^+ channels and may also activate mitogen-activated protein kinase pathways through Ras (Logothetis et al., 1987; Clapham and Neer, 1993; Reuveny et al., 1994; Wickman et al., 1994; Crespo et al., 1994; Faure et al., 1994). It has also been shown that one of the AC subtypes is activated synergistically by α- and βγ-subunit and another one is activated by α- , but inhibited by βγ-subunit (Iyengar, 1993). βγ-subunits can also activate PLCβ directly (Clapham and Neer, 1993; Smrcka and Sternweis, 1994).

Regulation of G Proteins
by Mood Stabilizers and Antidepressants

Effect of Mood Stabilizers on G Protein

Chronic administration of mood stabilizers (i.e., lithium, carbamazepine, and sodium valproate) is the primary treatment for BD, but the mechanisms of their therapeutic effects remain unknown. Since lithium was established as a mood stabilizer long before carbamazepine and sodium valproate, there is much more research on the molecular pharmacology of this agent than carbamazepine or sodium valproate. Lithium has been shown to exert many effects on the central nervous system (CNS) (Price et al., 1990; Odagaki et al., 1992). In animals, lithium increases the synthesis and turnover of 5-HT in presynaptic neurons, at least partly by increasing uptake of the 5-HT precursor tryptophan. The release of 5-HT into the synapse is also increased, particularly in hippocampus. Lithium diminishes the binding and function of postsynaptic 5-HT2 receptors, particularly in hippocampus, possibly reflecting compensation for primary presynaptic effects, and enhances electrophysiologic and behavioral responses mediated by postsynaptic 5-HT1A receptors. This compound has no consistent effect on β-adrenoceptor (βAR) binding, but it may alter βAR function as evidenced by decreased βAR-mediated stimulation of AC after lithium administration (Price et al., 1990). Many of these effects of lithium on the serotonergic and noradrenergic systems occur at lithium concentrations higher than those used therapeutically (Bunney and Garland-Bunney 1987; Wood and Goodwin, 1987; Risby et al., 1991). The lack of a clear elucidation of the effect of lithium on neurotransmitter systems and receptors together with the emerging understanding of the signal transduction system coupled to these receptors has resulted in a shift of focus toward postreceptor events and mechanisms. An increasingly convincing body of evidence has demonstrated that mood stabilizers exert their clinically relevant effects at G proteins and G protein-coupled effectors.

The earliest evidence of lithium's action on signal transduction pathways was findings that it could blunt receptor-activated AC activity. Forn and Valdecasas (1971) reported that lithium in vitro attenuated norepinephrine-stimulated cAMP accumulation. Following this study, much research demonstrated that in vivo treatment with lithium at therapeutic concentrations could blunt cAMP signaling following stimulation with various neurotransmitters and hormones. For example, chronic lithium administration increased basal cAMP and decreased noradrenaline and isoprenaline-stimulated cAMP production in rat brain, cultured renal epithelial cells, and human brain tissue (Forn and Valdecases, 1971; Goldberg et al., 1988; Mork and Geisler, 1989a,b; Masana et al., 1991, 1992). This effect is not limited to the noradrenergic system, but has also been shown for other neurotransmitter systems, such as dopamine and 5-HT, which signal through cAMP modulation (Mork and Geisler, 1989a,b; Newman et al., 1989b, 1991; Carli et al., 1994). The mechanism by which lithium increases basal AC activity has been hypothesized to be owing to stabilization of the inactive heterotrimeric form of G_i, thus reducing the inhibitory control of basal AC activity (Masana et al., 1992). A more direct effect of lithium on AC has been shown by Colin et al. (1991), in which the expression of AC types I and II was increased by chronic lithium treatment in rat brain.

Recently the actions of lithium have been attributed more specifically to effects at the level of the G protein. In an early study employing transformed rat pheochromocytoma (PC12) cells, Volonte (1987) found that lithium in vitro increased the density of [^{3}H]GTP binding. Chronic lithium treatment of rats was also shown to reduce GTP-dependent isoprenaline-activation as well as 5-HT inhibition of cAMP production, which further implied an effect of lithium on G proteins (Mork and Geisler, 1989a,b). Even more evidence is found in a much cited paper by Avissar and colleagues (1988), which demonstrated that lithium blunted both β-adrenergic and muscarinic receptor-agonist mediated increases in GTP binding in rat brain membrane, which they concluded was evidence that lithium could decrease G protein functionality. Inhibition of β-adrenergic receptor-G protein coupling by lithium may be related to its effects on G_s protein, whereas the muscarinic effects may be related to either a G_i isoform or G_q (Nathanson et al., 1978; Cockcroft and Gomperts, 1985). It is known that muscarinic receptors are divided into four subtypes (M1–M4) (Hammer and Giachetti, 1982; Barnard, 1988; Peralta et al., 1988). M2 and M3 inhibit AC activity, but M1 and M4 affect PI turnover (Gil and Wolfe, 1985; Barnard, 1988; Peralta et al., 1988). The M1 selective antagonist pirenzepine was 100-fold more effective than the M2 selective antagonist AD FX-116 in blocking carbamylcholine-induced GTP binding, which suggests that attenuation of lithium on muscarinic receptor agonist-mediated G protein is related to the M1 muscarinic receptor, and by extension that G_q may be involved (Avissar and Schreiber, 1989). This finding of a selective inhibition by lithium distal to M1 muscarinic receptors is also confirmed by electrophysiological studies on guinea pig hippocampal neurons in vitro (Muller et al., 1989).

Schreiber and coworkers (1991) found complementary evidence in patients with the report of increased βAR and muscarinic agonist-induced Gpp[NH]p (a nonhydrolyzable GTP analog) binding to lymphocyte membranes obtained from manic subjects compare to healthy controls. No such differences in agonist-induced [^{3}H]Gpp[NH]p binding were evident in euthymic lithium-treated BD patients compared with controls. Schrieber et al. (1991) thus interpreted these data to be consistent with hyperfunctionality of G proteins in BD, which could be reduced by lithium treatment. This is consistent with the findings of Risby et al. (1991), who found that lithium regulates basal-, guanine nucleotide-, and cesium fluoride-stimulated AC activity (the last two agents acting directly on G proteins) in platelets from healthy subjects. As will be elucidated later, these findings of an effect of lithium on G protein have been very important in establishing findings of G protein abnormalities in patients with BD.

Several studies have attempted to clarify the exact mechanism by which lithium decreases G protein function (Avissar et al., 1991; Colin et al., 1991; Lesch et al., 1991a; Li et al., 1991, 1993a). Avissar et al. (1991) suggested that lithium's specific site of interaction with G protein is related to competition with Mg^{2+} ions. Mg^{2+} ions interact with G protein at multiple sites and play a very important role in its functioning (Gilman,

1987). Magnesium abolished the inhibition by lithium of iso-prenaline- and carbamylcholine-induced increases in GTP binding to rat cerebral cortex membranes. Both Li et al. (1991) and Colin et al. (1991) found that chronic lithium treatment reduces rat cortical α_s and α_i mRNA levels, suggesting that lithium may also modify G protein functionality through the regulation of the genes expressing these G protein subtypes. This effect on G protein expression, however, may be quite complex and accompanied by compensatory posttranslational biochemical changes, since most studies have found no changes in α_s-, α_i-, α_o- or β-subunit levels in rat brain after chronic lithium treatment (Li et al., 1993a). Li and Jope (1995), however, found that chronic lithium treatment reduced the ability of NGF to increase G protein subunits (α_s, α_{o1}, α_{i1}, and β) in PC12 cells. In one study, carbamazepine had no effect on G protein mRNA or protein levels (Li et al., 1993a), and the effect of sodium valproate has not yet been determined. Since ADP ribosylation of G protein α-subunits has been shown to be important in regulating the function and possibly turnover of G protein α-subunits (Milligan et al., 1989; Milligan, 1993), a number of investigators have explored whether this process is a target of lithium. An earlier report on platelets from healthy subjects treated chronically with lithium showed that the pertussis toxin-stimulated ADP ribosylation of α_i was increased after chronic treatment with no changes found in α_i levels (Hsiao et al., 1992). Similar findings were also observed in platelets from lithium-treated subjects with BD (Manji et al., 1995a). More recently, Nestler et al. (1995) reported that lithium in vitro inhibits endogenous ADP ribosylation of α_s in rat brain homogenates. They also found that chronic administration of lithium to rats in vivo, under conditions that result in therapeutically relevant serum levels, increases endogenous ADP-ribosylation activity. In our laboratory (Young and Woods, 1996), we also found that chronic treatment with lithium markedly increased endogenous ADP ribosylation of both α_i and α_s in C6 glioma cells, whereas sodium valproate had an opposite effect and carbamazepine had little or no effect. Thus, to date there is evidence that lithium may alter G protein levels, and function both at the transcriptional and posttranslational levels, although a specific mechanism of action has not been completely established.

Effect of Antidepressants on G Proteins

Antidepressant drugs have been used in the treatment of depression for a long time, but the mechanism of their therapeutic effect is not clear yet. Most early studies focused on monoaminergic systems, which suggested that depression may be related to decreased monoaminergic release at CNS synapses. For example, tricyclic antidepressant drugs, such as desipramine and amitriptyline inhibit noradrenaline and 5-HT reuptake in nerve terminals, leading to increased neurotransmitter concentration in the synaptic cleft (Schildkraut, 1965; Coppen, 1967). It takes only a few hours for antidepressant drugs to inhibit monoamine reuptake and increase neurotransmitter concentration. However, a therapeutic effect is produced by antidepressant drugs only by chronic treatment for 2–3 wk. Further, other drugs, such as cocaine and amphetamine, have little antidepressant effect, but still inhibit monoamine reuptake (Zemlan and Garver, 1990). Numerous studies in animal models have shown that chronic treatment by tricylic antidepressants causes a downregulation of βAR and serotonin 5-HT2 receptors (Banerjee et al., 1977; Pandey et al., 1979b; Kellar et al., 1981). Although electroconvulsive treatment (ECT) downregulates βAR in rat brain, this treatment has been shown to upregulate rat brain serotonin 5-HT2 receptors (Bergstrom and Kellar, 1979; Kellar et al., 1981; Vetulani et al., 1981; Heal et al., 1989). Other agents with antidepressant properties do not downregulate βARs (Mishra et al., 1980). The relationship between the therapeutic effect of antidepressant drugs and receptor downregulation is still not well understood. These changes may be secondary to an increase of neurotransmitters induced by inhibition of monoamine uptake by antidepressant drugs.

In searching for a common mechanism of antidepressant action, more recent studies have focused on G proteins and G protein-coupled second messenger systems. Antidepressant treatment was reported to have decreased β-adrenergic agonist-stimulated production of cAMP (Vetulani et al., 1975, 1976; Mishra et al., 1980; Newman et al., 1987; Newman and Lerer, 1989). For example, Mishra et al. (1980) found that chronic administration of desipramine, mianserin, and zimelidine did not change basal cAMP level, but reduced the increase of cAMP levels induced by the β-adrenergic agonist isoproterenol. The B_{max} value of [³H]dihydroalprenolol binding was also inhibited by desipramine, but not by mianserin and zimelidine. Chronic treatment of rats with the antidepressant drugs imipramine, clomipramine, or clorgyline for three weeks did not effect basal [³H]GTP binding capacity, but partially inhibited increase of [³H]GTP binding capacity induced by isoproterenol. However, these drugs had no effect on muscarinic receptor agonist stimulated GTP binding (Avissar and Schreiber, 1992). ECT has also been shown to attenuate β-adrenergic and muscarinic receptor coupling to G proteins in rat cortex. For example, repeated ECT treatments were found to inhibit isoproterenol- and carbamycholine-induced increases in [³H]GTP specific binding to membranes prepared from rat cerebral cortex (Avissar et al., 1990).

More evidence of an effect on G proteins comes from the finding that chronic tricyclic, atypical antidepressant, or ECT treatment increased guanosine imidodiphosphate (Gpp[NH]p)-, sodium fluoride-, or forskolin-stimulated AC activity in membranes of rat brain (Menkes et al., 1983; Ozawa and Rasenick, 1989, 1991). These effects were confirmed by Newman et al. (1986, 1987, 1989), who found that antidepressant or ECT treatment enhances Mn^{2+} and forskolin stimulation of AC activity. Recently, Chen and Rasenick (1995a) found that in C6 glioma cells, chronic treatment with antidepressant drugs significantly increased Gpp[NH]p-stimulated AC activity. Furthermore, it is also reported that antidepressant drugs increased the GTPase activity of purified G_o and the binding of $[^{35}S]GTP\gamma S$ to the purified protein (Yamamoto et al., 1992). In in vitro animal studies, Ozawa et al. (1994) found that antidepressants, such as amitriptyline, clomipramine, desipramine, and mianserin, increase B_{max} and Kd of the hydrolysis-resistant photoaffinity GTP analog, P^3(4-azidoanilido)-P^1-5′ GTP (AAGTP) binding for each of the G proteins, especially $G\alpha_s$. Monoamine oxidase inhibitors, antipsychotics, and axiolytics had no effect. Lesch et al. (1991b) investigated G protein α-subunits, $G\alpha_s$, $G\alpha_i$, and $G\alpha_o$ levels in rat brain using an enzyme-linked immunosorbent assay, and found some brain region- and antidepressant class-specific changes in several G protein α-subunits. Two groups have not subsequently found changes in $G\alpha_s$, $G\alpha_i$, $G\alpha_o$, and $G\beta$ content detectable by immunoblotting after chronic antidepressant treatment in rat cerebral cortex (Chen and Rasenick, 1995b; Emamghoreishi et al., 1996). Although there was no change in the amount of G protein, antidepressant treatment increased the number of active $G\alpha_s$/AC complexes immunoprecipitated by anti-$G\alpha_s$ antibody (Chen and Rasenick, 1995b). These different results may be explained in part by the use of different techniques, and different species or strains of animals and cultured cells. The evidence described above suggests that signal transduction by G protein may represent one mechanism for the effects of these agents. Although antidepressants may not alter G protein levels, they may increase the coupling of G protein (particularly $G\alpha_s$) to effector enzymes, such as AC.

Summary

A consistent body of evidence has shown that lithium and antidepressant drugs regulate G protein function. Chronic treatment with antidepressants induces βAR desensitization while enhancing $G\alpha_s$-mediated stimulation of AC activity. Lithium may decrease G protein α-subunit gene expression, compete with Mg^{2+} ions for magnesium affinity sites of G protein, and increase pertussis toxin-catalyzed and endogenous ADP ribosylation, which may inactivate the α_i-subunit or may enhance α-subunit protein turnover. There is, however, little evidence that lithium treatment alters cellular G protein levels in animal models or directly in patients to support the clinical relevance of the effects of this drug. Much less is known about the effects of either carbamazapine or sodium valproate treatment on G protein or the cAMP signaling pathways, but available evidence suggests these agents may act differently than lithium.

Clinical Studies in Patients with BD and Major Depressive Disorder (MDD)

Complementing this body of evidence on the effects of antidepressants and mood stabilizers on signal transduction pathways are findings from clinical studies. Direct clinical evidence is found in studies on postmortem brain and in peripheral blood cells (i.e., platelets, leukocytes) obtained from subjects with mood disorders. As will be described, there are far fewer findings directly from patients on the mechanism of action of psychotropic drugs. In general, clinical findings support the conclusions drawn from studies described in the preceding section.

Studies in BD

Since lymphocytes have βARs coupled to AC through G_s, these cells have been commonly used as an analogous cellular model for clinical investigation of receptor sensitivity in BD (Hudson et al., 1993; Landmann et al., 1983). A number of groups found blunted norepinephrine- and isoproterenol-stimulated, but not PGE1-stimulated cAMP formation in mononuclear leukocytes (MNLs) from depressed patients compared with normal subjects (Pandey et al., 1979a; Extein et al., 1979; Siever et al., 1984; Mann et al., 1985; Halper et al., 1988). This appears to be one of the few relatively consistent biochemical changes identified in depression. These changes have been generally observed in unipolar patients, but have also been reported in BD (Hudson et al., 1993). Mann et al. (1985) also found this change in receptor sensitivity was not accompanied by changes in βAR density and affinity. They concluded that the blunted β-adrenergic cAMP response probably indicates a reduced responsiveness (desensitization) rather than a diminished number (downregulation) of βAR sites in depressed subjects (Mann et al., 1985; Halper et al., 1988). This notion is supported by recent findings suggesting elevated NE turnover in cerebral cortical, but not subcortical regions of BD postmortem brains (Young et al., 1994a) in the absence of changes in β-adrenoceptor densities (Young et al., 1994b) with controls. The lack of alterations in βAR number in transformed lymphocytes from BD subjects (Kay et al., 1993). This apparent desensitization was another of the findings that led a number of researchers to investigate possible alterations in postreceptor signaling in patients with BD.

Several studies have more directly examined G protein function and levels in patients with BD. Schreiber et al. (1991) reported increased agonist-stimulated [^{3}H]GppNHp binding in MNL membranes from manic patients compared with controls or euthymic lithium-treated BD subjects. Direct evidence indicating involvement of G protein dysfunction in the pathophysiology of BD comes from findings of elevated Gα_s immunoreactivity in postmortem brain from patients with an established life-time diagnosis of BD (Young et al., 1991, 1993). Compared with controls matched on the basis of age, postmortem delay, and brain pH, Gα_s immunoreactivity was significantly elevated in frontal, temporal, and occipital cortex but not in hippocampus, thalamus, or cerebellum. The lack of significant changes in Gα_o or Gβ immunoreactivities (Young et al., 1993; Avissar et al., 1996) illustrates the specificity of changes to Gα_s in this disorder. Early efforts to determine the functional correlates of the elevated Gα_s levels have revealed associated statistically significant increases in forskolin-stimulated AC activity in occipital and temporal cortical regions (Young et al., 1993). A significant correlation was also observed between forskolin-stimulated cAMP formation and

Gα immunoreactivity when examined across these cortical regions. These results suggest that increased amounts of Gα$_s$ and/or increased function of stimulatory G protein contributes to the pathophysiology of BD (Young et al., 1993).

It has also been reported that α$_s$ and α$_i$ immunoreactivity in MNL membranes was significantly increased in medication-free BD, but not in MDD patients when compared with age- and sex-matched healthy controls (Young et al., 1994b). These findings suggest that similar disturbances in the function of these G proteins occur in peripheral MNLs as in cerebral cortex of BD patients and also raises the possibility that they are trait-related, since the changes occurred in BD, but not MDD patients (Young et al., 1994b). Recently Manji et al. (1995a) also reported elevated MNL α$_s$ levels in drug-free and lithium-treated BD patients who were manic, depressed, or euthymic, which also supports the notion that these changes may be trait-related and are replicable. These studies suggest that disordered signal transduction at the G protein level, possibly linked to cortical noradrenergic postsynaptic receptors, is critical to the pathogenesis of BD.

A number of investigators have recently explored whether mutations might occur in the Gα$_s$ gene in BD patients and their family members, since such mutations have been identified as critical to the pathophysiology of several medical conditions. No mutations have been identified, however, in either the 5′ upstream or coding regions of the Gα$_s$ gene in several large pedigrees in both North America and Australia (Gejman et al., 1993; Le et al., 1994). Recent findings in our laboratory also show that Gα$_s$ mRNA levels are not altered in frontal, temporal or occipital cortex from subjects with BD (Young et al, 1996). Considering these data together, it seems plausible that other unidentified mechanisms which regulate Gα$_s$ levels (for example, protein turnover or membrane attachment) may be important in the pathophysiology of BD. To date there is no direct evidence to support this latter possibility.

Several more recent findings may lead one to reconsider whether the PI abnormalities are owing to a G protein-mediated defect. Although G_q levels were found to be increased in occipital cortex from subjects with BD (Mathews et al., 1996), GTPγS-stimulated [^{3}H]PI hydrolysis, but not PLC activity was decreased in this same region (Jope et al., 1996). The latter results were interpreted as evidence of an imbalance between the cAMP (overactive) and PI (blunted) signaling systems, which may be important to the clinical symptoms of BD and may be regulated in part by G proteins (Jope et al., 1996). Given the extensive body of evidence on the effects of lithium on the PI second messenger systems, it will be of interest to focus future investigations on this pathway in clinical studies with BD subjects.

Studies in MDD

Much of evidence cited in the section on clinical studies in BD patients derives from patients with MDD. As described in MDD subjects, desensitization rather than downregulation of βAR has been consistently repeated, with the findings of blunted β-adrenergic cAMP response and no change in βAR density (Mann et al., 1985; Halper et al., 1988). Studies have not consistently found abnormalities in G protein levels in subjects with MDD. Although Cowburn et al. (1994) observed that basal, GTPγS, or forskolin-stimulated AC activities were significantly lower in postmortem cerebral cortex membranes of depressed suicide victims, no significant differences in $G\alpha_s$ and $G\alpha_i$ levels were found. Ozawa et al. (1993) reported that binding of AAGTP, a hydrolysis-resistant photoaffinity GTP analog, to $G\alpha_{i/o}$ was increased in temporal and parietal cortices in depressive subjects. They also found that the ratio of $G\alpha_s/G\alpha_{i/o}$ AAGTP incorporation was significantly decreased in depressive patients. The decrease in the proportions of $G\alpha_s$ and $G\alpha_{i/o}$ suggests that hypofunction of the AC system may be one of pathophysiological changes in depression. Two independent studies have not found any differences in G protein levels in MNLs from MDD patients in contrast to patients with BD (Young et al., 1994b; Avissar et al., 1996). An unanswered, but interesting question is whether abnormalities further downstream to G protein, for instance in AC levels or a more distal mechanism, occur in MDD. In this regard, Shelton et al. (1996) very recently found that the activity of cyclic AMP-dependent kinase in cultured fibroblasts was lower in MDD patients than in healthy subjects, supporting abnormalities in G protein-coupled cAMP signal transduction system.

Summary

Despite the limited number of direct clinical studies in subjects with BD and MDD, and the limitations in extrapolating from findings in peripheral blood cells or postmortem brain with their inherent limitations, some consistency emerges on review of these data. G protein α-subunit function and possibly levels of the stimulatory α-subunit appear to be increased in blood cells and postmortem brain from subjects with BD based on findings from several independent laboratories. However, in depression G protein-coupled AC activity seems to be decreased without G protein level changes. The findings of altered G protein-coupled signal transduction in BD and MDD are very consistent with the body of evidence on the molecular pharmacology of mood stabilizers (especially lithium and antidepressant drugs) that also have a number of effects on G protein-coupled signal transduction pathways, often in the opposite direction to what was found in BD and MDD.

Conclusions

When neurotransmitters, modulators, hormones and other extracellular messengers bind to specific receptors at the cell surface, extracellular signals are transduced and amplified through G protein and G protein-coupled signal transduction systems. G proteins are a common pathway for a multiplicity of neurotransmitter receptor functions and play an important role in the regulation of postreceptor levels. Molecular abnormalities in G protein regulation may be important in mood disorders. There is already an interesting body of evidence to support the hypothesis that abnormalities in G protein-coupled signal transduction pathways are related to BD and MDD, and possibly to the response of this disorder to treatment with mood stabilizers. Increasing evidence has shown that currently available mood stabilizers and antidepressants involve regulation of G protein function. Findings on the effect of mood stabilizers and antidepressant drugs on G protein regulation may explain a diversity of biochemical effects of these drugs reported on various neurotransmitter systems, and may also explain both their antimanic and antidepressant clinical potency. These observations appear to provide strong support for the hypothesis that G protein-coupled pathways may be dysregulated in mood disorders. G proteins, as primary targets in the development of new drugs, are an attractive prospect for future studies in the treatment of mood disorders.

Acknowledgments

Supported by grants from NARSAD and the Medical Research Council of Canada. J. F. W. is a Canadian Psychiatric Research Foundation Fellow and L. T. Y is a Career Scientist of the Ontario Ministry of Health.

References

Avissar, S. and Schreiber, G. (1989) Muscarinic receptor subclassification and G proteins: Significance for lithium action in affective disorders and for the treatment of the extrapyramidal side effects of neuroleptics. *Biol. Psychiatry* **26**, 113–130.

Avissar, S. and Schreiber, G. (1992) The involvement of guanine nucleotide binding proteins in the pathogenesis and treatment of affective disorders. *Biol. Psychiatry* **31**, 435–459.

Avissar, S., Schreiber G., Danon, A., and Balmaker, R. H. (1988) Lithium inhibits adrenergic and cholinergic increases in GTP binding in rat cortex. *Nature* **331**, 440–442.

Avissar, S., Schreiber, G., Aulakh, C. S., Wozniak, K. M., and Murphy, D. L. (1990) Carbamazepine and electroconvulsive shock attenuate β-adrenoceptor and muacarinic cholinoceptor coupling to G proteins in rat cortex. *Eur. J. Pharmacol.* **189**, 99–103.

Avissar, S., Murphy, D. L., and Schreiber, G. (1991) Magnesium reversal of lithium inhibition of β adrenergic and muscarinic receptor coupling to G proteins. *Biochem. Pharmacol.* **41,** 171–175.

Avissar, S., Barki-Harrington, L., Nechamkin, Y., Roitman, G., and Schreiber (1996) Reduced β-adrenergic receptor-coupled Gs protein function and Gs immunoreactivity in mononuclear leukocytes of patients with depression. *Biol. Psychiatry* **39,** 755–760.

Backlund, P. S. Jr., Simonds, W. F., and Spiegel, A. M. (1990) Carboxyl methylation and COOH-terminal processing of the brain G protein gamma-subunit. *J. Biol. Chem.* **265,** 15,572–15,576.

Banerjee, S. P., Kung, S. L., Riggi, S. J., and Chanda, S. K. (1977) Development of β-adrenergic receptor subsensitivity by antidepressants. *Nature* **268,** 455, 456.

Barnard, E. A. (1988) Molecular neurobiology: separating receptor subtypes from their shadows. *Nature* **335,** 301–302.

Bergstrom, D. A. and Kellar, K. J. (1979) Effect of electroconvulsive shock therapy on monoaminergic receptor binding sites in rat brain. *Nature* **278,** 464–466.

Birnbaumer, L. (1993) Receptor-to-effector signaling through G proteins: roles for beta gamma dimers as well as alpha subunits. *Cell* **71,** 1069–1072.

Birnbaumer, L., Abramowitz, J., and Brown, A. M. (1990) Receptor-effector coupling by G proteins. *Biochem. Biophys. Acta* **1031,** 163–224.

Bunney, W. E., Jr. and Garland-Bunney, B. L. (1987) Mechanisms of action of lithium in affective illness: basic and clinical implications, in *Psychopharmacology: The Third Generation of Progress* (Meltzer, H. Y., ed.), Raven, New York, pp. 553–565.

Carli, M., Anand-Srivastava, M. B., Molina-Holgado, E., Dewar, K. M., and Reader T. A. (1994) Effects of chronic lithium treatments on central dopaminergic receptor systems: G proteins as possible targets. *Neurochem. Int.* **24,** 13–22.

Chen, J. and Rasenick, M. M. (1995a) Chronic treatment of C6 glioma cell with antidepressant increases functional coupling between a G protein (G$_s$) and adenylyl cyclase. *J. Neurochem* **64,** 724–732.

Chen, J. and Rasenick, M. M. (1995b) Chronic antidepressant treatment facilitates G protein activation of adenylyl cyclase without altering G protein content. *J. Pharmacol. Exp. Ther.* **275,** 509–517.

Clapham, D. E. and Neer, E. J. (1993) New roles for G protein βγ dimers in transmembrane signaling. *Nature* **365,** 403–406.

Cockcroft, S. and Gomperts, B. D. (1985) Role of guanine nucleotide binding protein in the activation of polyphosphoinositide phosphodiesterase. *Nature* **314,** 534–536.

Colin, S. F., Chang, H.-C., Mollner, S., Pfeuffer, T., Reed, R. R., Duman, R. S., and Nestler, E. J. (1991) Chronic lithium regulates the expression of adenylate cyclase and G$_i$-protein α-subunit in rat cerebral cortex. *Proc. Natl. Acad. Sci. USA* **88,** 10,634–10,637.

Coppen, A. (1967) The biochemistry of affective disorders. *Br. J. Psychiatry* **113**, 1237–1264.

Cowburn, R. F., Marcusson, J. O., Eriksson, A., Wiehager, B., and O'Neill, C. (1994) Adenylyl cyclase activity and G protein subunit levels in postmortem frontal cortex of suicide victims. *Brain Res.* **633**, 297–304.

Crespo, P., Xu, N., Simonds, W. F., and Gutking, J. S. (1994) Ras-dependent activation of MAP kinase pathway mediated by G protein βγ-subunits. *Nature* **369**, 418–420.

Emamghoreishi, M., Warsh, J. J., Sibony, D., and Li, P. P. (1996) Lack of effect of chronic antidepressant treatment on $G\alpha_s$ and $G\alpha_i$ subunit protein and mRNA levels in the rat cerebral cortex. *Neuropsychopharmacology* **15**, 281–287.

Extein, I., Tallman, J., Smith, C. C., and Goodwin, F. K. (1979) Changes in lymphocyte beta-adrenergic receptors in depression and mania. *Psychiatry Res.* **1**, 191–197.

Faure, M., Voyno-Yasenetskaya, A., and Bourne, H. R. (1994) cAMP and βγ-subunits of heterotrimeric G proteins stimulate the mitogen-activated protein kinase pathway in COS-7 cells. *J. Biol. Chem.* **169**, 7851–7854.

Federman, A. D., Conklin, B. R., Schrader, K. A., Reed, R. R., and Bourne, H. R. (1992) Hormonal stimulation of adenlyl cyclase through Gi-protein beta gamma subunits. *Nature* **356**, 159–161.

Forn, J. and Valdecasas, F. G. (1971) Effects of lithium on brain adenyl cyclase activity. *Biochem. Pharmacol.* **20**, 2773–2779.

Gejman, P. V., Martinez, M., Cao, Q. H., Friedman, E., Berrettini, W. H., Goldin, L. R., Koroulakis, P., Ames, C., Lerman, M. A., and Gershon, E. S. (1993) Linkage analysis of 57 microsatellite loci to bipolar disorder. *Neuropsychopharmacology* **9**, 31–40.

Gil, D. W. and Wolfe, B. B. (1985) Pirenzepine distiguishes between muscarinic receptor-mediated phosphoinositide breakdown and inhibition of adenylate cyclase. *J. Pharmacol. Exp. Ther.* **232**, 608–616.

Gilman, A. G. (1987) G proteins: transducers of receptor-generated signals. *Annu. Rev. Biochem.* **56**, 615–649.

Goldberg, H., Clayman, P., and Skorecki, K. (1988) Mechanism of Li inhibition of vasopressin-sensitive adenylate cyclase in cultured renal epithelial cells. *Am. J. Physiol.* **255**, F995–F1002.

Halper, J. P., Brown, R. P., Sweeney, J. A., Kocsis, J. H., Peters, A., and Mann, J. J. (1988) Blunted β-adrenergic responsivity of peripheral blood mononuclear cells in endogenous depression. *Arch. Gen. Psychiatry.* **45**, 241–244.

Hammer, R. and Giachetti, A. (1982) Muscarinic receptor subtypes: M1 and M2, biochemical and functional characterization. *Life Sci.* 2991–2998.

Heal, K. J., Butler, S. A., Hurst, E. M., and Buckett, W. R. (1989) Antidepressant treatments including sibutramine hydrochloride and electroconvulsive shock, decrease β_1 and β_2-adrenoceptors in rat cortex. *J. Neurochem.* **53,** 1019–1025.

Hepler, J. R. and Gilman, A. G. (1992) G proteins. *Trends Biochem. Sci.* **17,** 383–387.

Hsiao, J. K., Manji, H. K., Chen, G., Bitran, J. A., Risby, E. D., and Potter, W. Z. (1992) Lithium administration modulates platelet G_i in humans. *Life Sci.* **50,** 227–233.

Hudson, C. J., Young, L. T., Li, P. P., and Warsh, J. J. (1996) CNS transmembrane signal transduction in the pathophysiology and pharmacology of affective disorders and schizophrenia. *Synapse* **13,** 278–293.

Iyengar, R. (1993) Molecular and functional diversity of mammalian G5-stimulated adenylyl cyclases *FASEB J.* **7,** 768–775.

Jope, R. S., Song, L., Li, P. P., Young, L. T., Kish, S. J., Pacheco, M. A., and Warsh, J. J. (1996) The phosphoinositide signal transduction system is impaired in bipolar affective disorder brain. *J. Neurochem.* **66,** 2402–2409.

Kay, G., Sargeant, M., McGuffin, P., Whatley, S., Marchbanks, R., Baldwin, D., Montgomery, S., and Elliott, J. M. (1993) The lymphoblast β-adrenergic receptor in bipolar depressed patients: characterization and down-regulation. *J. Affect Disord.* **27,** 163–172.

Kellar, K. J., Cascio, C. S., Butler, J. A., and Kurtzke, R. N. (1981) Differential effects of electroconvulsive shock and antidepressant drugs on serotonin-2 receptors in rat brain. *Eur. J. Pharmacol.* **69,** 515–518.

Landmann, R., Burgisser, E., and Buhler, F. R. (1983) Human lymphocytes as a model for beta-adrenergic receptors in clinical investigation. *J. Receptor Res.* **3,** 71–88.

Le, F., Mitchell, P., Vivero, C., Waters, B., Donald, J., Selbie, L. A., Shine, J., and Schofield, P. (1994) Exclusion of close linkage of bipolar disorder to the G_s-α-subunit gene in nine Australian pedigrees. *J. Affec. Dis.* **32,** 187–195.

Lesch, K., Aulakh, C., Tolliver, T., Hill, J., and Woldzin, B. (1991a) Differential effects of long-term lithium and carbamazepine administration on G_s and G_i protein in rat brain. *J. Pharmacol. Mol. Pharmacol.* **207,** 355–359.

Lesch, K., Aulakh, C., Tolliver, T., Hill, J., and Murphy, D. (1991b) Regulation of G proteins by chronic antidepressant drug treatment in rat brain: tricyclics but not clorgyline increase $G\alpha_o$ subunits. *Eur. J. Pharmacol.* **207,** 361–364.

Li, PP., Young, L. T., and Warsh, J. J. (1991) Lithium decreased G_s, G_i-1 and G_i-2 α-subunit mRNA levels in rat cortex. *Eur. J. Pharmacol. Mol. Pharmacol.* **206,** 165, 166.

Li, P. P., Young, L. T., Y. K. Tam, D. Sibony, and J. J. Warsh. (1993a) Effects of chronic lithium and carbamazepine treatment on G protein subunit expression in rat cerebral cortex. *Biol. Psychiatry* **34,** 167–170.

Li, X. and Jope, R. S. (1995) Selective inhibition of the expression of signal transduction proteins by lithium in nerve growth factor-differentiated PC12 cells. *J. Neurochem.* **65,** 2500–2508.

Logothetis, D. E., Kurachi, Y., Galper, J., Neer, E. J., and Clapham, D. E. (1987) The βγ-subunits of GTP-binding proteins activate the muscarinic K$^+$ channel in heart. *Nature* **325,** 321.

Manji, H. K., Chen, G., Shimon, H., Hsiao, J. K., Potter, W. Z., and Belmaker, R. H. (1995a) Guanine nucleotide-binding proteins in bipolar affective disorder. *Arch. Gen. Psychiatry* **52,** 135–144.

Mann, J. J., Brown, R. P., Halper, J. P., Sweeney, J. A., Kocsis, J. H., Stokes, P. E., and Bilezikian, J. P. (1985) Reduced sensitivity of lymphocyte beta-adrenergic receptors in patients with endogenous depression and psychomotor agitation. *N. Engl. J. Med.* **313,** 715–720.

Masana, M. I., Bitran, J. A., Hsiao, J. K., Mefford, I. N., and W. Z. Potter. (1991) Lithium effects on noradrenergic-linked adenylate cyclase activity in intact rat brain and in vivo microdialysis study. *Brain Res.* **538,** 333–336.

Masana, M. I., Bitran, J. A., Hsiao, J. K., and Potter, W. Z. (1992) In vivo evidence that lithium inactivates G$_i$ modulation of adenylate cyclase in brain. *J. Neurochem.* **59,** 200–205.

Mathews, R., Li, P. P., Young, L. T., Kish, S. J., and Warsh, J. J. (1997) Increased Gq/11 immunoreactivity in postmortem occipital cortex from patients with bipolar affective disorder. *Biol. Psychiatry,* **41,** 649–656.

Menkes, D. B., Rasenick, M. M., Wheeler, M. A., and Bitensky, M. W. (1983) Guanosine triphosphate activation of brain adenylate cyclase: enhancement by long-term antidepressant treatment. *Science* **219,** 65–67.

Milligan, G. (1993) Agonist regulation of cellular G protein levels and distribution: mechanisms and functional implications. *Trends Pharmacol. Sci.* **14,** 413–416.

Milligan, G., Unson, G. C., and Wakelam, M. J. O. (1989) Cholera toxin treatment produces down-regulation of the -subunit of the stimulatory guanine-nucleotide-binding protein (Gs). *Biochem. J.* **262,** 643–647.

Mishra, R., Janowsky, A., and Sulser, F. (1980) Action of mianserin and zimelidine on the norepinephrine receptor coupled adenylate cyclase system in brain: subsensitivity without reduction in β-adrenergic receptor binding. *Neuropharmacology* **19,** 983–987.

Mork, A. and Geisler, A. (1989a) Effects of lithium ex vivo on the GTP-mediated inhibition of calcium-stimulated adenylate cyclase activity in rat brain. *Eur. J. Pharmacol.* **168,** 347–354, 1989.

Mork, A. and Geisler, A. (1989b) Effects of GTP on hormone-stimulated adenylate cyclase activity in cerebral cortex, striatum, and hippocampus from rats treated chronically with lithium. *Biol. Psychiatry* **26,** 279–288.

Moss, J. and Vaughan, M. (1988) ADP-ribosylation of guanyl mucleotide-binding regulatory proteins by bacterial toxins. *Adv. Enzymol.* **60,** 303–379.

Muller, W., Brunner, H., and Misgeld, U. (1989) Lithium discriminates between muscarinic receptor subtypes on guinea pig hippocampal neurons in vitro. *Neurosci. Lett.* **100,** 135–140.

Nathanson, N. M., Klein, W. L., and Nireneberg, M. (1978) Regulation of adenylate cyclase activity mediated by muscarinic acetylcholine receptor. *Proc. Natl. Acad. Sci. USA* **75,** 1788–1791.

Neer, E. J. (1995) Heterotrimeric G proteins: Organizers of transmembrane signals. *Cell* **80,** 249–257.

Nestler, E. J., Terwilliger, R. Z., and Duman, R. S. (1995) Regulation of endogenous ADP-ribosylation by acute and chronic lithium in rat brain. *J. Neurochem.* **64,** 2319–2324.

Newman, M. E. and Lerer, B. (1989) Post-receptor-mediated increases in adenylate cyclase activity after chronic antidepressant treatment: relationship to receptor desensitisation. *Eur. J. Pharmacol.* **162,** 345–352.

Newman, M. E., Solomon, H., and Lerer, B. (1986) Electroconvulsive shock and cyclic AMP signal transduction: effects distal to the receptor. *J. Neurochem.* **46,** 1667–1669.

Newman, M. E., Lipot, M., and Lerer, B. (1987) Differential effects of chronic administration of desipramine on the cyclic AMP response in cortical slices and membranes in the rat. *Neuropharmacology* **26,** 1127–1130.

Newman, M. E., Drummer, D., and Lerer, B. (1989) Single and combined effects of desimipramine and lithium on serotonergic receptor number and second messenger function in rat brain. *J. Pharmacol. Exp. Ther.* **252,** 826–831.

Newman, M. E., Ben-Zeev, A., and Lerer, B. (1991) Chloramphetamine did not prevent the effects of chronic antidepressants on 5-hydroxy-tryptamine inhibition of forskolin-stimulated adenylate cyclase in rat hippocampus. *Eur. J. Pharmacol.* **207,** 209–213.

Odagaki, Y., Koyama, Y., and Yamashita, I. (1992) Lithium and serotonergic neural transmission: a review of pharmacological and biochemical aspects in animal studies. *Lithium* **3,** 95–107.

Ozawa, H. and Rasenick, M. M. (1989) Coupling of the stimulatory GTP-binding protein Gs to rat synaptic membrane adenylate cyclase in enhanced subsequent to chronic antidepressant treatment. *Mol. Pharmacol.* **36,** 803–808.

Ozawa, H. and Rasenick, M. M. (1991) Chronic electroconvulsive treatment augments coupling of the GTP-binding protein G_s to the catalytic moiety of adenlyl cyclase in a manner similar to that seen with chronic antidepressant drugs. *J. Neurochem.* **66,** 330–338.

Ozawa, H., Gsell, W., Frolich, L., Zochling, R., Pantucek, F., Beckmann, H., and Riederer, P. (1993) Imbalance of the G_s and $G_{i/o}$ function in postmortem human brain of depressed patients. *J. Neural. Transm. (Gen Section)* **94,** 63–69.

Ozawa, H., Katamura, Y., Hatta, S., Amemiya, N., Saito, T., Ohshika, H., and Takahata, N. (1994) Antidepressants directly influence in situ binding of guanine nucleotide in synaptic membrane. *Life Sci.* **54,** 925–932.

Pandey, G. N., Dysken, M. W., Garver, D. L., and Davis, J. M. (1979a) Changes in lymphocyte beta-adrenergic receptor function in affective illness. *Am. J. Psychiatry* **136,** 675–678.

Pandey, G. N., Heinze, W. J., Brown, B. D., and Davies, J. M. (1979b) Electroconvulsive shock treatment decreases β-adrenergic receptor sensitivity in rat brain. *Nature* **280,** 234, 235.

Peralta, E. G., Ashkenazi, A., Winslow, J. W., Ramachandran, J., and Capon, D. J. (1988) Differential regulation of PI hydrolysis and adenyl cyclase by muscarinic receptor subtypes. *Nature* **334,** 434–437.

Post, R. and Ballenger, J. (1984) *Neurobiology of Mood Disorders.* Williams & Wilkins, Baltimore, p. 887.

Price, L. H., Charney, D. S., Delgado, P. L., and Heninger, D. R. (1990) Lithium and serotonin function: implications for the serotonin hypothesis of depression. *Psychopharmacology* **100,** 3–12.

Raymond, J. R. (1995) Multiple mechanisms of receptor-G protein signaling specifity. *Am. J. Physiol.* **38,** F141–F158.

Rens-Domiano, S. and Hamm, H. (1995) Structural and functional relationships of heterotrimeric G proteins. *FASEB. J.* **9,** 1059–1066.

Reuveny, E., Slesinger, P. A., Inglese, J., Morales, J. M., Iniguez-Lluhi, J. A., Lefkowitz, R. J., Bourne, H. R., Jan, Y. N., and Jan, L. Y. (1994) Activation of the cloned muscarinic potassium channel by G protein βγ-subunits. *Nature* **370,** 143–146.

Risby, E. D., Hsiao, J. K., and Manji, H. K. (1991) The mechanisms of action of lithium. II. Effects on adenylate cyclase activity and β-adrenergic receptor binding in normal subjects. *Arch. Gen. Psychiatry* **48,** 513–524.

Schildkraut, J. J. (1965) The catecholamine hypothesis of affective disorders: A review of supporting evidence. *Am. J. Psychiatry* **122,** 509–522.

Schreiber, G., S. Avissar, A. Danon, and R. H. Belmaker. (1991) Hyperfunctional G proteins in mononuclear leukocytes of patients with mania. *Biol. Psychiatry* **29,** 273–280.

Shelton, R. C., Manier, M. S., and Sulser, F. (1996) cAMP dependent protein kinase activity in major depression. *Am. J. Psychiatry* **153,** 1037–1042.

Siever, L. J., Kafka, M. S., Targum, S., and Lake, R. (1984) Platelet alpha-adrenergic binding and biochemical responsiveness in depressed patients and controls. *Psychiatry Res.* **11,** 287–302.

Smrcka, A. V. and Sternweis, P. C. (1994) Regulation of purified subtypes of phosphatidylinositol-specific phospholipase Cβ by G protein α- and βγ-subunits. *J. Biol. Chem.* **268,** 9667–9674.

Spiegel, A. M., Shenker, A., and Weinstein, L. S. (1992) Receptor-effector coupling by G proteins: Implications for normal and abnormal signal transduction. *Endocrinol. Rev.* **13,** 536–565.

Tamir, A. and Gill, D. M. (1988) ADP-ribosylation by chlera toxin of membranes derived from brain modifies the interaction of adenylate cyclase with guanine nucleotides and NaF. *J. Neurochem.* **50,** 1791–1797.

Vetulani, J. and Susler, F. (1975) Action of various antidepressant treatments reduces reactivity of noradrenergic cyclic AMP generation in limbic forebrain. *Nature* **247,** 495, 496.

Vetulani, J., Stawarz, R. J., Dingell, J. V., and Susler, F. (1976) A possible common mechanism of action of antidepressant treatments. Naunyn-Schmiedeberg's. *Arch. Pharmacol.* **293,** 109–114.

Vetulani, J., Lebrecht, U., and Pilc A. (1981) Enhancement of responsiveness of the central serotonergic system and serotonin-2 receptor density in rat frontal cortex by electroconvulsive treatment. *Eur. J. Pharmacol.* **76,** 81–85.

Volonte, C. (1987) Lithium stimulates the binding of GTP to the membranes of PC12 cells cultured with nerve growth factor. *Neurosci. Lett.* **87,** 127–132.

Wickman, K. D., Iniguez-Lluhi, J. A., Davenport, P. A., Taussig, R., Krapivinsky, G. B., Linder, M. E., Gilman, A. G., and Clapham, D. E. (1994) Recombinant G protein βγ-subunits activate the muscarinic-gated atrial potassium channel. *Nature* **368,** 255–257.

Wood, A. J. and Goodwin, G. M. (1987) A review of the biochemical and neuropharmacological actions of lithium. *Psychol. Med.* **17,** 579–600.

Yamamoto, H., Tomita, U., Mikuni, M., Kobayashi, I., Kagaya, A., Katada, T., Ui, M., and Takahashi, K. (1992) Direct activation of purified G_o-type GTP binding protein by tricyclic antidepressants. *Neurosci. Let.* **139,** 194–196.

Young, L. T. and Woods, C. M. (1996) Mood stabilizers have differential effects on endogenous ADP ribosylation in C6 glioma cells. *Eur. J. Pharmacol.,* in press.

Young, L. T., Li, P. P., Kish, S. J., Siu, L. P., Kamble, A., Hornykiewcz, O., and Warsh, J. J. (1991) Cerebral cortex $G\alpha_s$ protein levels and forskolin-stimulated cyclic AMP formation are increased in bipolar affective disorder. *J. Neurochem.* **61,** 890–898.

Young, L. T., Li, P. P., Kish, S. J., Siu, L. P., and Warsh, J. J. (1993) Postmortem cerebral cortex Gs alpha-subunit levels are elevated in bipolar affective disorder. *Brain Res.* **551,** 323–326.

Young, L. T., Li, P. P., Kish, S. J., and Warsh, J. J. (1994a) Cerebral cortex β-adrenoceptor binding in bipolar affective disorder. *J. Affect Disord.* **30,** 89–92.

Young, L. T., Li, P. P., Kamble, A., Siu, K. P., and Warsh, J. J. (1994b) Mononuclear leukocyte levels of G proteins in depressed patients with bipolar disorder or major depressive disorder. *Am. J. Psychiatry* **151,** 594–596.

Young, L. T., Asghari, V., Li, P. P., Kish, S. J., Fahnestock, M., and Warsh, J. J. (1996) Stimulatory G protein α-subunit mRNA levels are not increased in autopsied cerebral cortex from patients with bipolar disorder. *Mol. Brain Res,* **42,** 45–50.

Zemlan, F. P. and Garver, D. L. (1990) Depression and antidepressant therapy: receptor dynamics. *Prog. Neuropsychopharmacol. Biol. Psychiatry* **14,** 503–523.

G Proteins in the Medial Temporal Lobe in Schizophrenia

Fumihiko Okada

Introduction

Numerous controlled investigations by computed tomography and magnetic resonance imaging of living patients suffering from schizophrenia have found quantitative evidence of brain pathology in the form of enlarged third and lateral ventricles and increased cortical markings suggestive of reduced gyral mass or atrophy (Johnstone et al., 1976; Shelton and Weinberger, 1986; Suddath et al., 1989). In addition, controlled studies of postmortem brain tissue have found nonspecific, but objective evidence of abnormal brain structure in the periventricular limbic and diencephalic areas in schizophrenia (Lesch and Bogerts, 1984; Bogerts et al., 1985; Brown et al., 1986; Jakob and Beckmann, 1986; Roberts, 1990). Crow et al. (1989) showed asymmetrical enlargement of the temporal horn of the lateral cerebral ventricle in schizophrenia. These structural abnormalities occur in the absence of degenerative changes in the brain. It has been proposed on the basis of these studies that some anomaly of growth occurs during brain development, but little neurochemical evidence has yet been found that correlates with these structural findings.

Using samples from the same brain series as Crow et al. (1989), we measured the level of guanine nucleotide binding proteins (G proteins) of control and schizophrenic brains by two different methods: one was a ribosylation technique with

From: *Neuromethods, Vol. 31: G Protein Methods and Protocols*
Ed: R. K. Mishra, G. B. Baker, and A. A. Boulton Humana Press Inc.

a pertussis toxin or islet-activating protein- (IAP) catalyzed ADP ribosylation (Okada et al., 1990, 1991), and the other was a highly sensitive enzymatic immunoassay (Okada et al., 1994). We have found that the degrees of ADP ribosylation of G_i (the inhibitory guanine nucleotide regulatory protein) or G_o (another GTP binding protein of unknown function purified from brain) in the left putamen and hippocampus of schizophrenic brain were lower than those in controls (Okada et al., 1990, 1991), while the amounts of G_o determined by immunoassay were not changed (Okada et al., 1994).

In the present chapter, we have discussed the discrepancies between ADP ribosylation and immunoassay. Clarification of neurochemical characteristics based on the discrepancies may elucidate some aspects of the pathogenesis of schizophrenia.

Subjects and Methods

Postmortem Brain Samples

Postmortem brain samples were collected and stored as described previously (Okada et al., 1990). Patients met the criteria of Feighner et al. (1972) for a diagnosis of definite schizophrenia; controls were selected from previously healthy individuals without a history of neurologic or psychiatric disease, and were matched by sex and by age at death. Brains were bisected along the central sulcus. Left and right halves of the brain from each series were selected for anatomical examination or neurochemical studies at random.

Animals and Preparation of Tissue

The subjects were male Wistar-derived Donryu rats (200–230 g) kept under standard laboratory conditions (12:12 h light/dark cycle; free access to water and standard laboratory chow). The animals were injected IP once daily with chlorpromazine (10 or 20 mg/kg), haloperidol (1 or 2 mg/kg), or saline for 3 wk. Twenty-four hours after the final injection, the animals were decapitated, and the three areas (cerebral cortex, striatum, and hippocampus) were immediately removed from the brains.

Chemical Techniques

Tissues were homogenized, and crude membrane fractions were prepared as previously described (Okada et al., 1990). All membrane samples were stored at −80°C before analysis.

Radiolabeling of Membrane Proteins with [α-^{32}P]NAD and Their Separation by Polyacrylamide Gel Electrophoresis

The membrane preparation (100 μg of protein) was incubated with 25 μg/mL of IAP for 60 min at 30°C in 100 μL of 75 mM Tris-HCl buffer, pH 8.0, containing 2 μM [α-^{32}P]NAD (50 μCi/mL), 2.5 mM MgCl$_2$, 2.13 mM EDTA, 1 mM ATP, 1 mM dithiothreitol (DTT), 10 mM thymidine, and 25 μM GTP. The reaction was terminated by addition of 25 μL of a stop mixture containing 9.2% sodium dodecyl sulfate (SDS), 40% glycerol, 0.05% bromophenol blue, 20% β-mercaptoethanol, and 250 mM Tris-HCl. The contents were heated for 10 min in a boiling water bath. The labeled proteins were analyzed by subjecting the aliquots to sodium dodecyl sulfate-polyacrylamide gel electrophoresis (SDS-PAGE) as described by Katada and Ui (1982). Autoradiography was then conducted by exposure of the electrophoretogram to Kodak X-Omat film. Protein content was determined according to the method of Lowry et al. (1951) using bovine serum albumin as the standard.

Enzyme Immunoassays

The samples were solubilized with a solution containing 2% sodium cholate, 1 mM DTT, 1 mM EDTA, and 20 mM Tris-HCl at 0°C for 1 h. The homogenate was sonicated and then centrifuged at 4°C at 100,000g for 1 h, and the supernatant fraction was used for the analysis after dilution with 10 mM sodium phosphate (pH 7.0) containing 0.3M NaCl, 1 mM MgCl$_2$, 0.1% bovine serum albumin, 0.5% protease-treated gelatin, with 0.1% NaN$_3$, containing 20 μM AlCl$_3$, 6 mM

MgCl$_2$, and 10 mM NaF. The concentrations of Gα_o and Gα_{i2} in each sample were determined in triplicate by the enzymatic immunoassays prepared with antibovine Gα_o antibody and antibovine Gα_{i2} antibody, which were reported by Asano et al. in 1987 and 1989, respectively. Proteins were determined by the method of Schaffner and Weissmann (1973) with bovine serum albumin as the standard.

Results

Figure 1 shows the amount of [^{32}P]ADP ribose incorporated by IAP into the 39–41 kDa peptide in caudate head, putamen, and globus pallidus in controls and schizophrenics. Membranes from the left side of the putamen in schizophrenics showed significantly less intense radioactive labeling at 39–41 kDa than did the respective sides of controls. In this study, the total amount of G$_i$/G$_o$ proteins could not be measured directly, but the enzymatic activity of ADP ribosylation was estimated. The decrease in the IAP-labeled proteins amounted to 42% in the left-side putamen in schizophrenics when compared with respective controls. No significant differences were seen between the left and right caudate head or globus pallidus and right putamen when compared to their respective controls.

Figure 2 shows the amount of G$_i$/G$_o$ in five areas of the medial temporal lobe of control and schizophrenic brains utilizing pertussis toxin-catalyzed ADP ribosylation. Membranes from the left side of hippocampus in schizophrenics showed significantly less intense radioactive labeling at 39–41 kDa mol wt than did the respective sides of controls. The decrease in the IAP-labeled proteins amounted to 32% in the left side of hippocampus in schizophrenics when compared with respective controls. Although reductions were also seen in the left amygdala and parahippocampal gyrus, these did not reach significance. Insula and lateral temporal cortex were unchanged when compared to their respective controls.

Therefore, measurements of the level of G$_i$/G$_o$ in the schizophrenic brain utilizing pertussis toxin-catalyzed ADP

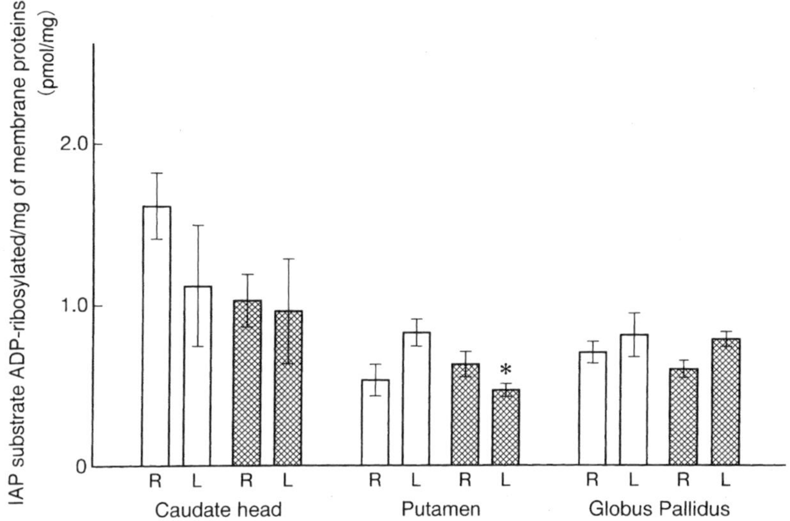

Fig. 1. The amount of [^{32}P]ADP ribose incorporated by IAP into the 39–41 kDa peptide in right (R) and left (L) side of the caudate head, putamen, and globus pallidus in controls (open column) and schizophrenics (shaded column). Each column represents the mean ± SEM of 5–12 subjects. *There was a significant decrease in the IAP substrate levels in the left side of the putamen in schizophrenics compared to the controls by analysis of variance (program BMDP2V), $p < 0.01$.

ribosylation showed that the amount of G_i/G_o was significantly decreased in the putamen and hippocampus of the left hemisphere in schizophrenics compared with the controls; unfortunately, these experiments with IAP could not distinguish multiple species of subunits of pertussis toxin-sensitive G proteins.

Figure 3 presents concentrations of the α-subunits of G proteins ($G\alpha_o$) using the highly sensitive enzyme immunoassay method. There were no significant differences in $G\alpha_o$ in the hippocampus and putamen of the left hemisphere in schizophrenic patients compared with controls, whereas there was a significant decrease in $G\alpha_o$ in the hippocampus and caudate head of the right hemisphere in schizophrenic patients.

Figure 4 shows concentrations of the α-subunit of G_{i2} proteins using a highly sensitive enzyme immunoassay method.

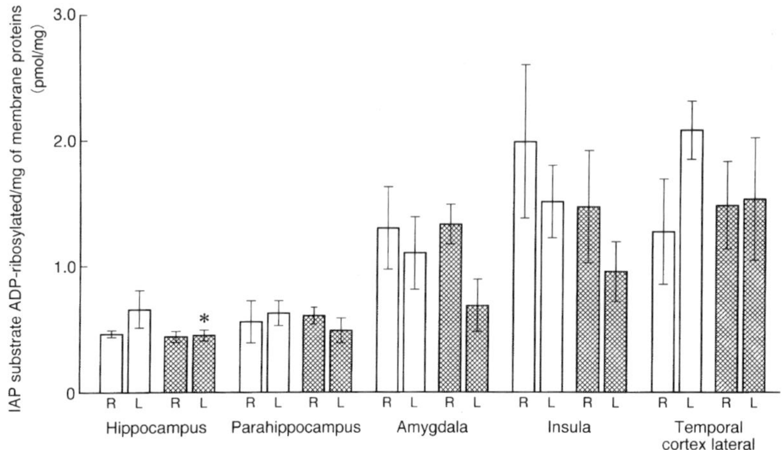

Fig. 2. The amount of [^{32}P]ADP ribose incorporated by IAP into the 39–41 kDa peptide in right (R) and left (L) side of the hippocampus, parahippocampus, amygdala, insula, and temporal cortex lateral in controls (open column) and schizophrenics (shaded column). Each column represents the mean ± SEM of 3–8 subjects. *There was a significant decrease in the IAP substrate levels in the left side of the hippocampus in schizophrenics compared to the controls by analysis of variance (program BMDP2V), $p < 0.05$.

The concentrations of Gα_{i2} did not differ significantly in any area.

Table 1 shows that there were no significant differences in the levels of ADP ribosylation among three areas, including striatum, of rat brain membranes from controls and those of groups receiving neuroleptic treatments.

Discussion

The family of G proteins plays a key role in the transmembrane signaling pathway in a variety of cells: G$_s$ and G$_i$ stimulate and inhibit, respectively, adenylate cyclase, and G$_o$ inhibits the voltage-dependent Ca^{2+} channel (Graziano and Gilman, 1987). Recently, Hepler and Gilman (1992) reviewed advances in G protein research; they indicated that the task of unraveling the complexity of the G protein-regulated cellular

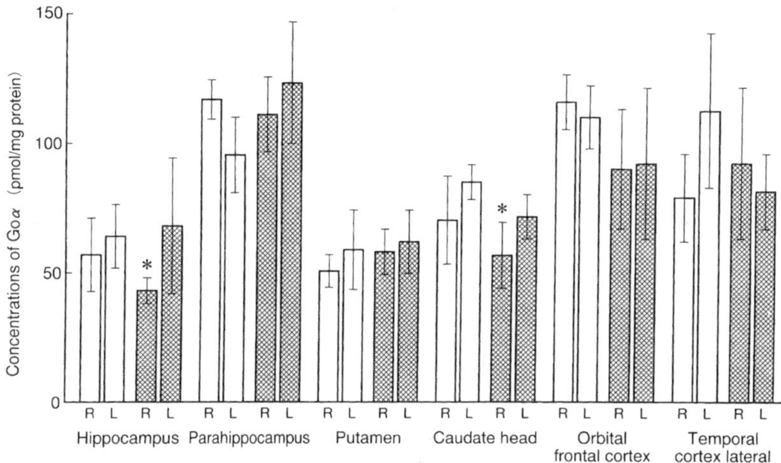

Fig. 3. Concentrations of Gα$_o$ in right (R) and left (L) sides of six brain areas in normal controls (open column) and schizophrenics (shaded column). Each column represents the mean ± SEM of 3–11 subjects. There was a significant decrease in Gα$_o$ in the hippocampus and caudate head of the right side in schizophrenic patients compared to controls by the ANOVA, $p < 0.05$.

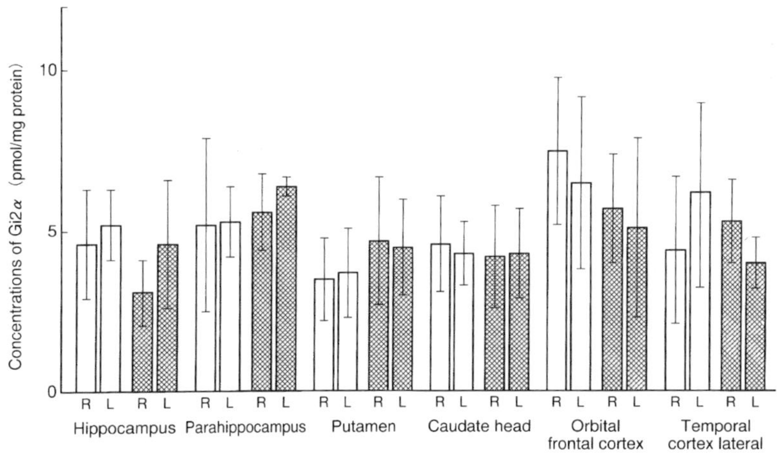

Fig. 4. Concentrations of Gα$_{i2}$ in right (R) and left (L) sides of six brain areas in normal controls (open column) and schizophrenics (shaded column). Each column represents the mean ± SEM of 3–11 subjects.

Table 1
ADP Ribosylation of G Proteins (G_i/G_o)
with IAP in the Cortex, Striatum, and Hippocampus
of Brains of Rats Treated with Neuroleptics for 3 Wk[a]

	Cerebral cortex	Striatum	Hippocampus
Control	100 (10)	100 (9)	100 (10)
CPZ 10 mg/kg	86 ± 8 (9)	112 ± 10 (10)	92 ± 6 (10)
CPZ 20 mg/kg	97 ± 11 (9)	112 ± 10 (9)	88 ± 11 (9)
HPD 1 mg/kg	84 ± 11 (8)	114 ± 12 (9)	88 ± 10 (9)
HPD 2 mg/kg	92 ± 10 (9)	116 ± 13 (9)	96 ± 11 (9)

[a]Data were quantified using a Fuji imaging analyzer (Fuji BAS 2000). They are given as percent of control, with the number of observations in parentheses. Each item shows the mean ± SEM. CPZ: chlorpromazine; HPD: haloperidol.

switchboard expands exponentially, as the number of identified G protein subunits continues to grow.

Measurements of the level of G_i/G_o in the schizophrenic brain utilizing pertussis toxin-catalyzed ADP ribosylation showed that the amount of G_i/G_o was significantly decreased in the putamen (Okada et al., 1990) and hippocampus (Okada et al., 1991) of the left hemisphere in schizophrenic patients compared with controls. Unfortunately, these experiments with IAP could not distinguish multiple species of subunits of pertussis toxin-sensitive G proteins. On the other hand, when we investigated the α-subunit of G proteins ($G\alpha_o$ and $G\alpha_i2$) using the highly sensitive enzyme immunoassay method, there were no significant differences in $G\alpha_o$ in the hippocampus and putamen of the left hemisphere in schizophrenic patients compared with controls, but there was a significant decrease in $G\alpha_o$ in the hippocampus and caudate head of the right hemisphere in schizophrenic patients (Okada et al., 1994).

Our study could not determine concentrations of $G\alpha_{i1}$ for methodological reasons; however, Asano et al. (1990) have reported that the amounts of $G\alpha_{i1}$ and $G\alpha_{i2}$ were about 36 and 8%, respectively, of that of $G\alpha_o$ in the hippocampus of rats. Therefore, the amounts of ADP ribosylated products reported previously in hippocampus or putamen were mainly those of the heterotrimeric structures of G_o and G_{i1} the level of ADP ribosylation of G_{i2} can be disregarded in this case.

One possible explanation for the discrepancies between ADP ribosylation and immunoassay was offered by Seeman et al. (1989, 1993). Dopamine receptors in schizophrenics could be locked in a high-affinity state. On the basis of their studies of dopamine receptors, Seeman et al. (1989, 1993) explained that normally G proteins rapidly switch dopamine receptors back and forth between states of high and low affinity, but in schizophrenics the receptor ceases to function. There is something wrong in schizophrenia with the receptors' ability to switch (Taubes, 1994).

When receptors couple with G proteins (or in the case of the dopamine systems, when dopamine-D2 or D4 receptors couple with G_i), the subunit equilibrium of heterotrimeric G_i ($\alpha\beta\gamma \rightleftarrows \alpha + \beta\gamma$) shifts to the right. Since ADP ribosylation by pertussis toxin occurs only when α has combined with $\beta\gamma$ in the tight trimeric structure, this reaction cannot be induced during dissociation of α from $\beta\gamma$. Therefore, when locked in a high-affinity state, ADP ribosylation is suppressed because of the dissociation of α from $\beta\gamma$. On the other hand, the trimeric structure is not necessary for reaction of the anti-$G\alpha_o$ antibodies against $G\alpha_o$. Therefore, Seeman et al.'s hypothesis explains the hitherto puzzling contradiction between our results using ADP ribosylation and those using the immunoassay (Okada et al., 1995). In turn, our results appear to support their hypothesis.

Another possible explanation of the discrepancies between ADP ribosylation and immunoassay could be a change in lipid modification of trimeric G proteins. All G protein α-subunits are modified at or near their N-termini by covalent attachment of the fatty acids myristate and/or palmitate (Wedegaertner et al., 1995). Myristoylation or, more specifically, N-myristoylation is the result of cotranslational addition of the saturated 14-carbon fatty acid myristate to a glycine residue at the extreme N-terminus after removal of the initiating methionine. A stable amide bond links myristate irreversibly to proteins. α-Subunits of the α_i family, including α_o, are myristoylated. Further, G protein γ-subunits are covalently modified by the 20-carbon isoprenoid geranylgeranyl or the 15-carbon isoprenoid farnesyl. Prenylation of γ is necessary for normal function of the βγ dimers. No lipid modification has been identified on G protein β-subunits. ADP ribosylation of $\alpha_o\beta\gamma$ is not supported, since nonmyristoylated α_o has a much reduced affinity for βγ and nonprenylated βγ shows a reduced binding affinity to α_o.

Abnormalities of functional regulation in G proteins, such as lipid modifications, may happen in the medial temporal lobe in schizophrenia even if amounts of G protein are not changed. Abnormalities in the enzymatic systems of neural cell membranes in this area may turn out to be pivotal in explaining many schizophrenic symptoms.

One more possibility, that decreased ADP ribosylation is owing to the effects of antipsychotic medications, was ruled out in rodent experiments after chronic drug treatments (Okada et al., 1996).

Summary

We have found that the degree of ADP ribosylation of G_o/G_i in the left putamen and hippocampus of schizophrenic brain was lower than that in controls, whereas the amounts of G_o determined by immunoassay were not changed. The discrepancies between ADP ribosylation and immunoassay could be explained by the following: (1) ADP ribosylation is suppressed because of dissociation of the α-subunit from the βγ-subunits when dopamine receptors in schizophrenics are locked in a high-affinity state; (2) ADP ribosylation of $\alpha_o\beta\gamma$ is not supported, since nonmyristoylated α_o has a much reduced affinity for βγ, and nonprenylated βγ, shows a reduced binding affinity to α_o. We have discussed (1), but there have been no investigations concerning (2). Abnormalities of functional regulation in G proteins, such as those of lipid modifications, may occur in the medial temporal lobe in schizophrenia even if amounts of G protein are not changed. Abnormalities in the enzymatic system of neural cell membranes in this area may turn out to be pivotal for explaining many schizophrenic symptoms. Another possibility, that decreased ADP ribosylation is owing to the effects of antipsychotic medications, was ruled out in murine experiments after chronic drug treatments.

References

Asano, T., Semba R., Ogasawara N., and Kato K. (1987) Highly sensitive immunoassay for the α-subunit of the GTP-binding protein Go and its regional distribution in bovine brain. *J. Neurochem.* **48,** 1617–1623.

Asano T., Morishita R., Semba R., Itoh, H., Kaziro, Y., and Kato K. (1989) Identification of lung major GTP-binding protein as Gi2 and its distribution in various rat tissues determined by immunoassay. *Biochemistry* **28,** 4749–4754.

Asano, T., Shinohara, H., Morishita, R., and Kato, K. (1990) Immunochemical and immunohistochemical localization of the G protein Gi1 in rat central nervous tissues. *J. Biochem.* **108,** 988–994.

Bogerts, B., Meertz, E., and Schonfeld-Bausch, R. (1985) Basal ganglia and limbic system pathology in schizophrenia: a morphometric study of brain volume and shrinkage. *Arch. Gen. Psychiatry* **42,** 784–791.

Brown, R., Colter, N., Corsellis J. A. N., Crow, T. J., Frith, C. D., Jagoe, R., Johnstone, E. C., and Marsh, L. (1986) Postmortem evidence of structural brain changes in schizophrenia. Differences in brain weight, temporal horn area, and parahippocampal gyrus compared with affective disorder. *Arch. Gen. Psychiatry* **43,** 36–42.

Crow, T. J., Ball, J., Bloom, S. R., Brown, R., Bruton, C. J., Colter, N., Frith, C. D., Johnstone, E. C., Owens, D. G. C., and Roberts, G. W. (1989) Schizophrenia as an anomaly of development of cerebral asymmetry. A postmortem study and a proposal concerning the genetic basis of the disease. *Arch. Gen. Psychiatry* **46,** 1145–1150.

Feighner, J. P., Robins, E., Guze, S. B., Woodruff, R. A., Winokur, G., and Munoz, R. (1972) Diagnostic criteria for use in psychiatric research. *Arch. Gen. Psychiatry* **26,** 57–63.

Graziano, M. P. and Gilman, A. G. (1987) Guanine nucleotide-binding regulatory proteins: mediators of transmembrane signaling. *Trends Pharmacol. Sci.* **8,** 478–481.

Hepler, J. R. and Gilman, A. G. (1992) G proteins. *Trends Biochem. Sci.* **17,** 383–387.

Jakob, H., Beckmann, H. (1986) Prenatal development disturbances in the limbic allocortex in schizophrenics. *J. Neural. Transm.* **65,** 303–326.

Johnstone, E. C., Crow, T. J., Frith, C. D., Husband, J., and Kreel, L. (1976) Cerebral ventricular size and cognitive impairment in chronic schizophrenia. *Lancet* **ii,** 924–926.

Katada, T. and Ui, M. (1982) Direct modification of the membrane adenylate cyclase system by islet-activating protein due to ADP ribosylation of a membrane protein. *Proc. Natl. Acad. Sci. USA* **79,** 3129–3133.

Lesch, A. and Bogerts, B. (1984) The diencephalon in schizophrenia: evidence for reduced thickness of the periventricular grey matter. *Eur. Arch. Psychiat. Neurol. Sci.* **234,** 212–219.

Lowry, O. H., Rosebrough, N. J., Farr, A. L., and Randall, R. J. (1951) Protein measurement with the Folin phenol reagent. *J. Biol. Chem.* **193,** 265–275.

Okada, F., Crow, T. J., and Roberts, G. W. (1990) G proteins (Gi, Go) in the basal ganglia of control and schizophrenic brain. *J. Neural. Transm. (Gen. Section)* **79,** 227–234.

Okada, F., Crow, T. J., and Roberts, G. W. (1991) G proteins (Gi, Go) in the medial temporal lobe in schizophrenia. Preliminary report of a neurochemical correlate of structure change. *J. Neural. Transm. (Gen. Section)* **84,** 147–153.

Okada, F., Tokumitsu, Y., Takahashi, N., Crow, T. J., and Roberts, G. W. (1994) Reduced concentrations of the α-subunit of GTP-binding protein Go in schizophrenic brain. *J. Neural. Transm. (Gen. Section)* **95,** 95–104.

Okada, F., Murakami T., and Tokumitsu, Y. (1995) Low levels of pertussis toxin adenosine diphosphate ribosylation in the schizophrenic brain. *Arch. Gen. Psychiatry* **52,** 319.

Okada, F., Ito, A., Horikawa T., Tokumitsu Y., and Nomura Y. (1996) Long-term neuroleptic treatments counteract dopamine D_2 agonist inhibition of adenylate cyclase but do not affect pertussis toxin ADP ribosylation in the rat brain. *Neurochem. Int.* **28,** 161–168.

Roberts, G. W. (1990) Schizophrenia: the cellular biology of a functional psychosis. *Trends Neurosci.* **13,** 207–211.

Schaffner, W. and Weissmann, C. (1973) A rapid, sensitive, and specific method for the determination of protein in dilute solution. *Anal. Biochem.* **56,** 502–514.

Seeman, P., Niznik, H. B., Guan, H.-C., Booth, G., and Ulpian, C. (1989) Link between D_1 and D_2 dopamine receptors is reduced in schizophrenia and Huntington diseased brain. *Proc. Natl. Acad. Sci. USA* **86,** 10,156–10,160.

Seeman, P, Guan, H.-C., and Van Tol, H. H. M. (1993) Dopamine D_4 receptors elevated in schizophrenia. *Nature (Lond.)* **365,** 441–445.

Shelton, R. C. and Weinberger, D. R. (1986) X-ray computerized tomography studies of schizophrenia: a review and synthesis, in *The Neurology of Schizophrenia* (Nasrallah, H. A. and Weinberger, D. R., eds.), Elsevier, Amsterdam, pp. 207–250.

Suddath, R. L., Casanoca, M. F., Goldberg, T. E., Daniel, D. G., Kelsoe, J. R., and Weinberger, D. R. (1989) Temporal lobe pathology in schizophrenia: a quantitative magnetic resonance imaging study. *Am. J. Psychiatry* **146**, 464–472.
Taubes, G. (1994) Will new dopamine receptors offer a key to schizophrenia? *Science* **265**, 1034, 1035.
Wedegaertner, P. B., Wilson, P. T., and Bourne, H. R. (1995) Lipid modifications of trimeric G proteins. *J. Biol. Chem.* **270**, 503–506.

G Protein Abnormalities
in Schizophrenia

*Naoki Nishino, Noboru Kitamura,
Chang-Qing Yang, Hideo Yamamoto,
Yutaka Shirai, Yasuo Kajimoto,
and Osamu Shirakawa*

Introduction

Since Emil Kraepelin coined the term dementia praecox, schizophrenia has been recognized as a group of illnesses of unknown etiology sharing common symptomatology with diverse clinical courses. Family, twin, and adoption studies indicate that genetic factors contribute to the etiology of schizophrenia. There has been evidence for involvement of many neurotransmitter receptor subtypes in the pathophysiology of schizophrenia. The champion of these is the dopamine (DA) receptor because DA-D2 receptors are targets of neuroleptics and have been found to be elevated in schizophrenia. The recent cloning of genes encoding five distinct DA receptor subtypes (D1–D5) and the discovery of a functional polymorphism within each gene have made it possible to investigate linkage of each DA receptor locus with genetic susceptibility to schizophrenia. To date there appears to exist no linkage between schizophrenia and the known DA receptors (Coon et al., 1993; Nöthen et al., 1994). Normal density or normal structure of a certain receptor subtypes does not always mean that the downstream partners of the receptor (G proteins, effectors, protein kinases, phosphatases) are normal.

From: *Neuromethods, Vol. 31: G Protein Methods and Protocols*
Ed: R. K. Mishra, G. B. Baker, and A. A. Boulton Humana Press Inc.

Therefore, in order to understand the neurochemical pathophysiology of schizophrenia, we must extend our research to transmembrane signaling systems. In this chapter we will review briefly basic aspects of G proteins and their abnormalities in specific disorders, and then present some data on G protein abnormalities in schizophrenia obtained in our laboratory using postmortem brain samples.

Role of G Proteins in Signal Transduction

G Proteins Act as a Transducer in Signaling Cascades

Heterotrimeric (α, β, γ) guanine nucleotide-binding regulatory proteins (G proteins) act as signal transducers by coupling seven-helix cell-surface receptors (serpentine receptors) to intracellular effector proteins. Receptor-mediated activation of a G protein results in the exchange of guanosine 5′-triphosphate (GTP) for guanosine 5′-diphosphate (GDP) on the G protein α-subunit, followed by dissociation of GTP-α from the receptor and from the noncovalently associated $\beta\gamma$-subunit complexes. Not only activated GTP-α, but also the released $\beta\gamma$-subunit complexes, alone or in cooperation with α-subunits, interact with effector molecules, such as adenylyl cyclase (AC) (Tang and Gilman, 1992), inositol phospholipid-specific phospholipase C (PLC) (Smrcka and Sternwein, 1993), and ion channels (Kurachi et al., 1989; Kim et al., 1989). The G protein α-subunits have intrinsic GTPase activity. Hydrolysis of the bound GTP terminates regulation of the effector, returning the α-subunit to the inactive GDP-bound form. Thus, the GDP/GTP exchange reaction serves as a molecular on-off switch in signal transduction cascades. Recent studies have solved the crystal structure of the G protein heterotrimer: the whole signaling complex does not act as a set of sequential bimolecular reactions, but rather like a "miniature (nano) machine" consisting of a lever (receptor), switch (Gα) and propeller (G$\beta\gamma$) (Clapham, 1996).

Receptor/G Protein/Effector Interaction

Many cellular responses are positively and negatively regulated by separate G protein-coupled pathways. Stimulation of receptors coupled with G_s, such as DA-D1 and β-adrenergic receptors, activates, whereas that of receptors coupled with G_i (G_{i1}, G_{i2}, and G_{i3}), such as DA-D2 and α_2-adrenergic receptors, inhibits AC, leading to bi-directional oscillation of intracellular adenosine 3',5'-cyclic monophosphate (cAMP). In general, the third intracytoplasmic loop of serpentine receptors appears to direct the interaction of the receptor with the appropriate G protein subtypes (Guiramand et al., 1995).

Most ACs are associated with the plasma membrane. cDNAs encoding eight ACs (types I–VIII) have been cloned. The most common motif of AC includes a short N-terminal region and two large cytoplasmic domains, punctuated by two hydrophobic stretches; each of the latter contains six transmembrane helices, a structure that is reminiscent of those that have been proposed for several transporters and channels (Krupinski et al., 1989). Based on sequence similarities, the isoforms have been classified into three subfamilies: the type I-like group (I, III, and VIII); the type II-like group (II, IV, and VII); and the type V-like group (V and VI). Functional properties are shared within these subfamilies (e.g., ACs of type I-like group are activated, whereas ACs of type V-like group are inhibited by Ca^{2+}/calmodulin). In mammalian brains, only three (I, II, and V) are expressed strongly, whereas two others (VII and VIII) are expressed at lower levels; each isoform has a distinct pattern of expression among brain regions (Mons and Cooper, 1995). The existence of different ACs in different cell types might allow for differential cellular responses to the same stimulus. All the mammalian ACs are activated by the α-subunit of G_s. AC is activated by G_s-linked pathways, but can be inhibited or stimulated further by βγ when other G protein G_i and G_o mediated pathways are activated simultaneously (Tang and Gilman, 1992). The exact mechanisms of inhibition of AC by the G_i protein are uncertain and may vary with the type of AC in question.

The increase in intracellular cyclic AMP (cAMP) level activates protein kinase A (PKA), which exists as a tetramer consisting of two regulatory subunits (R) and two catalytic subunits (C). The binding of cAMP to the inactive PKA R2C2 holoenzyme, much of which is anchored in the perinuclear region of the cytoplasm through membrane-associated anchoring proteins, releases the enzymatically active C subunit, which is then competent to phosphorylate its substrate proteins. A fraction of the C subunit population translocates into the nucleus, where it phosphorylates its nuclear substrates. One of the major nuclear PKA substrates is the transcription factor cAMP response element binding protein (CREB), which then binds to a palindromic response element cAMP-regulated enhancer (CRE) in cAMP-inducible genes. cAMP-inducible gene expression is crucial in a variety of cellular responses, including the establishment of long-term memory.

At least 30 serpentine receptors, such as the serotonin 5-HT1C and muscarinic m_3, activate PLC-β, which catalyzes hydrolysis of phosphatidylinositol 4,5-bisphosphate (PIP_2) to produce two second messengers, myoinositol 1,4,5-trisphosphate (IP_3) and *sn*-1,2-diacylglycerol (DAG). DAG is instantly phosphorylated by DAG kinase to form phosphatidyl acid (PA), which is then metabolized to phosphatidyl monophosphate and to phosphatidyl polyphosphates.

The most established path for activation of PLC-β (PLC-β1, PLC-β2, and PLC-β3) is through the pertussis toxin (PTX)-insensitive G protein (α-subunits of the G_q class subtypes, *see* Molecular Diversity and Specificity of G Protein). The PTX-sensitive G proteins (α-subunits of the G_i class, see below) that activate PLC-β are less well established, but there is evidence that both $G\alpha_o$- and $G\alpha_i$- as well as Gβ-subunits are involved (Neer, 1995). The α- and β-subunits appear to interact with separate domains of the PLC-β molecule. DAG increases sensitivity of protein kinase C (PKC) for Ca^{2+} and IP_3 triggers release of Ca^{2+} from the endoplasmic reticulum (ER) by binding to the specialized tetrameric ryanodine receptor/Ca^{2+} release channels (IP_3 receptor in nonexcitable cells) that span the ER membrane, thus synergistically activating PKC. Ryanodine receptor has six membrane-spanning segments, and is a member of the superfamily that includes the voltage- and second messenger-gated ion channels on the plasma membrane. The neuronal ryanodine receptor comprises three distinct types: skeletal, cardiac and brain types. In rabbit brains, the skeletal type is highly expressed in cerebellar Purkinje cells. The cardiac type is predominantly expressed throughout nearly the entire brain with markedly high levels in the olfactory nerve layer, layer VI of the cerebral cortex, the dentate gyrus, and cerebellar granule cells. The brain type ryanodine receptor expression is the least widely distributed throughout the brain, with high levels found in the hippocampal CA1 pyramidal layer, caudate nucleus, putamen, and the dorsal thalamus (Furuichi et al., 1994). In addition to voltage- or receptor-gated opening of Ca^{2+}-permeable plasmalemmal channels, the ryanodine receptor/Ca^{2+}-release channels can also play a crucial role in regulating Ca^{2+}-signals and compartmentalized function, such as excitability, neurotransmitter release, synaptic plasticity, gene expression, and apoptosis (Clapham, 1995).

PKC comprises a large family with multiple isoforms exhibiting individual characteristics and a distinct pattern of tissue distribution. Eleven isoforms of PKC have been identified so far in mammalian tissues, and the PKC family can be subdivided into three groups: "conventional" (Ca^{2+} plus DAG requiring) PKCs (cPKC: α, βI, βII, γ), "novel" (Ca^{2+}-independent, but DAG requiring) PKCs (nPKC: δ, ϵ, η, θ, μ), and "atypical" (which require neither) PKC (aPKC: ζ, ι). All require phosphatidylserine for activation (Nishizuka, 1995).

Thus, multicyclic enzyme cascades for signal transduction allow amplification, feedback, crosstalk, and branching of signals. Interference with any step of this signaling process can lead to aberrant cellular functions.

Molecular Diversity and Specificity of G Protein

In mammals, G protein α, β, and γ polypeptides are encoded by at least 16, 4, and 7 genes, respectively (Conklin and Bourne, 1993). Based on amino acid sequence similarity, α-subunits of G proteins can be subdivided into four classes (G_s, G_i, G_q, and G_{12}), and each class comprises specific isotypes: the G_s class includes α_s and α_{olf}; the G_i class includes α_{i1}, α_{i2}, α_{i3}, α_o, α_{gust}, α_{t1}, α_{t2}, and α_z; the G_q class includes α_{11}, α_{14}, α_{15}, α_{16}, and α_q; the G_{12} class includes α_{12} and α_{13} (Simon et al., 1991). The G_s class is modified by cholera toxin and the G_i class is PTX-sensitive (except α_z). Most α-subunits are widely expressed, except for α_t, α_{gust}, or α_{olf} (sensory organs), α_{16} (hematopoietic cells), and α_o (neurons). Some naturally occurring receptors can activate more than one class of G protein. A representative of these promiscuous receptors is the α_{2A}-adrenoceptor, which can activate G_i, G_s, G_z, and G_q (Conklin and Bourne, 1993).

Individual cells contain at least four or five types of G protein α-subunits. If all the α-, β-, and γ-subunits associated combinatorially and at random, there would be almost 100 different kinds of heterotrimers. There are over 100 G protein-linked receptors and many isotypes of effectors. How these G proteins encode specificity for receptors and effectors is unclear. The specificity exists both at the level of receptor/G protein and at the level of G protein/effector interactions, the structural determinants of which reside in α-subunits (Quick et al., 1994). Recent studies have documented contribution of the β and γ polypeptides to specificity at the level of receptor/G protein interactions as well. Four different β polypeptide sequences are known, and all have similar sequences (Simon et al., 1991). Because of the apparent sequence heterogeneity in the γ-subunits, functional differences of the $\beta\gamma$ complex have been attributed to the γ-subunits. Using microinjection of γ-selective antisense oligonucleotides into the rat pituitary tumor cell line GH_3, Kleuss et al. (1993) demonstrated that the muscarinic receptor is coupled to a G protein consisting of $\alpha_{o1}/\beta_3/\gamma_4$, and the somatostatin receptor to a G protein consisting of $\alpha_{o2}/\beta_1/\gamma_3$, but that both G proteins act to inhibit voltage-sensitive Ca^{2+} channels, suggesting that two G_o forms, distinct in all three subunits, discriminate between two distinct receptors, but functionally couple to the same effector.

Compartmentation of Receptor/G Protein/Effector Molecules

Receptor/G protein/effector interactions have been viewed as random collisions between the proteins freely floating in the membrane lipid bilayer. In this case, multiple receptor types can access the shared pool of G protein. Recent evidence suggests that in intact cells, some sets of receptors, G proteins, and effectors may be highly organized into specialized microdomains and not have access to other sets (Neubig, 1994), suggesting the existence and importance of stable assemblies of signaling proteins. It is now known that G protein α- or β-subunits associate with cytoskeletal proteins, including fodrin (nonerythroid spectrin).

Like erythroid spectrin, human fodrin is a heterodimer, consisting of α- and β-subunits (migrating at 240 and 235 kDa on SDS-PAGE, respectively). Each subunit exhibits an internal 106 amino acid repeating motif. The α-subunit is composed of 22 repeat units and contains the *src* homology 3 (SH3) domain, which is thought to be involved in the control of a small *ras*-like G protein. In the presence of Ca^{2+}/calmodulin, fodrin can undergo limited proteolysis by a Ca^{2+}-activated neutral protease, μ-calpain, to yield two fragments (each migrated at 150 kDa on SDS-PAGE). The β-subunit is composed of 17 repeat units (Dhermy, 1991). Near the C-terminus exists the pleckstrin homology (PH) domain. The PH domain has also been found in signaling molecules, such as *ras*-GTPase-activating protein, *ras*-guanine nucleotide releasing factor, β-adrenoceptor kinase (β ARK-1), PLC-γ, and dynamin (Haslam et al., 1993). Since the PH domain is thought to associate with the dissociated, prenylated, membrane-anchored G protein βγ-subunits and to be targeted to its membrane-bound receptor substrates, it is likely that the PH domain on β-fodrin serves as a recognition site for protein–protein interaction, whereby membrane-bound receptors and signal transduction complexes can be collected into topographic and functional ensembles (Lombardo et al., 1994).

Both the G protein α- and γ-subunits undergo posttranslational modification, resulting in the covalent attachment of specific lipophilic groups. The α-subunits are subject to N-terminal myristoylation (α_i, α_o, α_z) (Buss et al., 1987) or heterogenous acylation (α_t) at Gly 2 (Kokame et al., 1992), and most (except α_t) are regulated by reversible palmitoylation at Cys 3 (Wedegaertner and Bourne, 1994). In addition to functioning as membrane anchors, myristoylation of α_t subunits increases their affinity for βγ, and is required for interactions with effectors. Palmitoylation of α_s not only contributes to membrane association, but can regulate its cellular location and activity. On agonist activation of receptor, rapid depalmitoylation of α_s and membrane-to-cytosol translocation of α_s coincide, suggesting an additional mechanism of desensitization. The γ-subunits are farnesylated (γ_1) (Fukuda et al., 1990) or geranyl geranylated (γ_{2-4}) (Mumby et al., 1990) at the cysteine in the C-terminal CAAX motif. Although not required for βγ dimerization, this prenylation of γ-subunits is essential for interactions of βγ with α-subunits, receptors, and effectors. To date no lipid modification has been identified on G protein β-subunits. The primary role of the lipophilic modifications is to colocalize and orient the subunits on the membrane surface (Bigay et al., 1994).

Known G Protein Abnormalities in Human Disease

Naturally occurring mutations in G protein α-subunits have been identified as the cause of a substantial number of human disease (Spiegel et al., 1993) (Table 1). Multiple distinct heterozygous mutations of the $G\alpha_s$ gene, which disrupt either the protein or the synthesis of its mRNA, have been identified in affected family members with Albright Hereditary

Table 1
Abnormalities in G Protein-Coupled
Signal Transduction in Human Disease

Study	G protein mutation	Disease	Effects of mutation
Miric et al. (1993)	$G\alpha_s/Arg^{385}$	Pseudohypopara-thyroidism (PHP) type Ia	Loss of function
Iiri et al. (1994)	$G\alpha_s/Ala^{366}$	PHP type Ia with precocious puberty	Loss/gain of function
Weinstein et al. (1991)	$G\alpha_s/Arg^{201}$	McCune-Albright syndrome	Gain of function
Landis et al. (1989)	$G\alpha_s/Arg^{201}$ or Gln^{227}	GH-secreting pituitary tumor in acromegaly	Gain of function
Lyons et al. (1990)	$G\alpha_{i2}/Arg^{179}$	Ovarian and adrenocortical tumors	Gain of function

Osteodystrophy, whereas some missense mutaions have been identified, including one (Arg^{385}) shown to uncouple G_s from receptors, which should lead to impaired responsiveness of G_s protein on agonist activation of the receptor (Miric et al., 1993). GTPase-inhibiting mutations in the $G\alpha_s$ gene, which should lead to sustained activation of $G\alpha_s$ with increased cAMP production, have been identified in sporadic pituitary (Landis et al., 1989) and thyroid tumors (Arg^{201} or Gln^{227}) (Lyons et al., 1990), and McCune-Albright syndrome (Arg^{201}) (Weinstein et al., 1991). Activating mutations of $G\alpha_{i2}$ gene were identified in a minority of adrenal and ovarian tumors (Arg^{179}) (Lyons et al., 1990).

Abnormal function and expression of various G proteins have been implicated in a variety of psychiatric pathophysiologic states, such as schizophrenia, mood disorder, panic disorder, and chronic opiate/cocaine/ethanol use (Manji, 1992).

For example, in the caudate and the nucleus accumbens, a greater activation of AC by sodium fluoride (NaF) and a non-hydrized GTP analog $G_{pp}NH_p$ as well as a selective D1 agonist, SKF-38393, was observed in schizophrenic patients compared with controls (Memo et al., 1983), suggesting that the D1 receptors may be more efficaciously coupled to AC through G_s, at least in the striatum and the limbic system of schizophrenia. In the left parahippocampal gyrus and CA1 region of schizophrenic brains, Kerwin and Beats (1990) found an increase in forskolin binding thought to represent binding to the complex of $G\alpha_o$ and AC. Seeman et al. (1989) demonstrated that in the striatal tissues of control, Alzheimer, and Parkinsonian subjects, the DA-induced decrease in D2 receptor binding was inhibited by pretreatment with a D1-selective drug, SCH 23390, and that this D1-D2 link mediated by G proteins was missing in over half the striatal tissues from patients with schizophrenia and Huntington's disease.

Of the G protein subunits ($G\alpha_s$, $G\alpha_i$, $G\alpha_o$, and $G\beta$) examined, only $G\alpha_s$ immunoreactivity was elevated in the frontal and occipital cortex of patients with bipolar disorder (Young et al., 1991), suggesting that disturbances in G_s-mediated signal transduction may be involved in the pathophysiology of bipolar disorder. $G\alpha_s$ mRNA levels were increased in the hippocampus and the temporal cortex of patients with Alzheimer's disease (Harrison et al., 1991).

It is now known that various psychotropic agents can alter the amount of G proteins. Although lithium remains the treatment of choice for bipolar disorder, the mechanisms by which it attenuates the swings in mood between depression and mania remain unclear. In addition to inhibition of inositol monophosphatase, lithium might modulate the functions of the distinct G proteins. The chronic treatment of rats with lithium at therapeutically relevant serum concentrations decreased levels of mRNA and protein for the $G\alpha_{i1}$ and $G\alpha_{i2}$ and increased levels of mRNA and protein for AC type I and type II, in the cerebral cortex. It did not alter levels of $G\alpha_o$, $G\alpha_s$, and $G\beta$ (Colin et al., 1991). The chronic treatment of rats with cocaine, but not with several nonabused drugs, including haloperidol, produced similar changes compared to morphine in G proteins (decreased levels of $G\alpha_i$), AC (increased activity), and PKA (increased activity), in the nucleus accumbens (Nestler et al., 1990; Terwilliger et al., 1991). All these effects are observed on distinct G protein subtypes in a limited brain regions in spite of widespread occurrence and high abundance of G proteins in the brain, thus suggesting that G proteins may provide useful targets for new drug development.

G Protein Abnormalities in Schizophrenia

Aberrant Receptor/G Protein/Effector Coupling in Blood Cells from Patients with Schizophrenia

Human platelets provide convenient materials in the search for the biochemical pathology of schizophrenia, since they share similar functional proteins with neuronal cells. Of course, there have been continuing arguments regarding whether or not the data on platelets can represent functional states of the CNS, and we cannot totally neglect the influence of fluctuations of endogenous substances, such as blood catecholamines, on platelet function. Despite these drawbacks, there are some merits. We can evaluate signal transduction mechanisms in an intact or whole cell by applying receptor agonists and antagonists and enzyme activators or inhibitors. Moreover, by considering the time-point of sampling, we cannot only examine possible abnormalities in drug-naive patients, but also evaluate longitudinally the relation of data to clinical symptoms, treatment responsiveness, and prognosis. Table 2 summarizes the data on aberrant receptor–G protein–effector coupling in blood cells of schizophrenic patients.

Platelets

The plasma membrane of platelets, like that of neuronal cells, is equipped with prostaglandin E_1 (PGE_1) and α_2-adrenergic receptors, the activation of which causes an increase or a

Table 2
Aberrant Receptor/G Protein/Effector
Interaction in Blood Cells of Patients with Schizophrenia

Study	Subjects	Blood cells	Reported abnormalities
Kafka et al. (1979)	11M/9F drug-free	Platelets	Attenuated PGE_1-stimulated cAMP response in male patients
Rotrosen et al. (1980)	$N = 39$	Platelets	Attenuated PGE_1-stimulated cAMP response in acute, subacute, and chronic patients
Garver et al. (1982)	15M/5F	Platelets	Attenuated PGE_1-stimulated cAMP response in both sexes. No change in NaF-stimulated cAMP response
Kafka et al. (1986)	24M/8F	Platelets	Attenuated PGE_1-stimulated cAMP response
Kanof et al. (1986)	54M drug-free	Platelets	Attenuated PGE_1-stimulated cAMP response, regardless of prognosis of the illness
Kanof et al. (1987)	35M	Platelets	Attenuated PGE_1-stimulated cAMP response, negatively correlated with severity of positive symptoms
Kaiya et al. (1989)	15M/5F	Platelets	Enhanced thrombin-stimulated DAG production, in 6 out of 13 acute cases
Kaiya et al. (1990)	21M/14F	Platelets	Attenuated PGE_1- or forskolin-stimulated cAMP response, exclusively in male patients

Table 2 (*continued*)
Aberrant Receptor/G Protein/Effector
Interaction in Blood Cells of Patients with Schizophrenia

Study	Subjects	Blood cells	Reported abnormalities
Essali et al. (1990)	20M/23F	Platelets	Enhanced thrombin-stimulated inositol polyphosphates production, with no changes in nine drug-naive cases
Das et al. (1992)	3M/4F	Platelets	Higher PIP_2 level in seven treated patients
	15M/18F		Higher thrombin-stimulated generation of IP_3 in 18 on-drug and 9 drug-free patients, but not in 7 drug-naive patients lower thrombin-stimulated increase in intraplatelet Ca^{2+} level in 12 treated patients may be associated with desensitization of IP_3 receptors
Kanof et al. (1989)	47M drug-free	Leukocytes[a]	No changes in histamine-, isoproterenol-, or PGE_1-stimulated cAMP response
Ebstein et al. (1990)	8M/4F	Leukocytes[b]	No changes in PGE_1-, isopreterenol-, G_pNH_p-, or forskolin-stimulated cAMP response

[a]Polymorphonuclear leukocytes.
[b]EBV-transformed lymphocytes. M = male, F = female.

decrease in cAMP production through G_s or G_i, respectively. Pandey et al. (1977) reported that PGE_1-stimulated cAMP production in platelets is greater in four acute schizophrenic patients compared with five chronic patients or control subjects. In contrast, using a greater number of subjects, other investigators found that the PGE_1-stimulated cAMP responses are attenuated in schizophrenia, suggesting PGE_1 receptor hyposensitivity (Rotrosen et al., 1978; Kafka et al., 1979; Garver et al., 1982; Kanof et al., 1986). The attenuated PGE_1-stimulated cAMP response can be a marker of vulnerability to schizophrenia, since it is related neither to severity of symptoms, the age of onset (Kafka et al., 1983), nor to a marker of vulnerability of endogenous psychosis in general, since it is also observed in platelets from the depressed patients (Kanof et al., 1987).

Some of these researchers next studied whether the attenuation of PGE_1-stimulated cAMP response is owing to the decrease in the number of PGE_1 receptors or to the decrease in the amount or reactivity of G protein or AC coupled to PGE_1 receptors. NaF and $G_{pp}NH_p$ dissociate heterotrimeric G protein into α-subunit and $\beta\gamma$ subunit complexes, resulting in facilitated cAMP production. NaF also facilitates cAMP production through inhibiting GTPase. Forskolin directly stimulates the catalytic domain of AC, resulting in the cAMP production. Kafka et al. (1979) found the attenuated cAMP response in NaF-treated platelets from schizophrenics, but Garver et al. (1982) could not replicate this. Kafka et al. (1983) demonstrated an attenuated cAMP response in forskolin-treated platelets from male schizophrenics. They also found that the attenuation of PGE_1-stimulated cAMP response is reinstated by the application of $G_{pp}NH_p$ to the platelets of schizophrenics. These results suggest that the attenuation of the cAMP response is caused by abnormalities in either G protein or AC. In the brain, PGE_1 exerts an inhibitory control over DA release. Provided that the hyposensitivity of PGE_1 receptors in schizophrenic platelets reflects the hypofunction of PGE_1 receptors on the brain DAergic terminals, the latter could lead to an increase in DA release, a phenomenon that may well explain the DA theory of schizophrenia.

Kaiya et al. (1989) measured the thrombin-stimulated DAG production in platelets using tritiated arachidonic acid (AA) as a precursor. It was enhanced in 6 out of 13 acute schizophrenics, and the prognosis of such patients was better than that of the normal response group. They suggest an enhanced activity of PKC in schizophrenia. Thrombin stimulation of patients' platelets resulted in enhanced production of PA and inositol phosphates, suggesting increased platelet inositol phospholipid turnover caused by the increased PLC activity. These changes were found not only among patients who had been on neuroleptic treatment, but also among patients who had been drug-free for 11 mo. This thrombin-stimulated response did not differ between drug-naive patients and control subjects, suggesting that the changes are the result of neuroleptic levels that last for a long period after the drug discontinuation (Essali et al., 1990; Das et al., 1992).

Lymphocytes

Studies using human B-lymphocytes transformed by Epstein-Barr virus have merits in that one can distinguish contributions of environmental factors from those of genetic factors in cultured cells. The cAMP production in the particulate fraction of lymphocytes thus transformed did not differ between schizophrenic and control subjects, either in forskolin, Al/NaF, or $G_{pp}NH_p$ stimulation (Ebstein et al., 1990).

Polymorphonuclear Leukocytes

The mean life of polymorphonuclear leukocytes (PMNL) is shorter than lymphocytes. Therefore, studies using these cells have merits in that we can minimize the effect of circulating endogenous substance like catecholamines. The cAMP response in PMNL produced by PGE_1, β-adrenergic, or histamine receptor stimulation is normal in schizophrenic patients. It did not differ between patients in acute exacerbation and patients in remission (Kanof et al., 1986, 1989).

Aberrant Receptor/G Protein/Effector Coupling in Brains of Patients with Schizophrenia

G Protein Abnormalities in Temporal Lobe Structures in Schizophrenia

Brain imaging studies have revealed many structural and functional abnormalities in schizophrenia. Among these, the temporal lobe is one of the candidates for a site of schizophrenia, in addition to the frontal cortex and the thalamus. Moreover, there has been growing support for the notion that schizophrenia involves abnormal lateralization of cerebral function. For example, the presence of auditory hallucinations (Barta et al., 1990) and formal thought disorder (Shenton et al., 1992) in schizophrenia is associated with reduced volume of the left superior temporal gyrus. Schizophrenic hallucinations have been associated with decreased metabolism in the superior temporal gyrus (Cleghorn et al., 1992). There was a reversal of the normal asymmetry (left larger than right) in the planum temporale surface area of schizophrenics and severity of thought disorder in the patients correlated with planum asymmetry (Petty et al., 1995).

Since an abnormal lateralization of the temporal lobe exists in substantial cases of schizophrenia, aberrant second messenger systems could result. It seems that this is indeed the case. By using PTX-catalyzed ADP-ribosylation, Okada et al. (1990, 1991) measured the PTX-sensitive G proteins, and found that they are decreased in the putamen and hippocampus of the left hemisphere in schizophrenics compared with controls. However, in their more recent studies using an enzyme immunoassay, Okada et al. (1994) reported a decrease in $G\alpha_o$ but not $G\alpha_{i2}$ in the hippocampus and caudate head of the right hemisphere in schizophrenic patients compared with controls.

Using postmortem brain samples, we have studied possible changes in intracellular signaling systems in schizophrenia. The diagnosis for schizophrenia corresponded to DSM-IV categories. All schizophrenic patients were chronic cases whose duration of illness exceeded 20 yr. All had been on long-term neuroleptic treatment for at least 10 yr. Control subjects were selected from previously healthy individuals without a history of neurologic or psychiatric disorders and were matched by sex, by age at death, and by delay from death to autopsy. The methods used are as described previously (Nishino et al., 1993). A relative immunoreactivity of $G\alpha_s$, $G\alpha_i$, $G\alpha_o$, and $G\beta$ was measured in the superior temporal cortex (Brodmann's area 22, Wernicke's area, auditory association area), prefrontal cortex (Brodmann's area 9, frontal association area), and the entorhinal cortex (Brodmann's area 28, primary olfactory area) from schizophrenic patients and compared with control subjects. The G protein antisera used were AS/7, RM/1, GC/2, and SW/1, specific to $G\alpha_{i(1\ \&\ 2)}$, $G\alpha_s$, $G\alpha_o$ and $G\beta_{(1\ \&\ 2)}$ subunits, respectively.

$G\alpha_i$ and $G\alpha_o$ immunoreactivities in the crude membranes were decreased by about 30% in the left superior temporal cortex of schizophrenic patients compared to findings in the control subjects, on the left side, whereas $G\alpha_s$ and $G\beta$ immunoreactivities were unchanged between the two groups, on either side (Table 3). In the prefrontal cortex and entorhinal cortex of the left hemisphere, neither $G\alpha_s$, $G\alpha_i$, $G\alpha_o$, nor $G\beta$ immunoreactivities showed differences between the two groups (Table 4).

It is important to define the mechanism of such downregulation of the $G\alpha_i$ and $G\alpha_o$ in the left superior temporal cortex and to determine how it contributes to the pathophysiology of schizophrenia. We cannot totally exclude the possibility that the premortem use of neuroleptics affected our results. However, the left-to-right asymmetry, region specificity, and subtype specificity of changes in G protein immunoreactivities all argue against this possibility. Moreover, regulation of G protein and the cAMP system was not seen in response to chronic treatment with a neuroleptic drug haloperidol (Terwilliger et al., 1991). These results suggest little involvement of chronic neuroleptic treatment in our observations.

Decreased amounts of $G\alpha_i$ and $G\alpha_o$ can be induced by several different mechanisms: First, there is a possibility that the amounts of $G\alpha_i$ and $G\alpha_o$ are regulated at the level of gene

Table 3
Immunoquantification of G Protein
Subunits in the Superior Temporal Cortex[a]

G protein subunits	Left side ($N = 10$)		Right side ($N = 10$)	
	Controls	Schizophrenics	Controls	Schizophrenics
$G\alpha_s$				
52 kDa	100 ± 4.1	96.5 ± 4.1	100 ± 5.0	110.0 ± 6.0
45 kDa	100 ± 9.8	94.4 ± 5.6	100 ± 6.5	92.4 ± 5.9
$G\alpha_i$				
40/41 kDa	100 ± 9.0	70.1 ± 8.7[b]	100 ± 8.7	101.1 ± 8.8
$G\alpha_o$				
39 kDa	100 ± 7.8	74.2 ± 7.5[b]	100 ± 10.4	99.7 ± 7.6
$G\beta$				
35/36 kDa	100 ± 2.2	98.3 ± 3.4	100 ± 3.9	99.3 ± 4.4

[a]Values are the means $\pm$ SE. Densitometric readings were normalized to yield the average of the controls as 100%.
[b]Statistical significance compared with controls ($P < 0.05$, using Student's t-test, two-tailed)

expression in the left superior temporal cortex, that is, reduced transcription and/or translation of $G\alpha_i$ and $G\alpha_o$ genes. The promoter region of $G\alpha_o$ gene-regulating $G\alpha_o$ expression has been identified. Both transcriptional and posttranscriptional mechanisms are involved in regulating the expression of $G\alpha_o$. Transcriptional regulation is important for control of tissue-specific expression, whereas posttranscriptional mechanisms may be used to regulate the $G\alpha_o$ level in cells (Li et al., 1994). Our results suggest that both mechanisms may be involved. Second, in the left superior temporal cortex of the same patients, we have found accelerated proteolysis of fodrin α-subunits, as demonstrated by an increase in the ratio of the 150- to the 240-kDa fodrin immunoreactivities (Kitamura et al., in press). A 240-kDa fodrin α-subunit is proteolysed to a 150-kDa fragment by the activation of μ-calpain under conditions that favor the entry of Ca^{2+} through excitatory amino acid receptor-operated channels in the nerve cell membrane (Siman et al.,

Table 4
Immunoquantification of G Protein Subunits
in the Prefrontal Entorhinal Cortices of the Left Hemisphere[a]

G protein subunits	Prefrontal cortex		Entorhinal cortex	
	Controls (14)	Schizophrenics (10)	Controls (12)	Schizophrenics (11)
$G\alpha_s$				
52 kDa	100 ± 7.6	102.3 ± 7.2	100 ± 7.9	101.9 ± 6.4
45 kDa	100 ± 8.5	110.7 ± 13.2	100 ± 13.9	110.6 ± 16.5
$G\alpha_i$				
40/41 kDa	100 ± 14.7	88.5 ± 13.9	100 ± 17.9	108.7 ± 20.2
$G\alpha_o$				
39 kDa	100 ± 12.9	93.5 ± 7.8	100 ± 8.6	96.1 ± 11.2
$G\beta$				
35/36 kDa	100 ± 1.7	103.7 ± 2.7	100 ± 6.0	105.0 ± 5.9

[a]Values are the means $\pm$ SE. Densitometric readings were normalized to yield the average of the controls as 100%. Numbers in parentheses are the number of subjects.

1989), suggesting that the accelerated proteolysis of fodrin α-subunits may be the result of hyperglutamatergic activity via the *N*-methyl-D-aspartate (NMDA) receptors in the left superior temporal cortex of schizophrenics. Increased glutamate release can, in turn, enhance DA release via activation of NMDA receptors on the mesocortical DAergic nerve terminals. The chronic treatment of rats with cocaine, a DA uptake blocker, decreases levels of $G\alpha_i$ and $G\alpha_o$ (Nestler et al., 1990), and increases levels of AC and PKA (Terwilliger et al., 1991) in the nucleus accumbens, these findings being compatible with our results in the left superior temporal cortex. Therefore, the same adaptation mechanisms may operate to counteract the effect of DA in the target neurons. Finally, persistent activation of D2 receptors by DA may accelerate depalmitoylation of the $G\alpha_i$ and $G\alpha_o$ and hence translocation of $G\alpha_i$ and $G\alpha_o$ from membrane to cytosol, thereby reducing the membrane-bound $G\alpha_i$ and $G\alpha_o$. To prove this possibility, it is necessary to measure the $G\alpha_i$ and $G\alpha_o$ immunoreactivities in the soluble fraction as well as the particulate fraction.

Signal Transduction Utilizing cAMP as a Second Messenger

The regulatory (R) subunits of PKA are the only well-characterized cAMP receptor proteins. Therefore, we next measured the amount of PKA by monitoring specific ^{3}H-cAMP binding in the soluble fraction obtained from the superior temporal cortex of schizophrenic patients. The Scatchard analysis of the specific ^{3}H-cAMP binding revealed that B_{max} values from the schizophrenic group were significantly increased compared with control subjects, on the left side, but not on the right side. K_d values were unchanged between patients and controls, on either side, and between left and right superior temporal cortices in each group (Table 5) (Nishino et al., 1993).

In neural cell lines, exposure to dibutyryl cAMP increases tritiated cAMP binding and the amount of R subunits, but not the amount of C subunits or PKA activity. Moreover, this regulation was not observed in a glioma cell line. Therefore, an enhanced formation of cAMP would induce expression of PKA R subunits, leading to an increase in cAMP binding in the left superior temporal cortex of schizophrenic patients.

Signal Transduction Utilizing DAG and IP_3 as a Second Messenger

In the same postmortem brain samples, we measured the PKC and the ryanodine receptor/Ca^{2+} release channels (IP_3 receptor) in the left and right of the superior temporal cortex by using ^{3}H-4β phorbol 12,13-dibutyrate (^{3}H-PDBu) to label the regulatory domain of PKC (Nishino et al., 1989), and ^{3}H-IP_3 to label IP_3 receptors (Kitamura et al., 1989), respectively. We found no differences in the specific binding for PDBu or IP_3 between the schizophrenic and control subjects in either side of the cortex. No significant changes in PDBu or IP_3 binding were observed between left and right superior temporal cortices, in each group (data not shown).

Table 5
Binding Constants for Specific [³H]cAMP
Binding to Superior Temporal Cortex[a]

Binding constants	Left side ($N = 10$)		Right side ($N = 10$)	
	Controls	Schizophrenics	Controls	Schizophrenics
K_d (nM)	3.58 ± 0.36	3.43 ± 0.15	3.20 ± 0.48	3.53 ± 0.44
B_{max} (fmol/mg protein)	1200 ± 92	1591 ± 106[b,c]	1063 ± 100	1287 ± 89

[a]Values are the means $\pm$ SE. The concentration of free [³H]cAMP was 0.5–20 nM.

[b]$P < 0.05$ and

[c]$P < 0.05$ indicate statistical significance when compared to B_{max} values in controls on the left side, and the patients on the right side, respectively (Student's t test, two-tailed).

Concluding Remarks

In a series of experiments, we have demonstrated a decreased amount of $G\alpha_i$ and $G\alpha_o$, but not $G\alpha_s$ or $G\beta$, as well as an increased amount of R subunit of PKA in the left superior temporal cortex of schizophrenics. Thus, in schizophrenics, enhanced responsiveness of AC through a relative preponderance of stimulatory G protein over inhibitory G protein may facilitate cAMP production in the left superior temporal cortex. Dopamine neurons innervate the cerebral cortex in primates, and mRNA coding the DA-D1 receptor is expressed in the cortex in humans. D1 receptors may be more efficaciously coupled with AC through a relative preponderance of G_s over G_i in the left superior temporal cortex in schizophrenics.

We are only beginning to understand how G protein abnormalities relate to the pathophysiology of schizophrenia (DA hypothesis, glutamate hypothesis, structural abnormalities in the brain, and so on). With more extensive and elaborate exploration of the upstream (receptors) and downstream (intracellular signaling molecules) partners of G proteins, we may have new therapeutic approaches to schizophrenia.

References

Barta, P. E., Pearlson, G. D., Powers, R. E., Richards, S. S., and Tune, L. E. (1990) Auditory hallucinations and smaller superior temporal gyral volume in schizophrenia. *Am. J. Psychiatry* **147,** 1457–1462.

Bigay, J., Faurobert, E., Franco, M., and Chabre, M. (1994) Roles of lipid modifications of transducin subunits in their GDP-dependent association and membrane binding. *Biochemistry* **33,** 14,081–14,090.

Buss, J. E., Mumby, S. M., Casey, P. J., Gilman, A. G., and Sefton, B. M. (1987) Myristoylated α-subunits of guanine nucleotide-binding regulatory proteins. *Proc. Natl. Acad. Sci. USA* **84,** 7493–7497.

Clapham, D. E. (1995) Calcium signaling. *Cell* **80,** 259–268.

Clapham, D. E. (1996) The G protein nanomachine. *Nature (Lond.)* **379,** 297–299.

Cleghorn, J. M., Franco, S., Szechtman, B., Kaplan, R. D., Szechtman, H., Brown, G. M., Nahmias, C., and Garnett, E. S. (1992) Toward a brain map of auditory hallucinations. *Am. J. Psychiatry* **149,** 1062–1069.

Colin, S. F., Chang, H. C., Mollner, S., Pfeuffer, T., Reed, R. R., Duman, R. S., and Nestler, E. J. (1991) Chronic lithium regulates the expression of adenylate cyclase and Gi-protein α subunit in rat cerebral cortex. *Proc. Natl. Acad. Sci. USA* **88,** 10,634–10,637.

Conklin, B. R. and Bourne, H. R. (1993) Structural elements of G_α-subunits that interact with Gβγ, receptors, and effectors. *Cell* **73,** 631–641.

Coon, H., Byerley, W., Holik, J., Hoff, M., Myles-Worsley, M., Lannfelt, L., Sokoloff, P., Schwartz, J.-C., Waldo, M., Freedman, R., and Plaetke, R. (1993) Linkage analysis of schizophrenia with five dopamine receptor genes in nine pedigrees. *Am. J. Hum. Genet.* **52,** 327–334.

Das, I., Essali, M. A., de Belleroche, J., and Hirsch, S. R. (1992) Inositol phospholipid turnover in platelets of schizophrenic patients. *Prostaglandins Leukot. Essent. Fatty Acids* **46,** 65–66.

Dhermy, D. (1991) The spectrin super-family. *Biol. Cell* **71,** 249–254.

Ebstein, R. P., Bennett, E. R., Hadjez, J., Silver, H., Yedgar, S., and Lerer, B. (1990) Cyclic AMP second messenger signal generation in EBV-transformed lymphoblastoid cells from schizophrenic patients. *J. Psychiatr. Res.* **24,** 121–127.

Essali, M. A., Das, I., de Belleroche, J., and Hirsch, S. R. (1990) The platelet polyphosphoinositide system in schizophrenia: the effects of neuroleptic treatment. *Biol. Psychiatry* **28,** 475–487.

Fukuda, Y., Takao, T., Ohguro, H., Yoshizawa, T., Akino, T., and Shimonishi, Y. (1990) Farnesylated γ-subunit of photoreceptor G protein indispensable for GTP-binding. *Nature (Lond.)* **346,** 658–660.

Furuichi, T., Furutama, D., Hakamata, Y., Nakai, J., Takeshima, H., and Mikoshiba, K. (1994) Multiple types of ryanodine receptor/Ca^{2+} release channels are differentially expressed in rabbit brain. *J. Neurosci.* **14,** 4794–4805.

Garver, D. L., Johnson, C., and Kanter, D. R. (1982) Schizophrenia and reduced cyclic AMP production: evidence for the role of receptor-linked events. *Life Sci.* **31,** 1987–1992.

Guiramand, J., Montmayeur, J-P., Ceraline, J., Bhatia, M., and Borrelli, E. (1995) Alternative splicing of the dopamine D2 receptor directs specificity of coupling to G proteins. *J. Biol. Chem.* **270,** 7354–7358.

Harrison, P. J., Barton, A. J. L., McDonald, B., and Pearson, R. C. A. (1991) Alzheimer's disease: specific increases in a G protein subunit ($G_{s\alpha}$) mRNA in hippocampal and cortical neurons. *Mol. Brain Res.* **10,** 71–81.

Haslam, R. J., Koide, H. B., and Hemmings, B. A. (1993) Pleckstrin domain homology. *Nature (Lond.)* **363,** 309–310.

Iiri, T., Herzmark, P., Nakamoto, J. M., van Dop, C., and Bourne, H. R. (1994) Rapid GDP release from $G_{s\alpha}$ in patients with gain and loss of endocrine function. *Nature* **371,** 164–168.

Kafka, M. S., van Kammen, D. P., and Bunney, W. E., Jr. (1979) Reduced cyclic AMP production in the blood platelets from schizophrenic patients. *Am. J. Psychiatry* **136,** 685–687.

Kafka, M. S. and van Kammen, D. P. (1983) α-Adrenergic receptor function in schizophrenia: receptor number, cyclic adenosine monophosphate production, adenylate cyclase activity, and effect of drugs. *Arch. Gen. Psychiatry* **40,** 264–270.

Kafka, M. S., Kleinman, J. E., Karson, C. N., and Wyatt, R. J. (1986) Alpha-adrenergic receptors and cyclic AMP production in a group of schizophrenic patients. *Hillside J. Clin. Psychiatry* **8,** 15–24.

Kaiya, H., Nishida, A., Imai, A., Nakashima, S., and Nozawa, Y. (1989) Accumulation of diacylglycerol in platelet phosphoinositide turnover in schizophrenia: a biological marker of good prognosis? *Biol. Psychiatry* **26,** 669–676.

Kaiya, H., Ofuji, M., Nozaki, M., and Tsurumi, K. (1990) Platelet prostaglandin E1 hyposensitivity in schizophrenia: decrease in cyclic AMP formation and in inhibitory effects on aggregation. *Psychopharmacol. Bull.* **26,** 381–384.

Kanof, P. D., Johns, C. A., Davidson, M., Siever, L. J., Coccaro, E. F., and Davis, K. L. (1986) Prostaglandin receptor sensitivity in psychiatric disorders. *Arch. Gen. Psychiatry* **43,** 987–993.

Kanof, P. D., Davidson, M., Johns, C. A., Mohs, R. C., and Davis, K. L. (1987) Clinical correlates of platelet prostaglandin receptor subsensitivity in schizophrenia. *Am. J. Psychiatry* **144,** 1556–1560.

Kanof, P. D., Coccaro, E. F., Johns, C. A., Davidson, M., Siever, L. J., and Davis, K. L. (1989) Cyclic-AMP production by polymorphonuclear leukocytes in psychiatric disorders. *Biol. Psychiatry* **25,** 413–420.

Kerwin, R. W. and Beats, B. C. (1990) Increased forskolin binding in the left parahippocampal gyrus and CA1 region in post mortem schizophrenic brain determined by quantitative autoradiography. *Neurosci. Lett.* **118,** 164–168.

Kim, D., Lewis, D. L., Graziadei, L., Neer, E. J., Bar-Sagi, D., and Clapham, D. E. (1989) G protein $\beta\gamma$ subunits activate the cardiac muscarinic K^+-channel via phospholipase A_2. *Nature (Lond.)* **337,** 557–560.

Kitamura, N., Hashimoto, T., Nishino, N., and Tanaka, C. (1989) Inositol 1,4,5-trisphosphate binding sites in the brain: regional distribution, characterization, and alterations in brains of patients with Parkinson's disease. *J. Mol. Neurosci.* **1,** 181–187.

Kitamura, N., Nishino, N., Hashimoto, T., et al. (1997) Asymmetrical changes in the fodrin α subunit in the superior temporal cortices in schizophrenia. *Biol. Psychiatry,* in press.

Kleuss, C., Scherubl, H., Hescheler, J., Schultz, G., and Wittig, B. (1993) Selectivity in signal transduction determined by γ subunits of heterotrimeric G proteins. *Science* **259,** 832–834.

Kokame, K., Fukuda, Y., Yoshizawa, T., Takao, T., and Shimonishi, Y. (1992) Lipid modification at the N terminus of photoreceptor G protein α-subunit. *Nature (Lond.)* **359,** 749–752.

Krupinski, J., Coussen, F., Bakalyar, H. A., Tang, W-J., Feinstein, P. G., Orth, K., Slaughter, C., Reed, R. R., and Gilman, A. G. (1989) Adenylyl cyclase amino acid sequence: possible channel- or transporter-like structure. *Science* **244,** 1558–1564.

Kurachi, Y., Ito, H., Sugimoto, T., Shimizu, T., Miki, I., and Ui, M. (1989) Arachidonic acid metabolites as intracellular modulators of the G protein-gated cardiac K^+ channel. *Nature (Lond.)* **337,** 555–557.

Landis, C. A., Masters, S. B., Spada, A., Pace, A. M., Bourne, H. R., and Vallar, L. (1989) GTPase inhibiting mutations activate the α chain of the Gs and stimulate adenylyl cyclase in human pituitary tumors. *Nature (Lond.)* **340,** 692–696.

Li, Y., Mortensen, R., and Neer, E. J. (1994) Regulation of α_o expression by the 5'-flanking region of the α_o gene. *J. Biol. Chem.* **269,** 27,589–27,594.

Lombardo, C. R., Weed, S. A., Kennedy, S. P., Forget, B. G., and Morrow, J. S. (1994) βII-Spectrin (fodrin) and βIΣ2-spectrin (muscle) contain NH_2- and COOH-terminal membrane association domains (MAD1 and MAD2). *J. Biol. Chem.* **269,** 29,212–29,219.

Lyons, J., Landis, C. A., Harsh, G., Vallar, L., Grünewald, K., Feichtinger, H., Duh, Q.-Y., Clark, O. H., Kawasaki, E., Bourne, H. R. et al. (1990) Two G protein oncogenes in human endocrine tumors. *Science* **249,** 655–659.

Manji, H. K. (1992) G proteins: implications for psychiatry. *Am. J. Psychiatry* **149,** 746–760.

Memo, M., Kleinman, J. E., and Hanbauer, I. (1983) Coupling of dopamine D1 recognition sites with adenylate cyclase in nuclei accumbens and caudatus of schizophrenics. *Science* **221,** 1304–1307.

Miric, A., Vechio, J. D., and Levine, M. A. (1993) Heterogeneous mutations in the gene encoding the α-subunit of the stimulatory G protein of adenylyl cyclase in Albright hereditary osteodystrophy. *J. Clin. Endocrinol. Metab.* **76,** 1560–1568.

Mons, N. and Cooper, D. M. F. (1995) Adenylate cyclase: critical foci in neuronal signaling. *Trends Neurosci.* **18,** 536–542.

Mumby, S. M., Casey, P. J., Gilman, A. G., Gutowski, S., and Sternweis, P. C. (1990) G protein g subunits contain a 20-carbon isoprenoid. *Proc. Natl. Acad. Sci. USA* **87,** 5873–5877.

Neer, E. J. (1995) Heterotrimeric G proteins: organizers of transmembrane signals. *Cell* **80,** 249–257.

Nestler, E. J., Terwilliger, R. Z., Walker, J. R., Sevarino, K. A., and Duman, R. S. (1990) Chronic cocaine treatment decreases levels of the G protein subunits $G_{i\alpha}$ and $G_{o\alpha}$ in discrete regions of rat brain. *J. Neurochem.* **55,** 1079–1082.

Neubig, R. R. (1994) Membrane organization in G protein mechanisms. *FASEB J.* **8,** 939–946.

Nishino, N., Kitamura, N., Nakai, T., Hashimoto, T., and Tanaka, C. (1989) Phorbol ester binding sites in human brain: characterization, regional distribution, age-correlation, and alterations in Parkinson's disease. *J. Mol. Neurosci.* **1,** 19–26.

Nishino, N., Kitamura, N., Hashimoto, T., Kajimoto, Y., Shirai, Y., Murakami, N., Nakai, T., Komure, O., Shirakawa, O., Mita, T., and Nakai, H. (1993) Increase in [^{3}H]cAMP binding sites and decrease in $G_{i\alpha}$ and $G_{o\alpha}$ immunoreactivities in left temporal cortices from patients with schizophrenia. *Brain Res.* **615,** 41–49.

Nishizuka, Y. (1995) Protein kinase C and lipid signaling for sustained cellular responses. *FASEB J.* **9,** 484–496.

Nöthen, M. M., Wildenauer, D., Cichon, S., Albus, M., Maier, W., Minges, J., Lichtermann, D., Bondy, B., Rietschel, M., Körner, J., Fimmers, R., and Propping, P. (1994) Dopamine D2 receptor molecular variant and schizophrenia. *Lancet* **343,** 1301–1302.

Okada, F., Crow, T. J., and Roberts, G. W. (1990) G proteins (Gi, Go) in the basal ganglia of control and schizophrenic brain. *J. Neural Transm. (Gen. Section)* **79,** 227–234.

Okada, F., Crow, T. J., and Roberts, G. W. (1991) G proteins (Gi, Go) in the medial temporal lobe in schizophrenia; preliminary report of a neurochemical correlate of structural change. *J. Neural Transm. (Gen. Section)* **84,** 147–153.

Okada, F., Tokumitsu, Y., Takahashi, N., Crow, T. J., and Roberts, G. W. (1994) Reduced concentrations of the α-subunit of GTP-binding protein Go in schizophrenic brain. *J. Neural Transm. (Gen. Section)* **95,** 95–104.

Pandey, G. N., Garver, D. L., Tamminga, C., Ericksen, S., Ali, S. I., and Davis, J. M. (1977) Postsynaptic supersensitivity in schizophrenia. *Am. J. Psychiatry* **134,** 518–522.

Petty, R. G., Barta, P. E., Pearlson, G. D., McGilchrist, I. K., Lewis, R. W., Tien, A. Y., Pulver, A., Vaughn, D. D., Casanova, M. F., and Powers, R. E. (1995) Reversal of asymmetry of the planum temporale in schizophrenia. *Am. J. Psychiatry* **152,** 715–721.

Quick, M. W., Simon, M. I., Davidson, N., Lester, H. A., and Aragay, A. M. (1994) Differential coupling of G protein α subunits to seven-helix receptors expressed in *Xenopus* oocytes. *J. Biol. Chem.* **269,** 30,164–30,172.

Rotrosen, J., Miller, A. D., Mandio, D., Traficante, L. J., and Gershon, S. (1978) Reduced PGE1 stimulated ^{3}H-cAMP accumulation in platelets from schizophrenics. *Life Sci.* **23,** 1989–1996.

Rostrosen, J., Miller, A. D., Mandio, D., Traficante, L. J., and Gershon, S. (1980) Prostaglandins, platelets, and schizophrenia. *Arch. Gen. Psychiatry* **37,** 1047–1054.

Seeman, P., Niznik, H. B., Guan, H. C., Booth, G., and Ulpian, C. (1989) Link between D1 and D2 dopamine receptors is reduced in schizophrenia and Huntington diseased brain. *Proc. Natl. Acad. Sci. USA* **86,** 10,156–10,160.

Shenton, M. E., Kikinis, R., Jolesz, F. A., Pollak, S. D., LeMay, M., Wible, C. G., Hokama, H., Martin, J., Metcalf, D., Coleman, M., and McCarley, R. W. (1992) Abnormalities of the left temporal lobe and thought disorder in schizophrenia: a quantitative magnetic resonance imaging study. *N. Engl. J. Med.* **327,** 604–612.

Siman, R., Noszek, J. C., and Kegerise, C. (1989) Calpain I activation is specifically related to excitatory amino acid induction of hippocampal damage. *J. Neurosci.* **9,** 1579–1590.

Simon, M. I., Strathmann, M. P., and Gautam, N. (1991) Diversity of G proteins in signal transduction. *Science* **252,** 802–808.

Smrcka, A. V. and Sternweis, P. C. (1993) Regulation of purified subtypes of phosphatidylinositol specific phospholipase Cβ by G protein α and βγ subunits. *J. Biol. Chem.* **268,** 9667–9674.

Spiegel, A. M., Weinstein, L. S., and Shenker, A. (1993) Abnormalities in G protein-coupled signal transduction pathways in human disease. *J. Clin. Invest.* **92,** 1119–1125.

Tang, W.-J. and Gilman, A. G. (1992) Adenylyl cyclases. *Cell* **70,** 869–872.

Terwilliger, R., Beitner-Johnson, D., Sevarino, K. A., Crain, S. M., and Nestler, E. J. (1991) A general role for adaptations in G proteins and the cyclic AMP system in mediating the chronic actions of morphine and cocaine on neuronal function. *Brain Res.* **548,** 100–110.

Wedegaertner, P. B. and Bourne, H. R. (1994) Activation and depalmitoylation of $G_{s\alpha}$. *Cell* **77,** 1063–1070.

Weinstein, L. S., Shenker, A., Gejman, P. V., Merino, M. J., Friedman, E., and Spiegel, A. M. (1991) Activating mutations of the stimulatory G protein in the McCune-Albright syndrome. *N. Engl. J. Med.* **325,** 1688–1695.

Young, L. T., Li, P. P., Kish, S. J., Siu, K. P., and Warsh, J. J. (1991) Postmortem cerebral cortex Gs α-subunit levels are elevated in bipolar affective disorder. *Brain Res.* **553,** 323–326.

Index